MODERN TECHNOLOGY
IN THE
HEIDEGGERIAN PERSPECTIVE

MODERN TECHNOLOGY
IN THE
HEIDEGGERIAN PERSPECTIVE

Volume I

William Lovitt

and

Harriet Brundage Lovitt

Problems in Contemporary Philosophy
Volume 17a

The Edwin Mellen Press
Lewiston/Queenston/Lampeter

Library of Congress Cataloging-in-Publication Data

Lovitt, William, 1928-
    Modern technology in the Heideggerian perspective / William Lovitt
and Harriet Brundage Lovitt.
        p.   cm. -- (Problems in contemporary philosophy ; v. 17)
    Includes bibliographical references and index.
    ISBN 0-88946-345-X (vol. 17a). -- ISBN 0-88946-269-0 (vol. 17b)
    1. Heidegger, Martin, 1889-1976. 2. Techne (Philosophy)
3. Technology--Moral and ethical aspects. 4. Technology-
-Philosophy. I. Lovitt, Harriet Brundage, 1924-   . II. Title.
III. Series.
    B3279.H49L72   1995
    193--dc20   95-1886
                                                                    CIP

This is volume 17a in the continuing series
Problems in Contemporary Philosophy
Volume 17a ISBN 0-88946-345-X
PCP Series ISBN 0-88946-325-5

A CIP catalog record for this book is available from the British Library.

The Edwin Mellen Press   The Edwin Mellen Press
        Box 450   Box 67
Lewiston, New York   Queenston, Ontario
USA 14092-0450   CANADA L0S 1L0

The Edwin Mellen Press, Ltd.
Lampeter, Dyfed, Wales
UNITED KINGDOM SA48 7DY

Printed in the United States of America

IN MEMORY OF

CHARLES LEO LOVITT   STANLEY EVERET BRUNDAGE

AILEEN LINNEY LOVITT   SARA EMERY BRUNDAGE

CONTENTS
VOLUME I

ACKNOWLEDGEMENTS

This book is unlike many scholarly undertakings, in that we have neither sought nor received advice from others during its composition. The work that we have done, from the studying of Heideggerian materials to the writing and revising of our own lengthy text, has been carried on continually as a dialogue in which we have relied solely on one another for the sort of thought-furthering insight and corrective judgment that the bringing to bear upon written material of a mind or minds other than an author's own can so frequently provide. This has been an arduous and rewarding experience that has left us grateful each for the peculiar capabilities of the other and also for the sometimes surprising, shared resources of patience, perseverance and humor without which our project could never have been brought to a successful close.

This is not to say that we need acknowledge no help from others as we now present our work. On the contrary, much formative training and study lie for each of us behind the work that has here been done. In testimony to that all-important fact, we wish here to mention with deep gratitude a number of those who, in the course of our varied studies, by instruction and example taught us to endeavor always to let material considered speak first for itself, to think cogently regarding what was being said, to slight no detail while yet striving after a synthetic grasp, and, finally, to judge carefully, where judgment was appropriate, the character and adequacy of ideas and positions thus brought to light. Beyond this, among such teachers who early prepared us for the work now completed, several, especially those working in the biblical, historical and language fields, also opened to us the excitement and challenge of the kinds of exegetical and interpretive study that hold a prominent place in the pages to follow. We would render thanks, then, to the following: William T. Jones, Professor of Philosophy, Benjamin Bart, Professor of French, and John Gleason, Professor of History, of Pomona College; Ursula Niebuhr and Edmond La B. Cherbonnier, professors of Religion, Isabel Sterns and John Smith, Professors of Philosophy, and

Lorna McGuire, Professor of English, of Barnard College; Paul Tillich
and John Macquarrie, Professors of Theology, Reinhold Niebuhr and
David E. Roberts, Professors of Social Ethics and Philosophy of Religion
respectively, and James Muilenburg and Samuel L. Terrien, Professors of
Old Testament, of Union Theological Seminary, New York; John
Dillenberger and John Herman Randall, Professors of Religion and
Philosophy respectively, of Columbia University; Professors Jean Hering,
François Wendel, and Pierre Bourgelin, of the Faculty of Protestant
Theology of the University of Strasbourg; and Professors Wilhelm Szilasi
and Max Müller, of the Faculty of Philosophy, and Professor Walther
Rehm of the Faculty of German Language and Literature, of the Albert
Ludwig University, Freiburg im Breisgau.

In the practical sphere, we wish to acknowledge the awarding of
three separate grants of time and money by California State University at
Sacramento that during recent years very materially advanced our
pursuance of our work; we wish to express our gratitude to Marian
Tarbell, Lavon Johnson, Hannah Jaffe and Gusta Strummwasser for the
transcribing of many German texts into Braille; we wish to thank James
Bard, John Simon and Professor Bradley Dowden for helping variously to
launch us upon the work with the computers—one of which can speak—
that have played indispensible roles in our turning of our original hand-
written draft into this finished book; and we wish, finally, to thank Dr.
Clifford Browder for helpful editorial suggestions concerning our
completed manuscript, Anne Freitas for valuable proofreading, and Moira
Neuterman both for final proofreading and for extensive typing when our
project was young.

Harriet Brundage Lovitt & Charles William Lovitt
Fair Oaks, California
April, 1993

PREFACE

At the present time, less than twenty years after Martin Heidegger's death, the impact of his philosophical work has surely but begun to be felt. Without a doubt the influence of his thinking can confidently be expected to intensify and to spread ever more widely in coming years.

Heidegger's work, whose publication is, indeed, still in process, comprises as it stands a varied wealth of detailed analysis and proffered interpretation that has already made it a source of fructifying ideas over a wide spectrum of disciplines. Thanks, however, to difficulties inherent in that work, it has continually been approached by those concerned to study it very largely in a topical and fundamentally desultory way. The perspective from which Heidegger undertook his inquiries underwent a marked shift not long after he embarked upon his major work; continually he relied centrally on idiosyncratic uses of language for the conveying of his thought; and his thinking is frequently presented via isolated discussions, often without specific indications of inner relationships that exist among pivotally important explications and characterizations that lie scattered in them. These characteristics in particular have militated strongly against the consideration of Heidegger's extremely demanding body of thinking in its entirety. Their presence has, indeed, tended to lay that thinking open to interpretations that, because they have focused only upon one facet or portion of his thinking and have consequently been unable to reach to the depth and complexity resident there, have, regrettably, far too often remained to a considerable degree superficial or even misleading.

The present study has been written in the hope of contributing materially to the mitigation of this problem. We have found a single underlying outlook, a single way of understanding and portraying reality, to pervade Heidegger's philosophical work as a whole. We shall here be at pains to make evident the presence of that outlook as it shows itself systematically in diverse areas of his thinking, to the end that its disclosure may aid in rendering his thought accessible in a new way.

Beyond this, once again in the interest of genuine accessibility, we shall present an exposition of Heidegger's thinking that continually gives minute attention to his often highly individual uses of language, both through an explication of the meanings he intends and through a utilization of those meanings, so far as may be, in our own presentations. Any attempt truly to understand what Heidegger is saying requires of the student that he or she first take up a position within Heidegger's realm of discourse.   Anyone who would attempt to think with respect to Heidegger's thinking must be willing first literally to endeavor to think within it.   Our pervasive concern to set forth as faithfully as possible Heidegger's own way of speaking is intended to facilitate just such entry into the very difficult sphere of his thought.

In pursuing our study, we have centered our attention, as Heidegger centered his when speaking of our time, on modern technology, for to Heidegger that technology presented itself as peculiarly disclosive of the manner of being evinced by whatever *is* in this present age.   We undertake to consider the complex thought that is in play in Heidegger's understanding of technology and, on that basis, to view technology historically, i.e., in terms of its present character and scope and with respect both to its antecedence and to that which Heidegger sees to impend for it beyond the present.   We have found modern technology, when seen thus, to offer a theme remarkably well suited to focus toward clarity the inquiry into the depths and ranges of Martin Heidegger's thinking that is now so much needed among us in order that that thinking may be understood and assessed ever more accurately even as the scope and importance of its influence grows.

H. B. L.
C. W. L.

TOPICAL SUMMARY

VOLUME I

destining; the increasing pervasiveness of concealing and deception in history; disruptive divergence as informing the happening of every particular of what-is; the whiling of the particulars as presencing; deviance from whiling; self-assertion of the particulars; mere continuance; the origination of disruption in Being's happening as presencing; self-maintaining as the source of self-assertive drawing apart; the endangering of the Being of what-is; the surmounting of disruption in coming-together; the happening of Being as a handing-over keeping-in-hand; Being as a freeing, disposing gathering; the self-maintaining of what-is as itself; Being's unconcealing of itself always by way of some orderly manifold.

what-is upon man; man's answering self-assertive striving for security; inauthenticity; the transpiring of *Being*-there in everydayness; fallen-awayness as concomitant with the transpiring of *Being*-there in precariousness; concealing as the ultimate source of inauthenticity; the remaining awareness of transitoriness; anxiety; the awareness of thatness, utter contingency; the authentic, individually centered acknowledgement of uncanniness, of sheer thatness; the call of conscience; guilt; Being-toward-death; the perseverant self-opening of authentic *Being*-there: *Entschlossenheit*; the embeddedness of individual *Being*-there in its community, its transpiring as "historical"; freedom as letting-be; freedom as accordance with the happening of Being.

CHAPTER III: LANGUAGE 111
Language as constitutive of human existing.

Section 1 112
Discourse (*Rede*), as fundamental to the transpiring of human *Being*-there; spoken utterance (*Sprache*) as intrinsic to Being-in-the-world; meaningful interpreting, as constitutive in man's manner of existing in the world; assertion; the possibility of obscuring; communication as a sharing-with.

Section 2 117
Language as the self-unconcealing happening of the Being of what-is; man's language as an answering-to; language as a saying that shows; *Sage*, the primordial showing-saying that brings to meaningful appearance; the dialogical relation between primal Showing-saying and human apprehending; the eventuation of language in the spoken word; our saying-after; silence; the provision of meaningful disclosure via language; language as the carrying into play of Happening as be-waying (*Be-wegung*); *logos*, a gathering that lets lie forth together; naming, the disclosing of what-is; the deceptiveness of language, a manifestation of Being's happening as concealing; mere designating; words as disclosive of the manner of Being's happening.

Section 3 129
The provision, through language, of an open arena for human doing; thinking as an answering to Being's happening as Showing-saying, from out of a belonging together; the historical role of thinkers—the speaking forth of Being's happening; the dictates of Being's happening as unconcealing; *Dichtung* as that unconcealing, a name for thinking and art as dictated via language; poetry as revelatory; the art work as disclosive of Being's happening as Showing-saying; art and thinking; the constitutive role of art in on-going human life—the work as foundational for a historical people; the centrality here of language; change and concealing in the

taking place of the work of art; the changefulness of language as the vehicle of Being's ruling as disclosive happening; language as pivotal in the peculiar belonging together of Being and man.

belong—a new insight; the ruling of *Ereignis* as letting-belong; the immediacy of its encountering.

image of the divine; the crucial role of the heavens in the imaging forth accomplished through the poet; their imaging forth as a manifesting of the Being of what-is; poetic presenting as receiving and bringing to bear the genuine adjudging of whatever is; its bringing into play of the proper role of man, his pursuing of his way *in extremis* as the mortal; poetic presenting as a gift originating in *Ereignis*, enowning letting-belong together; poetic presenting as the completion of manifesting as such, which permits and accomplishes dwelling, through accomplishing the meaningful appearing of things; the grounding of all nurturing-building in poetic seeing and speaking; the immediate accomplishing of language in the poetic word. But is it the poetic or the techno-scientific that in fact plays a decisive role? Our time as one when no god appears.

The instrumental-anthropological defining of technology; the governance of technology from out of Being; technological using, the serving of some purposed end; proper usefulness and using, in contrast; the overmastering of self-presenting; the ubiquity of the technological approach; the evincing of the dominance of technology in language; the centrality of language for technology; informational language; the computer; "the language machine," a happening of Being; formalized vs. connotative language; our assertion of dominion via language.

Planning and calculating as fundamental to technology; technology's suppressing of uniqueness and promoting of likeness; the standing-reserve *(das Bestand)*, a mode of self-presenting; technology's central concern with available energy; technological setting-in-place-as-available-supply *(Stellen)*; such "setting" as a challenging forth; modern man's accomplishing of it as a setting in order *(Bestellen)*; technological setting-in-place-as-supply as a mode of Being's happening as revealing.

The meaning of *Wesen* ("essence"); *Wesen* as Being's happening as an enduring that disposingly gathers forth; the disclosing of any *Wesen* via the speaking of language; the *Wesen* of modern technology, *das Ge-stell*; the *Ge-stell* as the enframing summons ruling in every technologically motivated meeting of man and what-is; the *Ge-stell* as a happening of Being as primal Showing-saying; technology, as a bodying forth of this summons, as a strange mode of protective revealing; directing-toward and making-secure as primary tendencies of that revealing; a revealing that is a concealing; the withdrawing of presencing from view; technological revealing as a withholding from self-presenting; the decisive supremacy of technology.

Descartes, into a new way of being human, subjectivity; the dominion of subjectivity; its character as a despoiling; Being's ruling newly as will to will, in self-assertive subjectness and objectness.

Metaphysics as disclosive of the one destining ruling in the modern age; Descartes' authoritative words as providing the basis for modern science and technology; the founding of subsequent philosophy on his assumption that only in the sphere of the subjective could reality be validated; Leibniz: representing striving as definitive of Being; Kant: Being as established via objectifying; disclosing of the will to will; Hegel: Being as accomplished via reflexively structured thinking, the bringing of the Absolute to consciousness; the will to will as Nietzsche's will to power, the completion of the grounding disclosure provided through Descartes; the self-securing of the will to power as a will to mastery, in its reflexive forward movement as value positing; the self-conscious accomplishing of the will to power in man; Overman; the insurrection of man as subject; the bringing to nothing of Being as the appearing of that which appears from out of itself; the will to will as Being's self-negating; its self-deception with respect to its concealed nullity; emptiness the arena of all securing; the consequent insatiability of the will to will; the will's calculative character; its invincible insecurity; man's lordly standing as technological man; his assault on everything that is; the insatiability of technology; Enframing's ruling as will to will; the identity of the manner of happening of modern metaphysics and of modern technology; the grounding role of metaphysics vis a vis technology; the end, the completion, of philosophy; nihilism; the overcoming of metaphysics; beyond Nietzsche.

The meaning of "Greek".

Greek *technē*, a skillful knowing; the unfolding Twofold as *physis*; the administering via *technē*, as a masterful gathering forth, of its powerful self-disclosing; the widespread transpiring of *technē;* its accomplishment by the few; the wresting forth of a happening of Being as presencing; the belonging together of *physis* and *technē;* the ruling in both of Being as *logos*; belonging together as a counterposing in radical self-accomplishment; *technē* as a deed of daring; human apprehending as *Entscheidung*—an accomplishing of human *Being*-there; the radical fulfilling of the ruling happening and claim of Being; the exposure of the daring one to Being's happening as concealing; *technē* as a reckless knowing that opens itself to the extreme possibility, that of destruction.

of fitness-for; the precedence of *morphē;* the accomplishing of completion with respect to a particular entity; fitness-for as partaking of deficiency; the residing of any entity's Being in its *morphē;* the happening of *morphē* as *steresis*, its governing in absenting itself; the happening of *morphē* also apart from *hylē;* the superiority of *morphē;* original *physis* as distantly glimpsed in Aristotle's thinking; Aristotle's structuring of reality—the four *aitia*; the discernibility in the latter of Heidegger's primal Fourfold; the four *aitia* as manifesting the withdrawal of Being.

Aristotle's understanding of human utterance as the standard via which the Being of an entity is judged; language as a saying of something about something, not a locus of immediate disclosure; assertion as capable of disclosing Being; Aristotelian categories; stating, as at once assuming, declaring and fashioning the self-presenting of its subject; assertion as permitting human apprehending to grasp and secure Being for knowing; asserting as the locus of judgment as to whether something can or cannot be; truth as the correctness of assertion; Aristotle's logic, *logos* as assertion, a standard of judgment on which apprehending can implicitly rely; *morphē* as wholly knowable; Aristotle's establishing of the distancing of Being and man; the ultimacy of manifestness.

The bifurcating of human doing; the separating of philosophical knowing and *technē* as the province of craftsman and artist; the changed character of the primary comportment requisite upon man; the diminished role of *technē* and of its works; philosophical thinking as obscuring the immediate, uniquely particularized happening of what-is, in its pursuit of the secure; disruption throughout the happening of the Twofold; Platonic thought as the beginning of philosophy's end.

An historical delusion.

Latin translation, a transformation in the happening of Being; reality as a network of agency; *natura*; transformations in the naming of self-presenting; *forma* and *materia*; *causa*, working as an acting that effects; Aristotle's four "causes" as defined via agency; philosophical thinking and Christian theology; the obtrusion of agency in the defining of a thing; causality, the fragmenting happening of what-is, a self-belying of Being.

The impulse toward the discovery of grounds, as arising out of Being's happening as *logos*; *logos* as presencing and ground, and as human speaking; disruption in its happening; *ratio:* the obtruding of grounding; *ratio* as an adjudging reckoning; *ratio* as ground and as apprehending reason; *logos* as a relating-back (*relatio*); the increased dominance of human apprehending over Being as now shut from view in favor of what-is; the changed character of setting-in-place; *ratio* as accordant with Being's hidden happening as effecting causality.

Section 3   423
The modern bringing to light of the relating of thinking and what-is that is implicit in *ratio*; a fresh reliance on logic as disclosive of Being; Leibniz's making manifest of the manner of happening pertaining to *ratio*; his "Principle of Ground"; the rendering of the sufficient ground; a new decisiveness of causality; the correspondence of the real with that which logic discovers; the requiring of causality by the principle of ground; the "rendering" of the ground as a delivering-back of the object by the representing subject; the rendering of the sufficient ground, as establishing both knowledge and that which is known; efficient causality as the most thorough grounding possible; the knowing subject's securing of itself; the power of statement to confer or withhold Being; representing thinking as solely adequate to the determination of Being; verifiability via causality, the ground (*ratio*) presupposed in representing thinking; Leibniz's principle as manifesting the new manner of Being's happening.

Section 4   433
The presupposing of the validity of Leibniz's principle of ground by subsequent philosophical thinking; the grounding power of the principle in the spheres of science and technology; the happening of Being, now, as *ratio reddenda*; the transpiring in Leibniz's principle of the enframing summons (*das Ge-stell*) governing modern technology; the discrepancy between this *Ge-stell* and its Greek antecedent; deviant straying vis a vis the withdrawnness of Being; the consummating of Being's ruling as concealing; the age of the world picture; the adversarial encountering of man and what-is; the radical self-withholding of Being, as writ large in the happening of what-is; the evidencing of depleted happening in the meaning of working; energy, successor to *energeia*; the name "atomic age" as evincing the dominion of the principle of ground; the demand for grounding and groundedness as evidence of the self-withdrawal of Being; the extremity of that withdrawal—Being as value; prior puttings forward of adjudging, in response to the self-disguising of Being; Being as now all but forgotten; Being as imperiously ruling vis a vis man in his self-assertion.

Section 5   456
Heidegger's historical portrayals, technology included, as evincing the manner of happening of human existing as such; parallels between man's falling into

inauthenticity and human comportment in successive epochs; a disclosing of the link between the transpiring of technology and the structures belonging to human existing; a consistent portrayal of the happening of Being.

The inaccessibility to us of the wellsprings of the character of our age; the impossibility of humanly initiated change.

VOLUME II

Section 1   542

The self-presenting openness needful for man as a participant in the surmounting of
technology; the needed thinking; the needed openness as constituted from out of the
happening of Being; *Gelassenheit* (releasèdness) as the comportment of the needed
thinking; the antithesis of representing thinking; "waiting" thinking; Being as *die
Gegend* (the region)—connotations of openness and of letting remain; the relation of
waiting thinking to ruling openness; releasèdness as bestowed out of a prior
belonging; direct belonging, as wrought upon by concealing; a provisional
thinking—having and not having; the constitutive structure of human *Dasein*
manifested; accordant, waiting thinking as preparatory.

Section 2   552

Thinking, an answering to Being; complex fundamental thinking (*Gedanc*); *Gedanc*
as *Gedenken*, a responsive, gathering, mindful thinking; the latter's manifesting as
Gemüt, "disposition," a fervent mindfulness accordant with the Being of what-is;
Gedächtnis, memory, via which disposition comes to manifestation in specificity; its
gathering together of that which presents itself by way of time; primary mindful
thinking as obedience; the provisional character of the Twofold as the thought that is
the to-be-thought; primary mindful thinking as thanks; as a gift; pure Happening as
thinking's most critical concern (*das Bedenklichste*); the jeopardy of thinking; pure
thanks.

Section 3   565

Thinking, as primally understood, as always in play; its immediate transpiring as
genuinely waiting thinking; such genuine thinking as preparatory; its building of a
way, its own openness; reflecting (*Besinnung*); thinking's questioning (*Fragen*); the
dialogical character of genuine thinking.

Section 4   573

The relation of primary mindful thinking to language, in man's belonging to
Being—a transpiring in the domain of language; man as witnessing to the Twofold;
thinking's primacy, that via it language first comes into play; its openness as a
language-imbued abode for the happening of Being as the Being of what-is; its
fundamental role as a preparing—a building that bespeaks dwelling; its own
transpiring as a dwelling; its transpiring in our time; its constituting as an opening
"way"; the present need for thinking in releasèdness to bring to bear the
preparedness intrinsic to primary thinking.

of art; the fundamental unity, in distinction and interrelation, of thinking and art under the rule of *poiēsis*.

The difficulty of envisioning technology as transposed to another role; the poetic work of Johann Peter Hebel: friend to the "house of the world"—a "house" constituted via dwelling; the Fourfold therein; Hebel's concern with the structured world and with the world structure (*Weltgebäude*); his intention and comportment toward his reader; his circumspect, restrained speaking, as participant guest in the house of the world; Hebel's desire to make possible questioning concerning That— Being—which is announcing itself in the processes and conditions of "nature"; his acceptance of the scientific way of making nature known; his retrieving of scientific nature into naturalness; the declaring of the mysteriousness of the world by the on-going happening of nature; Hebel's thinking toward a more original dwelling; Hebel's work as instancing the role now possible for art.

The intensification now and the evidencing of the ruling of Enframing via calculability; the friend to the house of the world who is now wanting to us—his work; an implied locus for the bringing to pass of the surmounting turning within the happening of Being; how technology might happen if transformed; a positive word for technology; a glimpse of the transformed role possible for the technological here and now: "releasèdness toward things"; its implication for the future; how the work of the missing "housefriend" might take place in that future; "Only a god can save us," a declaration pointing to that turning; the radicality of the need for an inbreaking of saving happening; hints concerning possible loci for it; the other possibility: an extremity of technology's entrenchment; but one comportment as appropriate, self-forgetful preparing that says the present happening of Being and leaves technology free play.

Questions of interpretation and of assumed authority; matters of fact; obscureness.

Trenchant insights into the reality we know, the wisdom of immediate engagement from out of a presupposing of unitedness and interrelation; compelling focusings of attention; fruits of an inclusive perspective.

SYMBOLS

WORKS IN GERMAN:

AED   *Aus der Erfahrung des Denkens*, Pfullingen, Neske, 1954.

EHD   *Erläuterung zu Hölderlins Dichtung*, Frankfurt, Klostermann, 1951.

EM   *Einführung in die Metaphysik*, Tübingen, Niemeyer, 1953.

EN   *Der Europäische Nihilismus*, Pfullingen, Neske, 1967.

F   *Der Feldweg*, Frankfurt, Klostermann, 1953.

FD   *Die Frage nach dem Ding*, Tübingen, Niemeyer, 1961.

HHF   *Hebel der Hausfreund*, Neske, Pfullingen, 1958.

HW   *Holzwege*, Frankfurt, Klostermann, 1952.

ID   *Identität und Differenz*, in *Identity and Difference*, New York, Harper and Row, 1969. (Contains the English followed by the German.)

N I, II (German)   *Nietzsche* (Vols I and II), Pfullingen, Neske, 1961.

PL   *Platons Lehre von der Wahrheit mit eimem Brief über den "Humanismus,"* Pfullingen, Neske, 1954.

SD   *Zur Sache des Denkens*, Tübingen, Niemeyer, 1951.

SG   *Der Satz vom Grund*, Pfullingen, Neske, 1965.

SZ   *Sein und Zeit*, Tübingen, Niemeyer, 1953.

TK   *Die Technik und die Kehre*, Pfullingen, Neske, 1962.

US   *Unterwegs zur Sprache*, Pfullingen, Neske, 1959.

UK   *Der Ursprung des Kunstwerkes*, Stuttgart, Reclam, 1970.

VA   *Vorträge und Aufsätze*, Pfullingen, Neske, 1954.

VS   *Vier Seminare*, Frankfurt, Klostermann, 1977.

W   *Wegmarken*, Frankfurt, Klosterman, 1963.

WG   *Vom Wesen des Grundes*, in *On the Essence of Reasons*, trans.
          Terrence Malick, Evanston, Northwestern University Press,
          1969.   (Contains the German and the English on adjoining
          pages.)

WHD   *Was Heisst Denken?* Tübingen, Niemeyer, 1951.

ZS   *Zur Seinsfrage*, in *The Question of Being*, trans. William
          Klubach and Jean T. Wilde, New York, Twayne, 1958.
          (Contains the German and the English on adjoining pages.)

ENGLISH TRANSLATIONS:

BT   *Being and Time*, trans. John Macquarrie and Edward
          Robinson, New York, Harper and Row, 1962.

BW   *Basic Writings*, ed., David Farrell Krell, New York, Harper
          and Row, 1977.

DT   *Discourse on Thinking*; trans. John M. Anderson and E. Hans Freund, New York, Harper and Row, 1966.

EB   *Existence and Being*, ed., Werner Brock, trans. Douglas Scott and R. F. C. Hull, Chicago, Regnery, 1968.

EGT   *Early Greek Thinking*, trans. David Farrell Krell and Frank A. Capuzzi, New York, Harper and Row, 1975.

EP   *The End of Philosophy*, trans. Joan Stambaugh, New York, Harper and Row, 1973.

ER   *The Essence of Reasons*, trans. Terrence Malik, Evanston, Northwestern University Press, 1969.

HFH   "Hebel—Friend of the House," trans. Bruce F. Foltz and Michael Heim, in *Contemporary German Philosophy*, ed., Darrel E. Christensen et alia, University Park, Pennsylvania State University Press, 1983, Vol. 3, pp. 89-101.

ID   *Identity and Difference (Identität und Differenz)*, trans. Joan Stambaugh, New York, Doubleday, 1961. (The entire German text [pp. 23-76] follows the English [pp. 83-146].)

IM   *An Introduction to Metaphysics*, trans. Ralph Manheim, New York, Meridian, 1956.

K   "The Way Back into the Ground of Metaphysics," trans. by Walter Kaufmann of "Einleitung zu Was ist Metaphysik?" in Walter Kaufmann, ed., *Existentialism from Sartre to Dostoevsky*, New York, New American Library, 1975, pp. 265-279.

N I, II, III, IV (English) *Nietzsche*, (4 volumes), trans. Frank A. Capuzzi and David Farrell Krell, New York, Harper and Row.

P "The Pathway," trans. Thomas Sheehan, in Thomas Sheehan, ed., *Heidegger: The Man and the Thinker*, Chicago, Precedent Publishing, Inc., 1981.

Ph. "On the Being and Conception of *Physis* in Aristotle's Physics B, 1," trans. Thomas J. Sheehan, in *Man and World*, Martinus Nijhoff, The Hague, Volume 9, No. 3, August, 1976, pp. 219-270.

PLT *Poetry, Language, Thought*, trans. Albert Hofstadter, New York, Harper and Row, 1971.

QB *The Question of Being*, trans. William Klubach and Jean T. Wilde, New York, Twayne, 1958. (Contains both the German and the English on adjoining pages.)

QT *The Question Concerning Technology and Other Essays*, trans. William Lovitt, New York, Harper and Row, 1977.

TB *On Time and Being*, trans. Joan Stambaugh, New York, Harper and Row, 1972.

WL *On the Way to Language*, trans. Peter D. Hertz, New York, Harper and Row, 1971.

WT *What is a Thing?* trans. W. Barton and Vera Deutsch, Chicago, Regnery, 1962.

AUTHORS' PROLOGUE

The Way Before Us

AUTHORS' PROLOGUE

THE WAY BEFORE US

"Technology is a mode of revealing."[1] Thus Heidegger spoke in a lecture that inquired after the essence of modern technology. The statement is at once so simple and yet so outlandish that one might easily read it unthinkingly or, leaping to another extreme, might dismiss it as a flight of fancy undeserving of serious scrutiny. But still a third path invites us. These simple, unexpected words were in fact intended as a gateway opening into portentous and intricate ranges of thought. It is to those who would enter into the way there opened that the present study is addressed. It is our hope that by its means some clarity may be provided and some help given toward the acknowledging and understanding of what was for Heidegger the aspect of his thinking that most immediately brought to light the situation in which we of today in truth find ourselves in the midst of the reality that impinges upon us.

Technology is for us a major concern. Its processes and products are everywhere about us, and not one of us can fail to be aware to a significant degree of the extent to which its presence among us configures our ways of thinking and acting, determines the possibilities that lie open to us, and, in what has become one vast global context, structures often down to minute details the prevailing character of our lives. We cannot but welcome to a great extent the works of the technology of our time. What undreamed-of gifts and capabilities it has bestowed upon us! Yet nevertheless we fear it with a fear that sometimes breaks through our eager acceptance of its benefits and all but transfixes us with terror of the possibilities for self-destruction that technology has placed in our hands. How shall we combat the concentrations of depersonalizing power— economic, political, or ideational—that loom ever more threateningly around us? How shall we so marshall our resources as to save our environment, endangered now world-wide, from being catastrophically

---

[1] *Die Technik ist eine Weise des Entbergens.* "The Question Concerning Technology," VA 20, QT 12.

despoiled and polluted by the undertakings and effects of our technology? Or how shall we make our way back to sustainable safety from the brink of destruction to which we have come since the ultimate energy locked within the atom became ours to wield as a weapon against one another?

Yes, technology concerns us. Sometimes, indeed, it so looms as to fill the horizon of our thought. When this mood is upon us, then we are not far from the gateway to Heidegger's thinking. He too sees technology as the central, endangering phenomenon of our age. We tend to turn away from this insight. There is time. Solutions for problems will surely be found. Technology itself will open to us new possibilities. It is, after all, *our* technology. And we have many other sources of knowledge and understanding from out of which to orient ourselves and act. Although in serious ways technology daunts and threatens us, surely we shall, in the coming years, succeed in finding among ourselves intelligence and dedication sufficient to think through our difficulties and to master them.

Heidegger, for his part, is not so sanguine as we. He was deeply concerned to offer a portrayal of what he discerned as the true situation in which all of us find ourselves in this modern day, for to his mind there was profoundly lacking to us genuine knowledge of our age and of our place in it. In so doing, it was his wish to turn the attention of his fellows directly toward modern technology and to open to those who would thus look with him an understanding that would be disclosive for them of the true character of their time, of the import of their lives, and, beyond this, of the structures and significances pertaining to all their history, to what has been and to what might come to be.

Thinking must always begin where the thinker stands. It is for Heidegger always in an immediate encounter with that which encounters us that meaning is to be discerned. But when meaning is caught sight of, it is found to extend away into all ranges and aspects of a reality that, met thus in an immediate engagement of thinking, requires to be viewed according to long perspectives of time and according to subtly configured insights that consider it now from one vantage point, now from another. So it is with modern technology. The latter is a world-historical phenomenon, one indeed of preeminent import. For in it, the on-going Happening that must, for Heidegger, be the proper focus of any serious thinking, shows itself in a crucial way.

Everywhere technology, together with the science that is so often in one way or another allied with it, encounters us and determines our lives. But what are these encountering phenomena? Heidegger knew well our ordinary ways of thinking of such matters. He was exceedingly well informed regarding scientific work and technological achievement and was often doubtless far ahead of his hearers and readers in theoretical and practical knowledge in these fields. He was capable of characterizing science and technology in terms that the scientist or technologist could instantly recognize. This was for him a necessary starting point. But it was only that. Always he was moving to uncover fundamental meanings. Always he was inquiring after origins and implied relations. And always, as in those words "Technology is a mode of revealing," he was launching forth into a unique realm of discourse where both the ordinary listener and the trained specialist in any sphere—philosophical, scientific, or technological—must be hard put indeed to follow.

The study that we are here about to undertake is possessed of a peculiar character. Heidegger is fond of speaking of the pursuing of thinking as the pursuing of a "way," an identification that is intended to suggest a particularized forward movement in the direction prescribed by that which proffers itself to thinking as what demands to be thought. The image seems apt, if in a secondary sense, for our study as well. The way on which we shall embark is not, as Heidegger understands his own forays of thinking to be, originative, a path of primary disclosure. Nevertheless, it is, we believe, a vitally significant course. Certainly it is one that opens out before us, not as a pathway that we could ourselves hope to foresee and prescribe, but as one dictated by a unique subject matter that can be followed after and brought into view only through a thinking that is repeatedly under the necessity, as is the thinking that Heidegger essays and describes, of coming upon and taking up into itself some new insight, if it is properly to continue its course.

Needless to say, any significant thinking through of another's thought has something of this character. Often in such pursuings of thinking, we are beset by the danger that real or supposed similarities of outlook and terminology between our own thought and the thought that we intend to understand may blind us to the sort of thinking that is actually being demanded of us. In the case of Heidegger it is otherwise. His outlook is exceptional. His ways of speaking are strange. His work—in particular his understanding of modern technology with all that pertains to it—beckons us toward a very demanding course of thinking indeed. And yet this is a path that surely it behooves us to follow out. For it is a way that will lead through extended ranges wherein the most able thinker of our century strove to inquire into and bring to adequate utterance the reality in accordance with whose happening, to his mind, we *are*, all our ways are governed, and all that meets us is to be understood.

A formidable task lies before us. The work of Martin Heidegger is in many respects so difficult of access that the way to be followed here through his thinking will be long and intricate. Only a few salient indications of the course to be pursued as we seek to think our way in to Heidegger's thinking and of the demands that this undertaking lays upon writer and reader alike can here be given. But it is hoped that these will suffice significantly to mark out in advance the path to be taken, to suggest some of its turnings and the pitfalls that here and there lie along it, and hence to prepare it as an avenue of inquiry that will be both instructive and rewarding to any who undertake to pursue it.

The presentations that Heidegger offers to us are various. Sometimes he follows out particular themes with rigor, often to the disclosing of enormous intricacy within them. Sometimes, on the other hand, he does no more than let us catch glimpses of those themes, demanding thereby that we think for ourselves and bring to bear what we have learned elsewhere from his thought. His desire is, indeed, to engage his listener—his reader—directly in thinking, so that the listener may not merely hear about that which is under discussion but may himself take it up as the burden of his own thinking.

In a sense the present study is an unheideggerian enterprise. It involves a gathering together of Heidegger's words and thoughts on various themes and an explicating of them in a way that often presents a rounded portrayal where he himself gave only deft indications or proffered disparate depictions set forth now from one perspective, now from another. Yet in another sense this study is, we believe, not untrue to the spirit of Heidegger's work. For it arises out of long and arduous

thinkings-through of his words; and here a consistent effort has been
made not first of all to *speak about* what Heidegger says but to present
it—made surveyable, clarified, and amplified, it is true—in a direct
statement that will let it recommend itself to the reader in a manner not
unlike the immediacy belonging to his own speaking. In pursuance of this
undertaking, we have by no means sought to become conversant with the
whole corpus of Heidegger's writings. Rather, we have endeavored to
understand well such materials as, to our minds, bear most closely upon
our investigation and, drawing upon those materials in various ways, to
gather them together via an explication that can allow the reader access,
by way of peculiarly fruitful avenues of inquiry, to Heidegger's thought.

Here the reader will be kept continually in touch with Heidegger's
words themselves. Direct quotations will be frequent. In every instance
our primary concern has been to disclose as fully as possible the meaning
resident in Heidegger's German phrasing. To this end, all passages
drawn upon for this study have been considered in the German original,
and many have been freshly translated, sometimes in ways that depart
radically from existing translations. We have attempted, as far as
possible, to maintain consistency in our renderings of given words, so that
even where no reference is made to the original text, a careful reader will
be able to discern, with some specificity, that which lies behind what is
being said. Words of crucial import are often rendered not by single
words but by phrases—sometimes by slightly varied phrases, indeed, in
diverse contexts—in order that something of the wealth of connotation
belonging to them may be kept constantly in play in our discussion and
constantly before the reader's mind. In such cases the underlying German
words themselves are frequently sighted in the text. In addition, the
German texts on which expository passages are based are often given in
accompanying notes.

As further aids, we have placed at the beginning of the work a
topical summary of chapters, detailed section by section, and have given
in our notes both supportive citations and comments and numerous cross-
references to relevant material elsewhere in the study. Again, we have
included an index that, far from focusing on mere word occurrences, is
also primarily topical. Although it is by no means exhaustive, that index
can provide extensive access to the content of the book. It not only has

numerous words and occasional phrases as its primary entries and refers to passages in which their meaning is shown or in which they play a salient role in the discussion in progress, but it also contains among such entries a number of descriptive rubrics for which instancing references are likewise given. Every effort has been made throughout so to phrase the many sub-categories belonging to all entries as to ensure that they will instruct clearly regarding the points to which reference is being made. Finally, we have included a glossary of German words, when necessary giving the phrases used to display nuances in their meanings and, where a passage definitive of the meaning of the word exists, including a reference to that passage. We have avoided almost entirely the use of German words themselves—*Dasein*, for instance—to carry the weight of our discussion. For we ourselves are all too keenly aware how easily such words, even when their meanings have been discovered and are supposedly kept in mind, can be allowed to suffer erosion, imperceptibly losing the import intended for them and becoming mere opaque counters that impede instead of furthering understanding.

Although we have had the temerity to survey Heidegger's work and to draw together statements of his thinking that frequently stand in contexts widely disparate in the time of their composition and in the precise burden of their concern, we remain keenly alive to the fact that, before it can be lifted into juxtaposition with other Heideggerian presentations and viewed together with them, each particular formulation of thought must be understood from out of the specific extended discussion within which it inheres. Only when it is thus seen in its given context can the nuances of meaning bearing upon it and in play within it be properly discerned. With this in mind we have, in our notes, indicated not only the published volume but the specific work in which the reference in question is to be found. It is our belief that the common practice of citing in notes only entire volumes rather than individual writings serves seriously to impede, however unintentionally, proper access to Heidegger's thinking. Much of that thinking is wrought out in short pieces, essays of thought. No more than we would think of referring to the writings of Plato simply according to the titles of the volumes in which they appear should we allude to the writings of Martin Heidegger in this beclouding, uninstructive way.

The work that is here being undertaken is a quest of extensive scope. We shall be endeavoring to elucidate Heidegger's thinking with respect to a major theme that ramifies into all aspects of his work. It is our desire to let Heidegger speak for himself. Nevertheless, we must, ourselves, in such a study continually speak for him. Our perspective on his work and our insights into its meaning have been gained through long, pains-taking study, in pursuing which we have steadfastly sought to discover meaning, not to import it, and to let interconnections disclose themselves, not to impose them from out of our own preconceptions. We trust that in this we have succeeded to significant measure, but we also know well that that success cannot but be flawed. One's own preconceptions, some of which, indeed, may be unrecognized, are certainly not easy to lay aside. Beyond this, unquestionably it is out of perspectives on Heidegger's thinking already won that we have moved forward to consider new aspects of what he is saying; insights gained have become touchstones for insights sought; and meanings once ferreted out and accepted have become the basis for the discovering of other meanings in their turn. Such intrusions into what is intended to be unbiased interpretation can scarcely be wholly avoided and must not be gainsaid. Nevertheless, we have so often been forced by the material with which we have come to grips to shift our assumed bases of interpretation, to rethink, and to reformulate, that we cannot but believe that we have to a great extent discovered Heidegger's intended meaning and that in the succeeding pages we will very largely be bringing that meaning to light.

At points our discussion will, especially in the adducing of detail, range far beyond Heidegger's own. When this happens, an absence of annotation will alert the reader to the fact. In a few instances we attempt, if not to think the unthought in Heidegger's thinking, as he himself makes bold to do with other thinkers,[2] at least to say the unsaid. Always our discussion outruns his own, in that, where he contents himself with inquiry upon inquiry, we are attempting an overview that can allow often unconsidered interrelations of meaning to stand forth and far-flung patterns of thought to appear.

---

[2]WHD 71f; WCT 76f.

This is surely a sort of undertaking that would have made Heidegger uneasy. He would vehemently have resisted any attempt to "systematize" his thinking. And certainly that thinking cannot be presented as a logically coherent system. Heidegger could, when he chose, reason rigorously. But his formulations of his thinking consistently lie outside the accepted rational structures of thought. Reality was for him far more ample than anything those structures could encompass. And he insisted on according to it a spontaneity of self-presentation that must be matched by a spontaneity of response on the part of such critically significant thinking as he undertook to pursue. Nevertheless, consistency of perspective and congruence of descriptive formulation are to be found pervasively throughout his work. It is not too much to say, indeed, that that work is of a "systematic" character, in the sense that a single understanding of the determinative structures intrinsic to reality is to be found there. Often that unitariness of approach is concealed by the diverseness of theme and presentation that is constantly met with. In pursuing our study we have, however, found it to be abundantly evident. Through transcending and gathering together Heidegger's varied presentations we shall in fact, as one goal of our study, be endeavoring in these pages to make evident and to explicate this underlying understanding and the coherencies that attest it and make it manifest.

We have found the consistency in outlook and interpretation now in question to extend throughout Heidegger's work. We shall therefore draw in our study upon a wide range of his writings, from *Being and Time* to the late essays. We by no means share the view, long prevalent among scholars, that nearly all Heidegger's writings subsequent to *Being and Time* evince a sharp shift in his way of thinking. The whole weight of the evidence yielded by our study stands counter to such a supposition, and, indeed, Heidegger himself speaks tellingly against it. Thus, in commenting on "On the Essence of Truth,"—which was delivered as a lecture in 1930, three years subsequent to the publication of *Being and Time*—in an addendum published with the lecture in its second edition in 1949, Heidegger wrote:

> Every sort of anthropology and all subjectivity of man as
> subject is not only left behind, as it was already in *Being and Time*,
> and the truth of Being sought out as the ground of a changed historical
> fundamental-position, but the movement of the lecture sets out to think
> from out of this other ground (*Dasein*).[3]

Clearly, to his mind, the assuming of that new "ground," which assuming
he could in the same context call a "turning in the history of Being" began
with *Being and Time* and thereafter wrought itself out in his subsequent
work.[4]

Unquestionably, the perspective governing Heidegger's work did
change. In *Being and Time* his central concern was with the mode of
existing peculiar to man. Thereafter, his primary interest lay in depicting
the happening of Being. It can seem as though he forsook a
preoccupation with the defining elements and structures of human
existence and, turning away from man, directed his attention toward
various phenomena with a view to discerning their *Being*. In truth,
however, Heidegger's first, clearly stated concern in *Being and Time* was
with Being, and he averred that his inquiry regarding human existing
would open out toward a discovering of "the meaning of Being
generally."[5] His analysis of human existing was a portraying of it as the
openness via which Being brought itself to pass. More than this, his
centering of attention upon man meant equally for him a having-in-view
of the whole range of phenomena beyond man, albeit not as considered as
das Seiende (what-is), the companion happening happened by Being as
partner to itself that later received his attention. For man's characterizing
manner of existing was, precisely as such, Being-in-the-world. In setting
aside his work on *Being and Time*, Heidegger turned directly toward this
other pole of interest and concerned himself primarily with the Being of
what-is (*das Sein des Seienden*). In so doing, however, he did not lose

---

[3]"On the Essence of Truth," W 97, BW 141. For a statement by Heidegger to the
effect that, although in the course of its presentation over time, his thinking evinced
transformation, it did so out of an insight present from its inception in *Being and Time*,
see his letter to Father Richardson, which figures as the Preface to William
Richardson's *Heidegger: Through Phenomenology to Thought*, Martin Nijhoff, The
Hague, 1967, pp. xviff.
[4]W 97; BW 141.
[5]. . . *der Sinn von Sein überhaupt.* SZ 37; BT 61f. See also SZ 1; BT 1.

sight of man. The latter's role remained what it had originally been, that
of living out openness-for-Being. Heidegger did cease to scrutinize
human existing, in what we might call the detailedness of its taking-place,
as the primary locus where the happening of Being was to be discerned.
And, more importantly, he did not, as he had anticipated, portray Being
in terms of his analysis of that existing. But the insights he had gained in
that scrutiny continued to inform his thinking. In his numerous portrayals
of the happening of Being in the varied spheres of reality, as the Being of
what-is, it is, as we shall have ample occasion to see, the same
fundamental structuring—a structuring of happening by way of time—that
he consistently brings to view; and the specific manner of transpiring of
human existing—with its Being-configured constitutive aspects—wherein
that structuring is first evidenced and traced out can repeatedly be seen to
be in play.[6]

   In accordance with this, our understanding of Heidegger's work,
we shall not, as do so many interpreters, center our attention on *Being
and Time* as peculiarly or almost solely embodying that work as relevant
for interpretive thought. Rather, we shall consider such discussions in
Being and Time as bear directly on the matters with which our discussion
has variously to do. For it is our intention to bring what are for us salient
portions of its abundant   material into a presentation that must
preponderantly concern itself with later, often with much later, writings
belonging, like *Being and Time*, to a numerous corpus that demands
consideration *in extenso*.

   It is sometimes thought that, to a great extent, after *Being and Time*
for Heidegger incisive analysis gave way to quasi-mystical
pronouncement. At that juncture Heidegger did indeed move out beyond
his early formulations.   His direct concern with Being, which soon
became a concern with Being as presencing, opened to him vistas in
which complexity, depth, hiddenness, and innate mystery, elements by no
means absent hitherto, met him now in fresh ways. These he undertook
to set forth. During all the years of that later work, however, he never
repudiated the fundamental formulations that he had given in *Being and
Time*.[7] On occasion, indeed, he refers back to things there said. To his

---

[6]For support of this thesis in brief see Ch. I, p. 33, n. 36.

[7]Even a retraction of a specific point made in *Being and Time* is rare, but for such an

mind, clearly, his work as a whole was possessed of a unity, although—as would have seemed wholly fitting to him—of one wrought out by way of fresh discovery and an ever on-going pursuance of inquiring thinking.

It is this underlying unity of interpretation with respect to whatever phenomena may sustain the scrutiny of Heidegger's thinking that our discussion will presuppose. Forward movements in thought and insight and changed modes of characterization will at times be noted, but a decisive consistency of outlook will be assumed. The validity of this assumption will, we trust, be amply borne out by the ways in which, in instance after instance, the materials chosen for discussion will disclose the same patterns of thought.

In keeping with Heidegger's central concern with modern technology, specific allusions to technology are to be found widely scattered throughout the corpus of his works, particularly in later materials. Many are highly important. But only in two essays, "The Question Concerning Technology" and "The Turning," does technology become the main subject of discourse. These two, which were originally lectures delivered together in a series,[8] will necessarily form the core of our study, for other allusions can often best be properly understood only in their light.

Beyond overt discussion of modern technology, there lie for our purposes many areas of thought that must be explored. Heidegger's thinking on technology bears upon many other questions. At the same time, clearly, the fundamental tenets of his thinking are decisively in play in his understanding of technology. Accordingly, our inquiry will itself arise out of a consideration of primary understandings that show the governing orientation of Heidegger's thinking and will likewise open out into numerous avenues of investigation germane to our central theme.

Throughout our study our inquiring attention will be directed solely toward Heidegger's own formulations of his thought. No cognizance will be taken of interpretations other than our own. Having confidence in the rightness of many of our own insights and in the understanding of Heidegger's work out of which we speak, we have made bold not to call

---

instance see "Time and Being," SD 24, TB 23.

[8]The third lecture in the series of four, "Die Gefahr," which presumably also dealt extensively with technology, remains unpublished.

upon any other voice, for the offering of diverse interpretations might well entice the reader to step away from direct involvement in thinking steadily forward through what Heidegger is saying and might lead him or her simply to *think about* the particular point that was at issue. Again, no attempt will here be made to look beyond Heidegger's thinking itself and to ask after the origins of his ideas and modes of thought. Consideration of those origins could certainly open out into very fruitful and significant areas of study. But here it would be wholly out of place. It too might easily deflect attention from immediate encounter with the thinking that we are endeavoring to comprehend. And it might, in addition, lead, as such inquiries so often do, to a derogating of that thinking as merely derivative from one source or another. Heidegger is unquestionably heir to many traditions, both philosophical and theological, but his thinking is his own. In it contributing elements have been taken up and fused into aspects of a complex but consistent orientation of mind. Only by meeting and thinking together with Heidegger exclusively on his own terms can we hope to grasp at all what he is saying at any point in his work.

We are aware of having embarked on an almost presumptuous task. We would speak for Heidegger, undertaking to maintain a Heideggerian perspective and to explicate from out of it a wealth of complex thought. We would be the last to suggest that our presentation should in any sense supersede Heidegger's own work. That presentation, however adequately it may achieve its intended ends, cannot possibly be more than an imperfect, inadequate mirroring of Heidegger's wide-ranging, vastly complex thought, provided by persons who remain, avowedly, still but on the way toward a thorough understanding of the latter. Anyone who wishes truly to understand the thinking of Martin Heidegger can hope to do so only by coming to grips, out of a radical commitment to open-mindedness and thoroughness, with Heidegger's writings themselves, preferably, certainly, in the original German. Nevertheless, we believe that the present study can have a very important role to play. In it we shall endeavor to clarify Heidegger's thinking, opening many of its intricacies and inner structures to view, unfolding meanings and displaying often hidden connections of thought. Such work needs to be done. For the presentations of thought that Heidegger offers to us are often richly nuanced, filled with statements that presuppose disclosures made in far other contexts, and even shadowed by outright obfuscation on his part.

The fact that we are pursuing our work in English is, when seen from one perspective, a severe hindrance. By speaking always in translation we necessarily remain invincibly distant from the immediacy of Heidegger's own speaking. In this regard, indeed, Heidegger himself spoke words that should be taken with exceeding seriousness. "One can translate thinking," he said, "no more satisfactorily than one can translate poetry. At best one can circumscribe it. As soon as one makes a literal translation, everything is changed."[9] The changedness in question and the remoteness from the original utterance of thought that it unavoidably introduces should never be lost sight of by either translator or reader.

Viewed in another respect, however, this distancing can be seen to lend an advantage to our work. German is a language whose words abound in complex connotation. Heidegger avails himself masterfully of this character of his native language. To him words are well-springs of meaning, and clearly it is his settled intent to have in play every nuance in the words that carry the burden of his thought. Continually he uses families of words—e.g., *Ereignis, vereignen, eigentlich, eigens,* etc.[10]— intending that the various members of a group should resonate with one another. Deftly he places his key words, in context after context. For he continually shifts those words in their interrelation, thereby bringing to the fore one nuance or another. Sometimes he brings to prominence a word now archaic or all but forgotten in modern German. Or he may fashion for himself a word where none already exists. For the most part, however, the words that he uses with such skill are familiar ones. Frequently their fullness of connotation has been eroded by the subtly abrasive action of habitual use, but unquestionably he himself always takes their full force to be in play. Occasionally he hyphenates words of crucial import, thus clearly showing his intention. But for the most part he simply requires of those who seek to follow him in his thinking that

---

[9]"'Only A God Can Save Us,' *Der Spiegel*'s Interview with Martin Heidegger," *Der Spiegel*, Hamburg, Spiegel-Verlag, 31 May, 1976, issue No. 23, p. 217; *Philosophy Today*, Carthagena Station, Celina, Ohio, Winter, 1976, Vol. 20, No. 4, p. 282.
[10]For a discussion of this group of words, see Ch. IV, p. 149, n. 28.

they be alert to richness of meaning and to his own dedication to the bringing of it constantly to bear. Since the words that Heidegger uses are so dense with connotation, if one were to work solely in German it would, almost certainly, often be difficult, in seeking to explicate what he has written, to gain a vantage ground from which to see penetratingly into his language. Two problems exist. First, familiarity could veil, making it hard to sense in words any but the meanings that they ordinarily brought directly to mind. Beyond this, the very wealth inherent in the language could make it difficult to open out meanings and expose varied nuances to view through the use of words relatively simple in their connotation. English, for its part, is a language far wider in vocabulary than is German. Where German is rich in words fraught with multiple meanings, English frequently abounds in multiple words each of which carries distinctive import. Accordingly, although we must, in working in English, always remain, regrettably, significantly far from the fullness and immediate force of what Heidegger is saying, we can, at the same time, often analyze, name and work with crucial aspects of those meanings in such a way as to allow them to stand forth and inform our thinking as they could scarcely do if met with only in the original German words to which they intrinsically belong. This being true, we dare to hope, indeed, that our study may prove of value not only to the English-speaking student of Heidegger but to the German-speaking student as well, who may perhaps find in it the possibility of stepping out of the ambience of his own native tongue and of thereby gaining a purchase on Heidegger's language that might well elude him if he were confined to confronting the latter only as thoroughly familiar words with whose connotations he was already so conversant as to find it profoundly difficult to hear in them unwonted depths of meaning and to separate out, in so hearing, intricate implications and shades of significance.

Whatever the difficulties, we have consistently striven, in translation and exposition alike, after means of expression that would truly approximate the connotative richness and the wealth of nuanced juxtaposition that belong to Heidegger's own speaking. This endeavor eventuates frequently in very difficult English prose. Sometimes unorthodox phrases—"as which," for example, or "belongs into"—directly reflect Heidegger's own departures from usual German syntax.

In the vast majority of instances, however, such phrases have arisen out of our own need to bring to English utterance facets of meaning that often all but defy adequate presentation. Our practice of rendering crucially important words with phrases—indeed sometimes with varied phrases coupled together—repeatedly yields a text that makes stern demands on the reader's attention. We are well aware that, whereas a single trenchant word can speak immediately to the listening mind and permit thought to pass swiftly on, unfamiliar, even lengthy phrases characteristically impede apprehension, forcing the mind to pause and to search for what is being said. We can only ask and urge the reader to remember that solely via just such impeded thinking that must hesitate and concentrate itself to grapple time and time again with the unfamiliar can access be had to Heidegger's own utterance, burdened with implication and nuance as it is.

One salient characteristic of Heidegger's thinking is primarily responsible for the sort of difficulty of which we have just been speaking. This is the fact that to him that which presents itself to be thought upon does not do so as a congeries of discrete, inherently static items or elements among which thinking can range, surveying, comparing, and interconnecting at will. This is the course that we habitually follow in our conceptualizing thinking, and our words and syntactical structures ordinarily serve quite adequately for its pursuit, having indeed developed in just such service. Heidegger's reality, on the contrary, presents itself for thinking as an on-going, event-filled, shifting totality, far-flung yet single, whose most fundamental character is that of pure Happening.[11] Thought and language are for him therefore called upon to grapple with that which, if it is to be glimpsed and portrayed in and as itself, cannot properly be conceptually set forth and set fast, surveyed, dissected, manipulated, and, as it were, understood at leisure in terms of such connections and likenesses as would render it accessible to ordinary inquiring thought.

Our thinking moves characteristically in a simple linear way. For us, habitually, time, in which we live and think, moves uncomplicatedly forward, and central to our way of understanding is causation, which presupposes a forward movement from something to something else that

---

[11]Cf. Ch. I, pp. 23f.

follows from it. When we consider interrelationships, our thought moves linearly from one component of a relation to another. If the relationship is reciprocal, a like linear movement back toward the first component is undertaken. Heidegger's thinking does not move in this way; for, in accordance with his primary outlook, it must so move as to bring to light elements that are, while differentiated, intrinsically one and that must, if they are to be properly seen, be disclosed in a single juncture of time. For time is ever converging into the open present in which the given disclosure is being made. Again and again Heidegger requires of us that when we think of *this*, we must be thinking simultaneously also of *that*. The verbs that for him speak properly of relating are verbs such as gathering, disposing, or belonging-together. He does not speak of the causing of something by something else. Instead, he describes variously a letting-happen that bespeaks a mutuality of interplay among the participants in question.

It is, fundamentally, the necessity of displaying this unitiveness in multiplicity, this mutuality that must be laid hold of in simultaneity, that has forced us to present in the following study an outworking of Heideggerian thinking whose phrasing is often unusually complex and demanding. Indeed, the very sentences in which our discussion will be couched will, in their complexness and in their insistent gatherings-together of relevant aspects of themes into single, inclusive utterances, often tacitly evince this character of the thought with which we shall have here to do. Juxtaposed elements of that which is under discussion and juxtaposed nuances pertaining to those elements must constantly be presented together, and often the capabilities of language, and even of sentence structure, must be strained to achieve such presentation and to keep it in play in varied contexts.

Of crucial importance in this connection is the showing of the directionality that always obtains among the aspects and elements in question. Words such as "unto," "forth," and "hither" will be encountered with considerable frequency in these pages. Admittedly such words sound strangely in contemporary English utterance (a fact that doubtless evidences an impoverishment in our language). But their counterparts—consisting of prepositions, adverbs, and verbal prefixes—play a very significant role in the German idiom. Heidegger relies heavily upon these, often hyphenating words to disclose that reliance, and our following out of his thinking by way of formulations in English requires that dynamic prepositional and adverbial expressions of this sort be frequently in use.

That word "dynamic," itself but rarely used in our actual discussions of Heidegger's thinking, is perhaps the one that the reader should endeavor to keep most centrally in mind in seeking to grasp the character of the Heideggerian portrayals whose elucidation we are about to undertake. Heidegger's depictions—however intricate they may be— are continually astir with vitality. Our own are often lengthy and complex, but we have striven to keep that vitality decisively in play. Like Heidegger himself, we must often rely on a given context or, again, on the whole context of his thinking to lend the peculiar power of connotation that he requires to words that, if heard alone, might be quite other in import. While we shall continually seek to make needed meanings explicit and to keep fundamental configurations in view, we are keenly aware that only a steady purpose on the part of the reader to notice, remember, and expect movements and juxtapositions and gatherings that are untoward and ever underway can suffice to crown our efforts finally with success.

▧   ▧   ▧

We are about to enter upon a far-ranging course of thinking, in order that we may come to know what Martin Heidegger has to say concerning the salient phenomenon of our time that is modern machine technology. Radically unfamiliar thought couched in unfamiliar language that often strains to contain it—this is what Heidegger's writing offers us at every turn. He himself was keenly aware of the grave difficulties that must confront his readers. Yet he was in no way willing to swerve from his appointed path. Words that he wrote regarding one of his essays, "The Origin of the Work of Art," would surely apply to his writing as a whole:

The crisis remains unavoidable that at first, and even for a long time after, the reader, who naturally comes to the treatise from the outside, does not conceive and interpret matters from out of the secret realm wherein lie the wellsprings of what is to be thought. But for the author himself there remains the crisis that at each stage of the way he must speak, each time, precisely in language that is propitious and suitable.[12]

Thus stringent for Heidegger was the distressing necessity out of which he felt himself called upon to speak and his hearers to listen.

In order that our approach to our central theme, Heidegger's treatment of modern technology, may be made in a manner commensurate with the gravity and difficulty of the task before us, we shall not begin our study with a consideration of technology itself, as though the meanings that Heidegger would discover to us in that context could be found open at first sight to our inquiring thinking. Within and behind Heidegger's thinking on technology cluster the fundamental tenets and ruling orientation of mind out of which he thinks. These we must explicate and take to ourselves as elements in the indispensable perspective in accordance with which alone technology can show itself to us as it shows itself to him, namely, as the sphere by way of which modern man can hope most decisively to know himself and to experience the character of his age. Only when that perspective has been gained will we turn directly to technology, letting its character and scope, its beginnings, the possibilities gathered within it, and the claim that lies upon us in their light become the exclusive focus of our thought.

In all this, in whatever context, we shall be engaging first of all to think with Heidegger after the happening of Being, since that was the concern that met him *before* and *in* all else. Therein must lie our point of departure. And therein will always rest also, in the last analysis, our complex study's entire burden and course.

---

[12]"The Origin of the Work of Art," UK 100f, PLT 87.

PART ONE

Depictions of Being

CHAPTER I

THE BEING OF WHAT-IS

From the beginning the question of Being was Heidegger's fundamental concern.[1] All his thinking turns upon it. Therefore any attempt to understand that thinking must begin with a consideration of Being.

The phrase "the question of Being" may easily seem to us a formulation that has to do with nothing but an abstruse philosophical concept. Quite to the contrary, however, the question of Being is in fact one that, in the ordinary course of events, continually concerns us also. Constantly it matters vitally to us whether a thing or a state of affairs is or is not. All through our lives, in all manner of contexts, we care deeply whether persons or communities or all sorts of animate or inanimate objects do or do not have being. The word "being" serves in one of its uses as a designation for ourselves, as human beings, and can be used of other sorts of realities as well. But most characteristically, "Being," which is named and affirmed with forms of the verb to be, is spoken of and thought by us as a property belonging or not belonging to something or as a status or condition possessed by it. Accordingly, being is a "something"—a fundamental and valuable something—that is found among all the other somethings that we know, about which we think, and of which we speak.

This, our already pervasive concern with being, can provide us with a starting-point from which to begin to enter into Heidegger's thinking. But it can be literally only a point of departure. For Heidegger's understanding of the import of Being lifts the latter away from all else, seeing in it, radically, a primacy and uniqueness that is quite unfamiliar to us. We must follow him rigorously in this focusing of attention upon Being if we would discover the burden of his thought.

---

[1]SZ 1; BT 1. Cf. WG 96, ER 97, n. 59.

Only when we place ourselves at the center where this word is spoken can we hope to look perceptively into the various aspects of Heidegger's thinking, see them in perspective, and apprehend the relationships, often subtle and complex, that subsist among them.

1

To speak the word "Being" in isolation, as we have just been doing, is dangerous, for it could be misleading. Being is, for Heidegger, always and everywhere the Being of what-is, *das Sein des Seienden.*[2] Here "Being" should never be viewed as an abstraction that merely belongs to the realm of philosophical thinking. Being has indeed been

---

[2]"Postscript to 'What is Metaphysics?'" W 102, EB 385. Cf. WD 174, WCT 227; "On the Essence of Truth," W 96f, BW 140; and WG 26, ER 27. In the passage from the 1949 edition of "What is Metaphysics?" just cited, Heidegger avers in a characteristically oblique statement clearly intended to carry declarative force: . . . *wenn anders zur Wahrheit des Seins gehört, dass das Sein nie west ohne das Seiende, dass niemals ein Seiendes ist ohne das Sein* (". . . provided that it belongs to the truth of Being that Being never endures disposingly without what-is, that any being never is without Being"). This inclusive dual statement replaced, wholly without acknowledgement or explanation, an earlier one made in the same passage in 1943 that did not thus unequivocally couple Being and what-is. In that earlier version Heidegger wrote not . . . *Sein nie west ohne das Seiende* ("Being never endures disposingly without what-is"), but rather . . . *Sein wohl west ohne das Seiende* ("Being, indeed, endures disposingly without what-is"). The intent of Heidegger's original phrasing is by no means clear. It seems entirely probable, however, that with it he wished to point to the priority belonging to Being in its relation with what-is (see this chapter, pp. 30-32), and that he was asserting that the provisive happening of Being owed nothing to what-is while, on the other hand, nothing whatever that is could *be* apart from Being. In any event, Heidegger's later substitution of *nie* for *wohl* makes it clear that he wished to maintain unequivocally that Being, to be rightly thought, must be thought always as "the Being of what-is." The passage cited from WG at the beginning of this note, which is, in fact, a context in which Heidegger was concerned to speak of the "Difference" that he identified as "the *not* between what-is and Being" (WG 2, ER 3; cf. this chapter, p. 27)—an identification in which the aforementioned priority of Being with respect to what-is was ultimately at issue—well illustrates the fact that just this portrayal of an intrinsic mutuality of relation between Being and what-is that was thus insisted upon through emendation early appeared as central to his own thinking. On Heidegger's understanding of the verb *wesen,* which, as shown by our translation of *west* above, denotes for him an "enduring," that so happens as to provide and to dispose the provided into a particularity of taking place by way of time, cf. Ch. VI, pp. 252-257.

conceived and explicated in various ways in the Western philosophical tradition. That legacy, in fact, together with the preconceptions regarding Being that have arisen from it, prompts Heidegger to hesitancy in his use of the word "Being"[3] and undoubtedly contributes to a vast and careful intricacy in his every depicting of its meaning. Heidegger wishes to distance himself from all conceptualizing of Being, and instead to bring Being to direct statement and portrayal. For him, "Being is what is nearest [to man]."[4] It is powerfully and determinatively[5] astir in every entity whatever, including man himself.

And yet, on the other hand, Being is in no sense an entity beside all the entities that we know. Being is utterly unique. We can best say that it is a *happening*,[6] or perhaps better still that it is *pure Happening*. Here the word "happening" is to be taken in a fully verbal and not a nominal sense, as Heidegger shows through his constant use of the infinitive, the verb in its wholly indefinite form, to speak of Being.[7] Being is a Happening that manifests itself immediately in everything that in any way

---

[3]Cf., e.g, "A Dialogue on Language," US 93, 108f; WL 7, 19f. Cf. also ZS 74, QB 75.

[4]See the "Letter on Humanism," W 161f, BW 210.

[5]Whenever forms of "determine" are met with in this work, the reader should be careful to remember that the "determining" in question is not a relating of something to something else in terms of a causal structure. "Determine" reflects the German verb *bestimmen*, formed from *stimmen*, to attune, to incline, to predispose. The determining that is in question in Heidegger's portrayals of reality is quite other than the simple agency of cause and effect. It is to be thought, rather, as the playing of Happening upon happening. This interplay will, indeed, be the theme of our entire study. For an illustrative use by Heidegger of the notion of attuning, see "On the Essence of Truth," W 87ff, BW 130ff.

[6]This characterization of Being as "Happening" reflects, but is by no means solely dependent upon, Heidegger's use of the noun *Ereignis*, ordinarily meaning event, and the verb *ereignen*, to happen or come to pass. For an elucidation of the complex meaning that Heidegger finds for these words, see Ch. IV, pp. 143-170.

[7]Cf. EM 69f, IM 77. For a discussion by Heidegger of the infinitive to name Being, a usage that he subtly approves, while pointing to its difficulties, see the antecedent discussion, EM 57ff, IM 63ff, especially EM 59f, IM 66. Heidegger employs a great many infinitives in a great variety of contexts to speak of Being's manner of happening. Cf., e.g., EM 11, IM 12 for use of the infinitives *walten*, *aufgehen*, *entfalten*, etc., and *"Alētheia,"* VA 276, EGT 118, for *lichten*, *vorbringen*, *entbergen*, *verbergen*, etc.

is—in the vast complex of entities as a whole and in each entity seen in itself.

In order to speak of the Being of what-is, precisely in its immediacy, Heidegger uses an archaic German verb, *anwesen*, presencing.[8] Indeed "Being," he can say, "means the same as presencing."[9] Here is a word easier to comprehend than "Being," which yet opens out upon the latter. Anything that is truly present—in Heidegger's way of speaking, anything that presences—encounters us powerfully precisely from within itself. In thus confronting us out of its centeredness in itself, it confronts us as something mysterious. Anything that presences has what we might call its own in-itselfness, which we cannot penetrate. This mystery of that which presences, which as such shows us its Being, has nothing to do with any lack of information on our part. It belongs intrinsically to that which meets us. Whatever presences remains inviolable in its centeredness. It offers itself and yet keeps itself to itself.[10] From within itself, it imposes itself in its own way. In Heidegger's words, "presencing is lightingly opened self-concealing."[11]

Here we come upon a tenet that is most fundamental for Heidegger's thinking: the Being of what-is is, before all else, self-maintaining self-concealing (*Sichverbergen*). And this means that as self-concealing Being is also, and precisely as such, self-maintaining self-revealing, i.e., self-maintaining self-*dis*concealing (*Sichentbergen*).[12]

---

[8]The verb *anwesen*, no longer in use in German, is presupposed by the verbal adjective *anwesend* (present), and by the nouns *Anwesenheit* (presence) and *Anwesen* (property, real estate). Heidegger reinstates the verb, needing it to speak of Being.
[9]"Time and Being," SD 2, TB 2.
[10]Heidegger alludes ("*Alētheia*," VA 263, EGT 107) to the "governance of presencing" as discernible in "remaining sheltered and concealed in expectancy, in keeping-to-self (*an-sich-Halten*)."
[11]*Anwesen ist das gelichtete Sichverbergen.* "*Alētheia*," VA 263, EGT 108. On the meaning of *lichten*, see "The End of Philosophy and the Task of Thinking," SD 72, TB 65. On the light-bringing clearing (*Lichtung*) presupposed in the above statement, see this chapter, pp. 32f, and note 35. For details of the structure of Happening—the withholding-to-self which is, even as such, a moving forth as self-manifestation via an openness—that is variously bespoken by the words presencing and self-concealing, see this chapter, pp. 29-33, and n. 36.
[12]In this study, the verb *entbergen*, a word of Heidegger's own coinage, will regularly

be translated as to reveal or as revealing, and its companion noun *Entbergung* will be rendered as revealing. We have, however, felt it necessary in this context to approximate more closely the verb's literal meaning. *Entbergung* is formed from *bergen*, which means to hide, to rescue, to protect, to preserve, to harbor, and the prefix *ent-*, meaning from, out, away, dis-. *Bergen*, with the intensive prefix *ver-*, is the root component in *verbergen*, to conceal, and in *Verborgenheit*, concealment, concealing. Heidegger regularly uses *entbergen* in contexts in which a manifesting is first of all in question. Hence the translation "revealing." But that translation completely masks the verb's immediate kinship with its companion *verbergen* that speaks of concealing. Alternative translations of *entbergen*, all employing "harbor" (whose final syllable stems, in fact, from the same Indo-European root *koro-* as *bergen* [see *The American Heritage Dictionary of the English Language*, Boston, Houghton Mifflin, 1971, p. 1524]) will sometimes be used. But all translations can only partially overcome this inadequacy, and the reader must keep the full force of the underlying verb steadily in mind. Since, in addition to denoting a movement away to a new state—from, out, away—the prefix *ent-* can also denote the negating of the state that already obtains—dis- —and since, unquestionably, Heidegger habitually presupposes and draws upon all the connotations resident in the words that he employs, *entbergen* must also be taken to speak of a negating countering of concealing. Accordingly, although *entbergen* and *Entbergung* are not here usually so translated as to speak specifically of *unconcealing*, the notion of an unconcealing, a *dis*-concealing, certainly lies within these words and must be borne in mind when uses of *entbergen* and *Entbergung* are in question. In order to speak directly of unconcealing, Heidegger uses the noun *Unverborgenheit* and the adjective *unverborgen*. Both derive from a coupling of the negating prefix *un-* with the past participle of the verb *verbergen* (to conceal), built on *bergen*. Considered in themselves, they mean respectively unconcealment and unconcealed. *Unverborgenheit* could also be translated, more literally, as unconcealedness, but this rendering cannot suffice at all, as to a degree unconcealment can, to speak actively of the provisive happening of Being. Since *Unverborgenheit* always connotes provisive unconcealing, in our discussions "unconcealing," with its wholly active force, will often be employed to translate it or to indicate its presence behind the discussion in progress. *Un-* is not a verbal prefix in German. *Ver-* could serve that purpose, but it already serves as the intensifying prefix of *bergen* (perhaps with an implication of a negating of *bergen* as well) to speak of concealing. Hence no obvious verb that could clearly denote negating of the intensified preserving-sheltering that is concealing, i.e., no direct German parallel to English unconceal, is possible. And *entbergen* must be understood to fulfill this role precisely in fulfilling that which is, contextually speaking, the more obvious for it, namely that of naming a manifesting, a revealing, that is a harboring-forth that, as such, protects and preserves. All such revealing is a surmounting of concealing, as a self-surmounting happening of concealing itself. *Bergen* and all the words built upon it, except for the adjective *unverborgen*, unconcealed, speak only of the happening of Being. *Entbergen* and *Entbergung* do so with particular force since, with the use of *ent-* they in themselves speak of the happening-away as which Being comes to pass as the Being of what-is. Cf. this chapter, pp. 30-33. For examples of Heidegger's

Indeed it must be so. Heidegger can ask: "What would a self-concealing be if it did not restrain itself," if it did not keep maintainingly to itself, i.e., "in its turning toward arising," in its turning toward going revealingly forth?[13] A pure self-concealing that simply shut itself in upon itself would utterly perish.

The notion of self-concealing that is here in question is one that must seem quite paradoxical to us. This is a self-concealing, an implosive keeping-to-self, that is, as such, a dynamic going-forth. It is an utter self-containedness that yet opens itself in disclosure—a poised yet puissant "Stillness" that brings itself to pass as an all-pervading concealing-fraught unconcealing,[14] via which alone whatever is, i.e., whatever presences from out of its keeping-to-self, is brought to be.

It is as this self-manifesting, self-maintaining self-concealing that Being is, and indeed must be, the Being of what-is—if we will understand "must" to allude not to any external compulsion playing upon Being but to the "need" inherent in its own most intrinsic character.[15] As self-concealing, Being so goes forth revealingly as to preserve itself and yet conceal itself in its very self-manifestation.

Being is the pure Happening that meets us in whatever is. But Being and what-is are not two separate somethings that stand in external relation to one another. To think thus would be to think only of what-is and not of Being. Neither is Being definingly to be understood, as so often it has been in the philosophical tradition, as the "ground" of what-is, i.e., as that on which whatever is—however it is—ultimately depends.[16] Such thinking can never reach to Being *as* Being. It can

---

closely similar but nuanced use of *entbergen* and its derivatives and of *Unverborgenheit* and *das Unverborgene*, see "On the Essence of Truth," Sections 4 and 6 (W 83ff, 89ff; BW 126ff, 132ff).

[13]"*Alḗtheia*," VA 271, EGT 114.

[14]On concealing as the heart of unconcealing, the "place of Stillness (*Ort der Stille*)," see "The End of Philosophy and the Task of Thinking," SD 74f, TB 67f. On the role that Heidegger finds for that "Stillness," see Ch. III, p. 121, Ch. IV, pp. 146-148 & 160f, and Ch. V, pp. 180-182 & 197f.

[15]"The Anaximander Fragment," HW 329, EGT 43. On this "need," see this chapter, p. 53, and n. 107.

[16]"The End of Philosophy and the Task of Thinking," SD 62, TB 56. Cf. "The Onto-

remain only in the sphere of the isolable, moving solely via conceptualization and still doing no more than concern itself exclusively, if loftily, with what-is.[17]

Heidegger early spoke of Being and what-is as happening after the manner of a unique differentiation. This, the "Ontological Difference (*ontologische Differenz*)," is identified by him as "the *not* between Being and what-is."[18] Recognition of it is mandatory for the safeguarding of the uniqueness of Being. It is no less requisite, also, for any understanding of the interrelation that obtains between Being and what-is. The disclosedness of the two ever instances but a single achieving of unconcealing from out of a singleness of transpiring; and yet by reason of this Difference, every such disclosing happens in a dual way.[19] The Difference in question, which is in his later work often intricately probed if not necessarily named by Heidegger, inheres ever at the center of his thinking.[20]

For Heidegger Being and what-is are possessed of twoness, but it is a twoness of a peculiar kind. For him the phrase "the Being of what-is" and its seldom used correlate "what is in Being" speak of the unique "Twofold [Zwiefalt] of Being and what-is," which as such is yet one.[21] Here "what-is" does not mean any particular entity or being, or even any

---

Theo-Logical Constitution of Metaphysics," ID 121, 54.

[17]Cf. Ch. III, pp. 130f. On grounding see Ch. XI, p. 417.

[18]Cf. WG 2; ER 3; WG 26 & 22, ER 27 & 23. Regrettably the title of the English translation of this work, "On the Essence of Reasons," is poorly chosen. The German title, *Vom Wesen des Grundes*, can best be translated "The essence of Ground." *Grundes* is singular, and it does not refer primarily to "reasons" but to "ground" as that which grounds. The lengthy, complex discussion has to do with grounding as broadly considered. "Reasons" are in question, but only as one instancing of grounding. See further, Ch. XI, p. 415, n. 52.

[19]WG 26; ER 27. With reference to the unconcealment in question (see WG 18-20, ER 19-21) Heidegger there speaks of the one "necessarily bifurcated manner of enduringly happening (*Wesen*) of truth (*Wahrheit*)," where "truth" names that unconcealment itself. On truth as unconcealment, see this chapter, p. 38, and n. 49. On the meaning of *Wesen*, cf. Ch VI, pp. 252-257.

[20]Cf. especially, this chapter, pp. 30-35, 44f, & 52-55; Ch. II, p. 77f; Ch. III, pp. 119f; and Ch. IX, 336f.

[21]"*Moira*," VA 240, EGT 86; WD 174, WCT 227.

mere aggregate of intrinsically separate entities. Rather, what-is is a unitary manifold of particulars, within whose totality every entity whatever belongs, not as a member of an aggregation but as a participant in what is first of all a single, vastly intricate happening: what-is as a whole (*Seiendes im Ganzen*).[22] And indeed, when particular entities are in view in their particularity, they can rightly be caught sight of only when seen to belong inherently within that manifold.[23]

The simple uninflected present participle *seiend* (being), the translation of the Greek present participle *eon*, most clearly carries for Heidegger the meaning of this Twofold.[24] These participles are as such possessed both of nonspecific verbal meaning and of nominal force. This use of language is instructive. It indicates that, as the Being of what-is, Being—which we have called pure Happening—maintains itself as a dual unity comprised of Happening as such and of particularized happening that can be named as this entity or that. We may perhaps call this latter "shaped happening." We must, however, be careful not to think of it primarily in terms of our category of substance as the latter is often understood. That is, we must not envision it as something that, as an entity, is possessed of specificity but that remains otherwise undefined. For substance in this sense is not definitively thought of as partaking of activity, as our context requires. Heidegger is speaking of that *as which* pure Happening takes shape, i.e., particularizes itself, in the manifold of entities that we know; hence his use of the inflected participle *Seiendes*—a verbal noun made specific now by inflection—to allude to that multitudinously shaped manifold. But the latter remains *shaped happening*. It is not, in whole or in part, fundamentally other than

---

[22]Cf. "On the Essence of Truth," W 87f, BW 130f, and "The Word of Nietzsche, 'God is Dead,'" QT 55, n. 5. For a particular perspective on what-is as happening in entirety, namely as "world," vis a vis human existing as Being-in-the-world, cf. Ch. II, pp. 61-63.

[23]See "On the Essence of Truth," W 92, BW 136, where Heidegger uses, characteristically, *Seiendes im Ganzen* for "what-is as a whole" and *Seiendes als eines solchen* for "a particular being." For a discussion of this passage, see Ch. II, p. 92, n. 125.

[24]"*Moira*," VA 240, EGT 86. Cf. also "*Logos*," VA 227, EGT 76.

Happening as such. It is *happening* shaped and particularized with the concreteness and specificity that in fact give rise to our ordinary notion of substance. All manner of entities—gods, men, trees, houses, concepts, interconnections—belong to the manifold that Heidegger names *Seiendes* (what-is). Everything is, then, whatever specific character it may display, intrinsically *happening*. Moreover, the structuring that belongs to Being as Happening and permits this to be true belongs likewise to what-is as a whole, as happening, and to whatever centeredly shaped particulars comprise its manifold. We must cling to this insight, difficult though it be, if we are properly to apprehend what Heidegger is saying with the words "the Being of what-is."

The pure Happening that is named with the word Being is, as Heidegger understands it, intricately nuanced in its bringing of itself to pass. It is, when most deeply considered, as self-concealing "that which closes itself in a way that veils."[25] And yet since, even as such self-closing, that Happening "must" as Being preserve and maintain itself, it opens itself and comes to self-manifestation as the Being of what-is. In so doing, it brings itself to pass as the Same (*das Selbe*), i.e., as a unique unitariness that is, precisely as such, a uniting-sundering that is at once a holding apart and a drawing of the parted toward one another.[26] Bringing itself to pass as this unique differentiating, this Happening opens itself as a "between (*Zwischen*)" wherein Being and what-is—themselves modes of the one Happening here in question—can happen reciprocally just in happening from out of and indeed *as* that unitingly differentiating Same[27]

---

[25]*des sich verhüllend Verschliessenden.* "The Onto-Theo-Logical Constitution of Metaphysics," ID 133, 65.
[26]"The Onto-Theo-Logical Constitution of Metaphysics," ID 132, 64. See also "Language," US 24-26, PLT 202f, where the character of this *Unter-Schied* as a uniting sundering is stressed. There, in a discussion some years anterior to that with which we are now concerning ourselves, in which particular instantiations of the Twofold as "world" and "thing" are under consideration as sunderingly united, Heidegger emphasizes that the *unter* of *Unter-Schied* must be understood as corresponding to Latin *inter* and must be seen as bespeaking an intimacy of the sundered. Cf. Ch. V, pp. 180-183. *Unterschied* ordinarily means difference, distinction.
[27]ID 132f, 65.

in a manner of transpiring that carries unitariness into accomplishment as underlyingly, determiningly in play in every distinguishing and every differentiation by way of which the single Twofold brings itself to pass.[28]

The character of Being as the provisive bringing into play of this intricacy of Happening is shown us in bold relief when Heidegger undertakes to elucidate this unique interrelating[29] of Being and what-is in their unity. He writes: "Being of what-is means: Being that *is* what-is" (italics ours). But this *is* does not have the meaning we would expect. For Heidegger adds, "'is' speaks here transitively, transitioningly."[30] It

---

[28]Cf. EM 106, IM 117.

[29]Here and elsewhere in this study the reader should be careful rigorously to separate "relate" from its usual meaning. Here "relate," in contexts where the Being of what-is is in any way under consideration, does not speak at all of a connecting of inherently discrete elements, one of which is to be subordinately referred to the other. The relating that is now in question is a *holding-toward-one-another* of participants that, while they are distinct from one another, are ultimately one. For we are using "relate" to speak of the happening of the single Twofold. See further, Ch. IV, p. 145, n. 10.

[30]*Das "ist" spricht hier transitiv, übergehend.* "The Onto-Theo-Logical Constitution of Metaphysics," ID 132, 64. See also "A Heidegger Seminar on Hegel's *Differenzschrift*," translation by William Lovitt, *The Southwestern Journal of Philosophy*, Vol. XI, No. 3, Fall, 1980, p. 38. Heidegger there elaborates on the import of "is" as transitive, attributing this notion to Meister Eckhart, although the allusion to Meister Eckhart does not appear in the English translation of the seminar, since it was not present in the text of the transcript that was presented for translation from the French. For the complete German rendering of the passage, which is a translation from the reworked French original (*Question IV*, Paris, Gallimard, 1976, p. 258), see VS 63. See also "On the Essence of Truth," W 96, BW 140.

This use of "is" as transitive points up in the most emphatic way a fact that must never be lost sight of in any consideration of Heidegger's treatment of the Happening named as Being in its character as provisive, namely, that when initiating is in any way ascribed to that Happening as such, that initiating is in no sense to be thought of in terms of what we understand as causation. It has nothing whatever to do with the achieving of a "result" through the acting of one something, understood as agent, upon another. Here it is not sufficient to think of an absence of cause. We must, rather, have in mind a "Happening" that is a happening-away-toward-another which is wholly other than "causing." Heidegger early raised this question of the distinction between the causal and a Happening that provides as an origin but is not causal, in a discussion in *On the Essence of Reasons* specifically concerned with human freedom in which, in a manner characteristic of his early work, he continues to use a familiar philosophical term—here "cause" —but endeavors to infuse into the word fresh meaning. (WG 102-104; ER 103-105.) For Heidegger "causality" as it is regularly understood is to be seen as a disguising of the Happening as which Being must be intrinsically thought.

shows us Being "in the manner of a passing-over, a transition, to what-is." But Being is in no sense something that quits its place and goes to meet another something, namely what-is, that is already there. No, Being is sheer Happening that, as such, happens itself over to that which is itself intrinsically *happening* but which, without the initiatory "movement" of Being, is not at all, has no reality whatever. Heidegger continues with an unusual transitive use of the verb "pass over" or "go (*gehen*)," reinforcing the extraordinary use of *is* above. With "what-is" as implied object, he writes: "Being passes [that] over thither, revealingly (*entbergend*) comes powerfully upon [that] which, through such coming-upon (*Überkommnis*), first arrives (*ankommt*) as that which is unconcealed from out of itself."[31]   What-is *happens* in its own right; for the unconcealing, the happening of Being, that comes upon it, i.e., that centers itself into the shaped happening as which it happens, rules as intrinsic to it, bringing it to appearance as whatever it is.[32]   Like the Happening, Being, that happens it, its happening self-unconcealedly is a self-preserving, a self-maintaining of that which has a distinctiveness and inviolability of its own; for what-is and Being are one and the same Happening and possessed of the same character.   And yet they are a Twofold, wherein priority of initiating rests with Being. The self-

---

Cf. Ch. III, pp. 130f, and Ch. XI, pp. 409-411.

[31]*Sein geht über (das) hin, kommt entbergend über (das), was durch solche Überkommnis erst als von sich her Unverborgenes ankommt.* "The Onto-Theo-Logical Constitution of Metaphysics," ID 132, 64. Heidegger's use of *kommen . . . über* is difficult. The verb *kommen* with *über* and an object in the accusative ordinarily means to come upon or to get over or surmount. But the verb *überkommen* with inseparable prefix means to get possession of, to receive, to seize, to overcome, to befall, to be transmitted. Heidegger here uses *kommen . . . über*, but he repeatedly employs the noun *Überkommnis*, a word of his own construction. And this latter usage makes it most probable that he intends connotations of *überkommen* to be heard in his use of *kommen . . . über*. Nevertheless, direct connotations of *coming* also remain, as may be seen from the fact that to *kommt . . . über* here corresponds *ankommt*, meaning "arrives," literally, "comes unto." Hence the translation "comes powerfully upon."

[32]For an affirming in a particular instance, that of the work of art, of just this happening of Being as unconcealing immediately in and as that which specifically is, cf. Ch. III, p. 135, [especially] n. 123. The reference there is to *Wahrheit* (truth), which for Heidegger is to be understood as unconcealing. See this chapter, p. 38, and n. 49.

manifesting of what-is is secondary. What-is achieves its "arriving," which for it is "to harbor itself in unconcealment,"[33] only because Being, happening *as* unconcealment, comes revealingly upon it.   Pure Happening, bringing itself to pass as Being, happens itself away as a partner to itself. Therewith it allows itself to maintain itself, for it allows itself to happen shapedly, in self-maintaining particularity, so as to bring itself into play as unconcealing.

Being and what-is happen in sunderedness from one another precisely in their happening toward one another in coming-upon and arriving.   As the uniting-sundering Same that in holding apart achieves this happening-toward, pure Happening is, with respect to both Being and what-is in their happening thus distinctively a "Difference *(Differenz)*" that transpires as an accomplishing carrying-out *(Austrag)* that at once reveals, in preservingly harboring forth, and preserves, in harboring protectingly.[34]   In that carrying-out there rules a clearing, a light-

---

[33]*sich bergen in Unverborgenheit*. ID 132, 64. The motif of *preserving* is present more strongly at this point in Heidegger's discussion than an English translation can adequately show.   In our translation, we glimpse it only in the thought that what-is preserves itself in unconcealment. But it must be remembered that the verb *bergen*—to hide, to rescue, to harbor, to protect, to preserve—provides the central element for the words translated as revealingly *(entbergend)*, unconcealed *(Unverborgenes)*, and unconcealment *(Unverborgenheit)*.   Being comes upon what-is in such a way as to reveal it, and that *revealing* is, more literally, a preserving away, a harboring-forth. Furthermore, the unconcealing that is therewith vouchsafed to what-is and informs its self-happening is the counterpart to that self-concealing *(Sichverbergen)*, which, as self-maintaining as inviolable, threatens Being with extinction and "necessitates" that Being as self-preserving self-concealing happen simultaneously as self-preserving self-unconcealing, in going revealingly forth as the Being of what-is. See this chapter, pp. 24-26, and n. 12.

[34]*entbergende-bergende Austrag*.   (Heidegger italicizes.) "The Onto-Theo-Logical Constitution of Metaphysics," ID 133, 65. The noun *Austrag* ordinarily means carrying-out, issue, decision. It directly parallels the meaning of *Differenz* from Latin *differe*.   The presence for Heidegger of strong connotations of differentiating in the word is shown by his use in the essay *"Logos"* of *Austrag* and *austragen* in parallel with *diapheromenon* (differing or carrying out) in a passage where differentiating is in question. See *"Logos,"* VA 221, EGT 71. At the same time *Austrag*, and with it here *Differenz*, is primarily defined for Heidegger by the meaning he accords to *Unter-Schied* as a severing that is a uniting; and, as the phrase cited at the beginning of this note shows, the carrying-out in question, as a carrying to issue, is a bringing to full culmination of the bringing together that fulfills that distinguishing sundering. See

permitting opening up (*Lichtung*) as which the Happening that is intrinsically self-closing brings itself to pass self-unconcealingly.[35] That puissant self-opening originatingly "bestowes the away-from and toward one another of coming-upon and arrival," permitting the twofold unitary happening as which Being and what-is come reciprocally to pass as one. Via it pure self-closing Happening brings into play its own self-differentiating self-relating.[36]

"Language," US 25, PLT 202.

[35]ID 133, 65. On the meaning of *Lichtung* (clearing), see "The End of Philosophy and the Task of Thinking," SD 72, TB 65. The self-opening of self-concealing Happening as this light-permitting clearing is variously invoked by Heidegger as the provisive manner of self-bringing-to-pass via which that Happening's self-unconcealing is accomplished. See especially, this chapter, pp. 19-22, Ch. II, pp. 58-60, Ch. IV, pp. 146-148 & 151, and n.7; Ch. V, pp. 178f & 174f; and Ch. IX, p. 353; and Ch. XIII, pp. 505f.

·[36] This structure in the manner of happening of the Being of what-is which shows us the manner of self-bringing-to-pass of pure Happening per se—namely, a going-forth from out of a self-maintaining keeping-to-self, which happens via a separating-away that is an opening-up and which, just in so happening, is simultaneously a gathering-toward returning-to-self—was first elaborated by Heidegger through an analysis of the existing (*Dasein*) of human beings seen as the time-structured locus wherethrough that Happening can be directly discovered. (See Ch. II, pp. 58-60, 61f, & 68-75. For a succinct affirmation of human existing as disclosive of the manner of happening of Being, see Ch II, p. 77, n. 78.) In *Being and Time* Heidegger anticipated inquiring into "the meaning of Being generally" by way of his inquiry into the human existing wherein Being could be centrally discerned. (SZ 37; BT 61f. Cf. Authors' Prologue, p. 9). This he never overtly did, and neither did he allude, in his portrayals of the Being of what-is, to the seminal role that belonged to his understanding of the manner of that existing. Nevertheless, returned from that analysis, the fundamental structure of happening intrinsic to every particular aspect via which man's existing is in *Being and Time* presented as structured forth stands as central to all his subsequent diverse depicting of the ever differentiative, unitary happening of Being and of whatever is, i.e., of man and of all else. The presence of this structure as underlying portrayals that will follow here a propos of Being, of what-is, of man in his pursuing of his existing as such and his accomplishing of the varied undertakings pertaining to it—apropos, indeed, of all happening whatever—will often be noted. This is the primal structuring that is, for Heidegger, determinative of Happening as such. It should constantly be kept immediately in mind. Beyond our present discussion, cf. in particular: (of what-is as a manifold), Ch. I, paragraph below; (of Being's happening as time), Ch. I, pp. 37-40, and n. 50; (of Being's happening of what-is into deviance), Ch. I, pp. 45f, and n. 77; (of presencing), Ch. I, pp. 51f, and n. 104; (of the happening of Being as handing-over keeping-in-hand), Ch. I, pp. 53-55, and n. 107; (of what-is as what presences), Ch. I, p. 55, and n. 112; (of human existing as

At the same time, that happening as Difference likewise permits pure Happening to manifest and maintain itself in happening as the manifold as which that Twofold brings itself to light; for the differentiating thus brought into play ever ramifies throughout the happening forth of the Twofold. Disclosure of what-is, is disclosure of what-is *in its particularity*. Precisely there the differentiating that permits disclosure at all fulfills itself in the distinctiveness, the happening in specificity, that we ever find to pertain to whatever *is*. Again the oneness of the Happening of which we are speaking shows itself in that what-is, as which that Happening shapes itself forth, happens precisely after the manner of pure Happening itself. Here this means that, even as Being and what-is happen from out of a sundering that is fulfilled in a responsive arriving-into, so what-is transpires by way of a differentiating in which the differentiated—its particularized entities—just as differentiated, happen also toward one another in their belonging to the unitary manifold that is named "what-is *(das Seiende)*." Indeed, in the context in which Heidegger first speaks of "the Being of what-is," which is in fact the context in which his discussion moves to the proclaiming of the Ontological Difference obtaining between the two, he portrays all distinguishing of whatever is, and all identifying of whatever is

---

"transcending" as Being-in-the-world), Ch. II, pp. 61-62 & 72-75, and n. 19, and Ch. III, p. 112f; (of existing as directly instancing Being's happening), Ch. II, pp. 72f & 74-76; (of existing as inauthentic and authentic), Ch. II, pp. 89-110, n. 176; (of the interpreting displayed via language), Ch. III, pp. 113-115, and n. 15; (of *Ereignis*), Ch. IV, pp. 144-146, and n. 12, (of *Ereignis* that is *Enteignis*), Ch. IV, pp. 163-165, and n. 89; (of pure Happening), Ch. IV, pp. 146-148, and n. 25; (of poetic measuring-seeing), Ch. V, pp. 200f, and n. 117; (of the transpiring of technology), Ch. VI, pp. 228f, and n. 12; (of the transpiring of science), Ch VII, pp. 268-275, and n. 10; (of Descartes' representing thinking), Ch. VIII, pp. 298-300, and n. 24; (of will to will), Ch. VIII, pp. 304f, and n. 40; (of Being's ruling as subjectiveness and objectiveness), Ch. VIII, pp. 304f, and n. 42; (of subsequent metaphysical thinking), Ch. VIII, pp. 306-310, and n. 61; (of Nietzsche's will to power), Ch. VIII, pp. 310-313, and n. 63; (of *technē*), Ch. IX, pp. 334f & 338, and n. 35; (of *ratio*), Ch. XI, pp. 417-419, and n. 66; (of the thinking presupposed and required in Leibniz's principle of ground), Ch. XI, pp. 425-428, and n. 92; (of thinking carried out in releasèdness), Ch. XIV, pp. 544f, and n. 22; (of "disposition"), Ch. XIV, pp. 554-560, and n. 59; and (of the viewing of art), Ch. XVI, pp. 663f and n. 40.

distinguished through the manifesting of the "what" and "how" of its appearing, as stemming from the happening of Being, whose self-disclosing as that "what" and "how," via the Difference by way of which it happens vis a vis what-is, permits any disclosing of what-is at all.[37] It is the self-sundering of pure Happening, by way of the happening-into-Being that is thus permitted, that lies behind or, more properly, inheres in, the distinctiveness, the variedness in specificity of character, that constitutes what-is precisely as a manifold and that therewith lets the Twofold unfold as the Twofold that it is. This disclosing of what-is in its multitudinous particularity is ever the sine qua non of the accomplishing of self-maintaining that pure Happening, as self-concealing that unconceals itself, consummates in its forth-happening as the Being of what-is.[38]

This self-maintaining is a lasting. But it is not the sort of lasting that we ordinarily think of, a mere continuance that stretches through linear time from one now-point to another. This is the unique self-maintaining as which pure Happening brings itself to pass in its carrying out of itself in differentiation by way of shaped happening. The Being of what-is, as which that Happening maintainingly accomplishes itself, so happens as to remain.[39] And, in virtue of that initiatory remaining, whatever is, whatever presences, happens lastingly.

To speak of the Happening in question as such initiatory remaining, Heidegger uses the verb *anwähren*, formed from *währen*, to endure.[40] This is an enduring that must, once again, be thought transitively. For it is as this enduring that Being, accomplishing pure Happening, comes initiatingly upon what-is, allowing the latter to present

---

[37]WG 20, 26; ER 21, 27.
[38]The all important particularizing disclosure of what-is often comes to the fore as such in Heidegger's thinking. It is shown us from various perspectives. Cf. especially this chapter, pp. 44-46, 49-51, & 54f; Ch. III, pp. 112f & 135; and Ch. V, pp. 185f.
[39]*bleiben*. Cf. "The Turning," TK 38, QT 38f. It seems worth noting that in everyday German speech the verb *bleiben* is very often employed to speak of the being of something with respect to its whereabouts.
[40]"The Onto-Theo-Logical Constitution of Metaphysics," ID 132, 64.

itself as unconcealed. This enduring is nothing other than pure Happening as the latter brings itself to bear as presencing (*Anwesen*). The verbs *wesen* and *währen* are, Heidegger repeatedly points out, the same in etymology and meaning.[41] The unique, initiatory enduring as which, as Being, pure Happening maintains itself is an enduring-unto (*Anwähren*), i.e., unto *us*, the human beings who are met by the presencing of whatever presences.[42] And in presencing, *An-wesen*, the import of that self-maintaining enduring that, happening initiatorially, permits arrival into self-maintaining enduring, always sounds.[43] The presence that proffers itself to our awareness must never be thought of as inherently apart from duration. For Heidegger "presence (*Anwesenheit*) means: tarrying that constantly approaches and concerns men, is reached out to them and has reached them."[44] In the happening of the single Twofold of Being and what-is, as Being comes revealingly upon what-is and what-is arrives into and harbors itself in the unconcealment as which Being thus rules, Being, in happening as presencing, i.e., as enduring-unto, is letting what-is presence, letting it endure unto men. And that enduring-unto (*Anwähren*) is precisely what it means for what-is "to be what-is."[45]

The Being of what-is is the enduring—the constituting enduring—of what endures, an enduring that is inherently directed toward man. Yet "enduring" is a word that suggests to us "time" much more than "Being." Indeed, Heidegger's thinking on Being opens immediately into his thinking on time, for to him the two are intrinsically one.

---

[41]Cf. e.g., "The Question Concerning Technology," VA 38, QT 30, and "Science and Reflection," VA 50, QT 161. It is *Wesen* that holds the foremost place by far in Heidegger's portrayal of the happening of Being as provisive enduring. For a detailed discussion of *Wesen* see Ch. VI, pp. 252-257.

[42]"Time and Being," SD 13, TB 12.

[43]Heidegger takes the prepositional prefix *an-* to combine in itself the meanings "toward" and "into": *Die deutsche Preposition "an" bedeutet ursprünglich zugleich "auf" und "in."* WHD 143. This sentence is omitted from the English translation of *Was Heisst Denken?* (*What Is Called Thinking?*).

[44]*Anwesenheit besagt: das stete, den Menschen angehende, ihn erreichende, ihm gereichte Verweilen.* "Time and Being," SD 13, TB 12.

[45]"The Onto-Theo-Logical Constitution of Metaphysics," ID 132, 64.

2

To follow Heidegger's portrayal of time we must lay aside our customary notions of what time is. For Heidegger time is not first of all a succession of discrete instants, let alone of days or years—spaces of time ceaselessly passing away. Time is not some sort of non-spatial container in which entities exist and events occur. Neither is it the "fourth dimension" of a universe that is to be conceived basically in spatial terms.

Time, for Heidegger—genuine time[46]—is precisely that "opening clearing of self-concealing" by way of which Being, happening as self-*unconcealing*, in accomplishing the uniting intrinsic to it as self-

---

[46]Our use here of the word "genuine" is intended to point to time when considered in itself, i.e., as it is happened via the happening of Being. It points to time, that is, as considered quite apart from any ordinary understanding of it, which can only be secondary and which must appear as a misunderstanding of time as Heidegger wishes to portray it. Behind our use of "genuine" lies Heidegger's use of *eigentlich* (proper, real, actual, intrinsic), a word that for him bespeaks the manner of happening that pertains to anything as properly its own (cf. Ch. II, p 85, n. 102), and that must, in contexts referable to his thinking in its fully developed scope, be taken to suggest the Bringing-to-pass, *Ereignis*, in play as provisive in every phenomenon (cf. ch. IV, pp. 144-151, and pp. 154-162, and n. 28). Heidegger himself uses the phrase *eigentliche Zeit* in speaking of the provision of time as such (cf. "Time and Being," SD 18, 16; TB 17, 16). To our regret, we have found no viable English translation for *eigentlich* that can immediately display the notion of that which is one's own. The word "proper" would be the obvious choice, were it not for the fact that "proper" inevitably suggests a sort of morally tinged ascription of rightness and wrongness that Heidegger would utterly have rejected. Throughout our study, with the exception of uses of *eigentlich* (and *uneigentlich*) to speak of human existing, where "authentic" will be employed (cf. Ch. II, pp. 85f and n. 102), we shall consistently use "genuine" as a direct translation of *eigentlich*. Beyond this, when in our discussion we ourselves wish, from within the perspective of Heidegger's thinking, to speak of that which is under consideration as it is in itself, apart from other conceptions of it, we shall employ the word "genuine" for that purpose. The decision to use two different words for the translating of *eigentlich* has been a difficult one and in one respect remains questionable, since the consistency of Heidegger's usage all but compels a like consistency in our own. It is because of the very centrality of *eigentlich* and *uneigentlich* for Heidegger in his characterizing of human existing that we have chosen not to use "authentic" for other renderings of *eigentlich* or for the suggesting of its presence behind our own words. For we have concluded that the very centrality of "authentic" in the characterizing of man could too easily result in the immediate association of every phenomenon spoken of as "authentic" first of all with man and not, as must be the case, with the pertaining of that phenomenon to the Happening that is named as the Being of whatever is.

differentiating, brings itself to pass as the presencing of what presences.[47] Heidegger can call time (*Zeit*) "the first name of the truth of Being,"[48] where truth (*Wahrheit*), with a meaning drawn from Greek *a-lētheia*, means the unconcealment (*Unverborgenheit*) as an opening-up that discloses as which Being happens.[49]

There is between "Being" and "time" no simple identity, but these words speak from different perspectives of the one happening-forth as disclosure of the Being of what-is. Time is the name for the self-opening as which pure Happening, as self-unconcealing, happens as the Twofold and appears enduringly to man as the presencing of what presences. As that opening, i.e., as time, pure Happening happens as on-goingness, and that on-goingness allows it to happen as self-preserving. For it permits particularizing; it permits the unfolding of the Twofold as the Being of the vast and varied manifold of what-is. *Time* is *Being* seen as on-goingly opening itself that it may, as the Being of what-is, bring itself to pass as unconcealing.[50]

---

[47]"Letter to Father Richardson," pp. xx - xxi, in William J. Richardson, S, J., *Heidegger: Through Phenomenology to Thought*, The Hague, Nijhoff, 1967.

[48]"The Way Back into the Ground of Metaphysics," W 205. For the English translation by Walter Kaufmann, see Walter Kaufmann, *Existentialism from Dostoevsky to Sartre*, New York, Meridian Press (expanded edition), 1975, p. 273. Hereafter the translation will be referred to as "K," as indicated in the list of symbols on p. xiii.

[49]The word *alētheia* is comprised of *lēthē*, forgetfulness, and the privative prefix *a-*. Heidegger's regular equivalent for *alētheia* is *Unverborgenheit*, unconcealment. This is an initiatory unconcealment that is inherently a protective preserving. Like time, Being's happening as unconcealment, i.e., as truth, is identified with the light-permitting clearing (*Lichtung*) that preservingly opens Being's happening as self-concealing, therewith permitting presencing. ("On the Essence of Truth," W 96, BW 140; "The End of Philosophy and the Task of Thinking," SD 79, TB 71) On the connotations resident in *Unverborgenheit*, cf. this chapter, pp. 24f, n. 12. For Heidegger, *Wahrheit*, the ordinary German word for truth, carries connotations of disclosive protecting. See this chapter, p. 54, n. 109.

In "On the Essence of Truth," Heidegger, starting from the notion of truth as the correctness of a statement with respect to that which is spoken about, arrives at "truth" as meaning ultimately just this happening of Being as unconcealment. (W 96; BW 139).

[50]In the discussion of time that follows, the same structuring will be found as that obtaining in the happening of Being as the Being of what-is, i.e., an opening in differentiation that is simultaneously a coming-toward of the differentiated. See this

The time of which Heidegger speaks is precisely the time that we know—the time of immediacy of presencing, of remembrance, and of anticipation. Yet here again Heidegger's way of thinking about time is markedly different from our own. We do indeed speak of "time to come." But if we say that the future will become the present and the present must become the past, is it not because we find time to be ever advancing so as to leave itself and its contents behind as past and irrevocably gone, while its leading edge, the present, moves steadily forward? For Heidegger, however, time is Being, as the presencing of what presences, happening as on-going self-opening. As such, time is to be thought always as an on-goingness that moves *toward* us. For presencing is an enduring unto men.

Time is a ceaseless interplay of three "dimensions" or reachings-forth *(Reichen)*. As that interplay, it reaches forth presencing.[51] "Arriving"—corresponding to our "future"—ever moving forward, reaches forth "having-been"—our "past"—before it, moving the latter away through its approach. Concomitantly, "having-been" accommodates itself to on-coming "arriving," moving away before it and permitting its advance. The two never meet. Rather, their mutual relating reaches forth the present between them.[52]

---

chapter, pp. 29-33, and n. 36.

[51]"Time and Being," SD 14f, TB 14. The exact translation of *Reichen* would be "reaching." Since the use of a simple "reachings" would not here be meaningful, we have used "reachings-forth," which is in fact the equivalent of *Erreichen*. This translation accords, as "reaching" could not, with the meaning that Heidegger gives for "dimension." The latter "is not thought here only as the area for possible measurement, but as extending throughout, as reaching-out that lightingly clears and opens up." (*Dimension wird hier nicht nur als Bezirk der möglichen Abmessung gedacht, sondern als das Hindurchlangen, als das lichtende Reichen.*) "Time and Being," SD 15, TB 15.

[52]"Time and Being," SD 14f, TB 14f. The words that Heidegger here uses for the "dimensions" of time point up his view that in speaking of time one is ultimately speaking of Being. Instead of using only the usual German word for the future, *Zukunft*, he joins with it the noun *Ankunft* (arrival) to carry the main burden of his portrayal of the future. (Here Heidegger's use of *kommen . . . über, Überkommnis*, and *ankommen* as pivotally important words in his portrayal of the happening of what-is from out of Being should be recalled, since they suggest the transpiring of that happening by way of time that we are now considering. (Cf. this chapter, pp. 31f, and

The three reachings-forth of presencing—arriving, having-been, and present—are constantly allowed to approach each other and as constantly held apart by a "nearing" that is, as a drawing of the differentiated toward one another—and most importantly as a bringing into play of the primordially provisive, self-sundering Same that rules as a uniting-sundering—ever intrinsically in play in the self-bringing-to-pass of the originative Happening named as Being.[53] Its distancing of the three even as it accomplishes their approaching provides an openness.[54] And that openness, wherein time's reachings-forth are ever gathered on-goingly toward one another, is what we know in our experiencing of time.[55] Indeed we, as human beings, are called to live out time's openness.[56] For it is unto man that Being, by way of time, brings itself to pass as initiatory enduring. Man is "needed" to receive Being's interrelating self-reachings-forth by way of time, in order that he may allow to Being the accomplishing of its self-opening and hence the fulfilling of its happening as unconcealing.[57]

Here the words "openness" and "distancing" must not be thought

---

n. 31.) Instead of the German noun that ordinarily denotes the past, *Vergangenheit*, he uses the past participle of the verb *sein* (to be), *gewesen*, which can best be translated as "been," or less literally as "has-been" or "having-been"; or he employs a noun formed from *gewesen*, *Gewesenheit* ("has-beenness"). For the present he uses the noun *Gegenwart*, which regularly means both "the present" and "presence," and he then draws out these two aspects of meaning together, linking *Gegenwart* immediately with *Anwesenheit*, a noun likewise meaning presence, which, like its kindred *Anwesen*, speaks specifically of Being.

[53]"Nearing" here translates *die Nähe*, literally "nearness." *Nähe* consistently speaks for Heidegger of a ruling mode of happening that provides. This is "the Nearing that nears." "The Turning," TK 47, QT 49. Accordingly, we have rendered *Nähe* whenever possible with the gerund that can suggest an active accomplishing of relating, and not with the inescapably static noun. Heidegger does not specifically discuss the relation subsisting between Nearing and the self-sundering Same. He does, however, display it. See, e.g., "The Nature of Language, US 106, WL 214, and Ch. IV, pp. 147f. On the ruling of originative Happening as Nearing, cf. further Ch. IV, pp. 147f & 160f; Ch. V, pp. 183f, 189f; & 197f; and Ch. XIII, pp. 519f.

[54]"Time and Being," SD 16, TB 15.

[55]Cf. SZ 328f, 326; BT 377, 374, and cf. Ch. II, pp. 69f.

[56]Cf. "Time and Being," SD 16, TB 17. On the complexity of this living out, see Ch. II, pp. 61-110.

[57]Cf. SD 12f, TB 12.

of in spatial terms. Heidegger is not concerned in this context with any measureable "extent" of time lying between particular time-points. Rather, he is concerned with an opening-up to possibility, to appearing and disappearing, to supersession and change. And he is concerned, therewith, with the permitting, from out of the Twofold's happening by way of time, of particularization within undergirding unitariness—a particularization which he sees, indeed, as ceaselessly and ubiquitously in play from the presencing of readily specified entities or the individual existing of persons to the self-carrying-out of humankind as distinctive peoples or the epochal differentiation of identifiable extendings of time.[58]

The Being of what-is, indeed, happening by way of the complex on-goingness that the opening-up of time permits, happens ever as an arising.[59]   As such, it comes to pass as the Being of that which ever newly comes which is, as happened via time, ever that which is passing away.[60]   This is Being's happening as that initiatory enduring that meets us in the presencing, the enduring-unto, of that which presences toward us.   Seen now more amply, that presencing again encounters us as mystery—now the mystery of an on-going self-presenting that, as partaking inherently of arriving and departing, imbues the encountered with an ungraspability, an unpossessibility, that attest the ultimately inviolable centeredness of that which, in its transiency, from out of some unplumbable inwardness, withdraws itself even in its self-presenting.[61]

Beyond this, indeed, lies the presencing of what-is as a whole, within which and from out of which the mystery-fraught presencing of any such particular happens. Therein resides mystery in the ultimate sense. Happened on-goingly by way of time, what-is as a whole—as an intricate host of arriving, departing particulars—so presents itself in its ever-transpiring totality as wholly to withhold itself from simple

---

[58]Cf. this chapter, pp. 49f & 47-49, and Chapter III, p. 137.
[59]"*Alētheia*," VA 268, EGT 111.
[60]"The Anaximander Fragment," HW 322f, 337; EGT 37, 41f. Cf. this chapter, pp. 49f.
[61]Cf. HHF 28, HFH 97. Cf. Ch. XVI, p. 657. On mystery as manifest in presencing as such, cf. this chapter, p. 24.

immediate meeting. It is, as such, inaccessible to all grasping of its proffering of itself. Its disclosing of itself can but disclose its unreachability and unpossessibility, the sheer evincing of mystery as such.[62]

The on-going self-opening that allows time to reach forth presencing toward us centers in the present. Whatever is, i.e., whatever presences, meets us as in some way present in the present. It may possess immediate presence in the present. On the other hand, it may itself be absent, either as something still future or as something already past, and yet it may, nevertheless, encounter us powerfully in the present (*Gegenwart*). The presencing (*Anwesen*) of what presences can, then, be either immediate—present-time—presence (*Gegenwart*)[63] or presencing as absenting (*Abwesen*).[64] In happening itself away by way of the openness that time provides, Being makes room for its happening of itself as this diverse presencing.

For Heidegger Being is a plenitude possessed of inexhaustible riches.[65] But plenitude belongs to Being in its happening as concealing and unconcealing. As such, it is in no sense a fullness predicated upon a timelessness that must preclude change. As the plenitude of that Happening that, as self-concealing, must needs self-maintainingly differentiate and particularize itself as the Being of what-is, so as to happen as self-unconcealing, it is, rather, a fullness that requires its own self-accomplishing by way of time and the change and diversity that time permits. Therefore, we might say that "prior to" its self-differentiating as the Being of what-is, Being opens itself as time, i.e., as the on-going openness wherein, as Being unfolds itself as the Twofold, the possible can

---

[62]Cf. "On the Essence of Truth," W 89ff, BW 132ff.

[63]The dual meaning of *Gegenwart* (present) is here in question. See this chapter, p. 39, n. 52. Cf. "Time and Being," SD 10f, TB 10f.

[64]"Time and Being," SD 13f, TB 13.

[65]"Science and Reflection," VA 70, QT 181f. Cf. EM 74, IM 82f. This theme of plenitude shows itself in Heidegger's portrayal of the divine as ruled from out of the holy (*das Heilige*)—from *Heil*, "wholeness, wellbeing"—or from out of the serenely joyous (*das Heitere*)," both modes of the originative Happening that is named as Being, as that Happening holds sway provisively as wholeness and as gladsome abundance. See Ch. V, pp. 206f, and n. 142, and Ch. XIII, p. 524.

occur and the new take place. The words "prior to" do not here have a temporal meaning. They speak, rather, of Being which, purely as the Happening that it is, *does* provide for itself the open clearing that is time, as a carrying out of its very happening as unconcealing. They speak of Being at what we might call the roots of its happening, where, as the coming provisively into play of pure initiating, "before" happening toward men—by way of time—as the presencing of what presences, it happens purely as itself.

3

The diversity that, in its self-maintaining, Being allows to itself in its unfolding as the Twofold is not limited to the variety that is accomplished through the appearing and disappearing or the immediate presencing and absenting of the host of particulars belonging to the unitary manifold of what-is. The structuring according to which Being carries itself out self-differentiatingly by way of time, as the Being of what-is, remains constant: happening as "Difference," Being comes revealingly upon what-is, and what-is, itself happening from out of this "coming-upon," preserves itself in unconcealment, in Being. But, as Being happens on-goingly, the *manner* of its happening revealingly—and hence of the self-unconcealing happening of what-is—is in its every aspect marked by changefulness and even by discrepancy.

The Being of what-is happens as self-unconcealing self-concealing. When this Happening brings itself to pass as unconcealing, it surmounts itself as concealing.[66] The two modes of the one Happening counter one another, whether in the intricate presencing of given particulars viewed in small compass or in the unfolding of the manifold of what-is as seen in sweeping perspective. For they counter one another in the happening of what-is as a whole. Each tendency can be named and considered in itself, but the two are always in play as one. When we speak of one, we must be alert to the counter happening of the other.[67]

---

[66]Cf. this chapter, pp. 24-26.
[67]"On the Essence of Truth," W 89ff, BW 132ff.

Heidegger shows strikingly this all-pervasive duality in the Happening that is Being through the use of the German verb *scheinen* for that Happening. *Scheinen* means both to shine and to seem and hence names Being equally as a genuine and a spurious disclosing.[68] The two poles of this duality belong likewise to the self-manifesting of what-is as that which Being lets appear. For this happening of what-is the verb *erscheinen* (to appear), which carries the same ambiguity, is regularly employed.[69] In the unfolding of the Twofold clear disclosure and mere seeming, genuine appearing and deceptive semblance, happen together in patterns small and great, in the on-going changefulness as which the Being of what-is brings itself to pass by way of time. And beyond this, in all Being's happening of what-is, unconcealing both arises out of concealing and evinces the latter in hiddenly blocking from view the very concealing that it disclosingly transcends.[70]

Thus, in the Happening named as the Being of what-is, concealing and unconcealing happen in one. The ever-shifting happening of the two is, however, not equipoised. For Being happens first of all as a concealing that, in happening as unconcealing, is ever surmounting itself. And always and everywhere this its fundamental character is pervasively decisive.

Even in Being's constitutive bringing of itself to pass self-sunderingly as the Being of what-is from out of the "Difference" as which pure Happening rules, the "priority" of concealing tells in its happening. The concealing in question is a withdrawing-to-self. Accordingly, in happening what-is into the light-bringing clearing—time—where it can appear, Being retires self-concealingly behind what-is.[71] In the happening of what-is as a whole, in the transiency and ever-shifting kaleidoscopic self-presenting of its particulars in interrelation with one another, Being as concealing transcended by unconcealing immediately rules. Yet such happening, although ever primordial, ever also falls into

---

[68]EM 76; IM 85.
[69]EM 76; IM 85. Cf. "The Way to Language," US 252f, WL 122.
[70]"On the Essence of Truth," W 89ff, BW 132ff.
[71]"The Anaximander Fragment," HW 310f, EGT 26f.

eclipse. The particularity belonging to what-is is so brought to pass as to conceal it. Pure Happening, as concealing that unconceals, does not let itself be seen, by reason of the particularized manifold of *shaped* happening—what-is—that appears. And what-is, likewise keeping itself self-concealingly to itself, draws away from Being. In its very self-maintaining, it insistently presents itself as though alone in its appearing,[72] while in its manifold, in like manner, particular configurations and particular centerings so stand forth as to hide from view the whole, in its Being, within which they inherently belong.[73]

This tension between concealing and unconcealing is constantly in play within the single happening of the Twofold. It is not, however, a tension that holds Being and what-is in an unvarying relation. "Being," as Heidegger says, "turns evasively away" even as, in happening so as to unfold itself as the Twofold, "it reveals itself into what-is,"[74] i.e., harbors itself forth disclosively into the shaped happening as which it— unitary Happening—happens as partner to itself. Not only does Being conceal itself behind what-is; but as self-maintaining self-concealing centering in itself, Being, as it were, in a disrupting of the primordial unity of its happening, turns tangentially away from what-is in permitting it unconcealment and likewise disruptingly happens what-is tangentially away (*beirrt . . . das Seiende*) from itself. Being so happens what-is, in permitting to it unconcealment, that the latter—at once happened and happening itself—goes divergingly astray from Being.[75] And, ruled by such disruption of unity, what is, in so diverging, suffers a disruptedness

---

[72]HW 310; EGT 26.
[73]The above paragraph explicates a passage in "The Anaximander Fragment" (HW 310, EGT 25f), as that passage has been understood in the light of Heidegger's chronologically prior discussion in "On the Essence of Truth" (W 89ff, BW 132ff) of the fact that concealing as unconcealing, i.e., the happening of the Being of what-is, is manifested first of all in the self-presenting of what-is as a whole and is lost sight of as that happening of what-is is obscured from within itself.
[74]*Das Sein entzieht sich, indem es sich in das Seiende entbirgt.* Heidegger stresses this point by stating it as a separate paragraph that is repeated verbatim subsequently in his discussion. "The Anaximander Fragment," HW 310f, EGT 26.
[75]See HW 310, EGT 26. *Dergestalt beirrt das Sein, es lichtend, das Seiende mit der Irre.*

that masks its intrinsic wholeness with a fragmented, deviant self-presenting of its particulars.   What-is "disruptively diverges about (*umirrt*) Being," skirting around it.   And therewith, as the bringing to manifestation in concreteness and specificity of the happening of Being as self-withdrawing self-concealing, what-is "founds," i.e., through its happening brings into play, the deviance-fraught domain (*Irrtum*) of such disruptive divergence (*Irre*) from Being;[76] and concomitantly it evidences that divergence in the vast on-going complex of mis-prizing and mis-dealing that we know as erring and that we find so frequent in our encountering of what-is.[77]

Thus once again it is via time that Being allows the diversity belonging to its way of happening to transpire—here a diversity born specifically of its happening as a disruptive diverging, in its coming to pass as self-differentiating.   We do not speak now of time, the self-opening-up of self-concealing, viewed with respect to Being as the latter reaches itself forth via time's dimensions as the presencing of the particulars of what-is that present themselves as such in an accomplishing of particularized self-disclosure as a surmounting of keeping-to-self. Rather, we speak of time as seen with respect to Being in its happening as

---

[76]Heidegger's sentence runs in full: *Das Seiende ist in die Irre ereignet, in der es das Sein umirrt und so den Irrtum (zu sagen wie Fürsten- und Dichtertum) stiftet.* HW 310; EGT 26. The meaning of *die Irre* as shown by Heidegger in "On the Essence of Truth" (W 91f, BW 135f) is presupposed in the argument of this paragraph to this point.

[77]Cf. "On the Essence of Truth," W 92, BW 136. In the portrayal of the happening of the Twofold that we have just been detailing, the basic structure that we have found for the Happening named as the Being of what-is, namely a going-forth via a separating-away that is yet a coming-together, appears again; for Being reveals itself into what-is and what-is arrives into the unconcealment as which Being permits it to happen. See this chapter, pp. 29-33, and n. 36. But that structure has now a particular guise that shows us the presence of a strong element of distortion and diminution within it. The depleting alienation-from-self that takes place when the deviant divergence here in question is in play is, as is the primary structure itself, a manner of Being's happening that Heidegger first described in his delineating of human existing. Cf. Ch. II, pp. 89f. For discussions of such transpiring as pertaining to man in history and to the manifold of what-is see Ch. VIII, pp. 304f, and n. 42, and pp. 303f; Ch XI, pp. 436-440; and Ch. XII, pp. 473f.

an initiatory providing that, as a surmounting of evasive self-withdrawing, governs inclusively throughout vast ranges of the manifold of what-is, by way of extensive openings-up of time. This means that we speak of time as the milieu of the historical.

To Heidegger "history (*Geschichte*)" is no mere succession of events to be understood as a congeries of causes and effects. What we call "history" may appear thus to the science of historiography. But to Heidegger reality as history is far richer in the complex interconnections of its on-going takings-place. It is that transpiring complex as humanly lived out and as understood always according to some identifying mode of happening that renders it meaningful to those who take their way via its course.[78] History is a transpiring of happening (*Geschehen*) that is accomplished through a human questioning into reality and through a resolute confronting of the latter that brings it to light, i.e., through undertakings that allow the Being of what-is to be caught sight of and therewith allow the meaning as which Being brings itself rulingly to pass to be disclosed.[79]

Heidegger concerns himself with such history as the on-going manifold of what-is seen as ordained from out of the Happening named as Being that, in a vast intricacy of ever-shifting particularization, in happening it, sends (*schickt*) it on its way.[80] Viewed thus, that manifold is Western history. For it is in the West, beginning among the Greeks, that Being is named and that, as the Being of what-is, it happens self-revealingly, making itself known variously to the thinking and accomplishing of Western man.[81] Seen in relation to what-is as "history" in the fullest sense, Being's "movement" in coming powerfully and revealingly upon what-is is its self-sending-forth; it is an ordaining destining (*Geschick*) that gathers and disposes into the specificity of time

---

[78]Cf. SZ 375, BT 427f, and "On The Essence of Truth," W 85, BW 129.

[79]EM 34; IM 36f. Cf. "On the Essence of Truth," W 84f, BW 128f, and EM 109f, IM 120f. On the happening of Being as "meaning (*Sinnen*)," see "*Alãtheia*," VA 276, EGT 118.

[80]"Science and Reflection," VA 63, QT 175.

[81]Cf. "Time and Being," SD 8f, TB 8f. Cf. also Ch. III, pp. 130-138.

and circumstance the multitudinous particulars as which what-is takes place.[82]

Being rules after the manner of a self-sending-forth, but this self-sending-forth is itself likewise a self-withholding, a happening of Being as self-withdrawing self-concealing. Accordingly, Being's holding sway as destining is "epochal" in character. Here "epochal" does not speak directly of periods of time but of self-withholdings of Being in accordance with which particular destinings, ever bringing themselves to bear via the apprehending and life of specific human communities, variously govern what-is as the latter is provided by way of particular lengthy openings-up of time.[83]

Suddenly, repeatedly, Being's self-sending-forth happens anew as self-withholding. Again and again, at particular junctures in the widely extended opening-out of time, it changes, therewith changing its way of ordaining and disposing what-is.[84] And in this facile self-transforming, the happening of the Twofold in disruptive divergence rules. Accordingly, Being as self-sending-forth tries now this possibility and now that. And ever in so doing, happening on-goingly, it withdraws and withdraws, therewith happening what-is "historically" as fraught with deception, as a realm where misdirection and misunderstanding occur, and occur to an ever-increasing extent.[85] In its primordial happening as concealing, as the Being of what-is as a whole, Being conceals itself in happening as an unconcealing that lets what-is stand away from it and lets the particulars of what-is present themselves in a semblance of enduring and a semblance of nearness.[86] And in history, wherein this manner of

---

[82]In speaking of *logos*, the happening of the Being of what-is as selecting-gathering, Heidegger writes, *Inwiefern ist aber der Logos das Geschickliche, das eigentliche Geschick, d.h. die Versammlung des Schickens, das Alles je in das Seine schickt? Die lesende Lege versammelt alles Schicken bei sich, insofern es zubringend vorliegen lässt, jegliches An- und Abwesende auf seinen Ort und seine Bahn zuhält und alles ins All versammelnd birgt.* "Logos," VA 222, EGT 72. Cf. Ch. III, pp. 124, where the import of this passage is detailed.

[83]"Time and Being," SD 9, TB 9, and "The Origin of the Work of Art," UK 41-45, PLT 41-44.

[84]"The Turning," TK 37f, QT 37; SD 9, TB 9.

[85]"The Anaximander Fragment," HW 310f, EGT 26.

[86]Cf. "On the Essence of Truth," W 89ff, BW 133ff.

Being's ruling happening is extended via time, it evinces itself in ever more undisclosive modes of that presencing as which Being, configuring its ruling by way of destining upon destining, gatheredly, enduringly happens what-is as what presences, i.e., as what endures toward men.[87]

The disruptive divergence that thus informs the happening of the Twofold as history immediately informs, likewise, the happening of what-is, as so viewed, in every one of its particulars. Indeed, the historical aside, in accordance with the priority of the happening of Being as concealing that must ever be transcended in any unconcealing, that divergence is discernible in their presencing as such.

What-is, happening as a manifold, presences, i.e., endures toward us, in that it passes over the open region (*Gegend*) belonging to unconcealing, which happens as the present (*Gegenwart*).[88] That is, happening in particularity and by way of the particularity of time, what-is "whiles" (*weilt*).[89] Whiling, the particulars as which it happens follow one another into and out of immediate presence in the present, yielding to one another in orderly interrelation. Their very coming is, indeed,

---

[87]Unlike other fundamental aspects of the happening of the Being of what-is which rest back upon Heidegger's early portrayals of the human manner of existing, *Dasein* (cf. this chapter, p. 33, n. 36, and p. 46, n. 77), this characterization of that happening as governed to an ever increasing extent by concealing is found only in his depiction of history. It has no parallel in his discussions of *Dasein*.

[88]For another use of *Gegend*, see Ch. XIV, pp. 546f.

[89]"The Anaximander Fragment," HW 322f, EGT 37. For Heidegger what-is, as what presences, is what whiles (*das Weilende*) or what whiles in particularity (*das Je-weilige*). The adjective *jeweilig* usually means actual, for the time being. Heidegger's hyphenating of it in using it as a substantive stresses the note of particularity resident in *je*, "at all times," "at any given time," "each." Our portrayal of what-is in terms of the "particulars" as which it happens rests on this usage.

The discussion that follows here is based upon Heidegger's elucidation of the Anaximander fragment. Heidegger sees that fragment as the earliest word of Greek philosophy in which Being as the Being of what-is, the presencing of what presences, brings itself to utterance. Hence his intricate discussion of the fragment is no antiquarian exercise but an elucidation of the manner in which for him Being shows itself to happen. And the portrayal of Being's happening that there emerges does not serve merely to tell us what Heidegger believes Anaximander intended to say. It tells us, rather, what Heidegger, thinking in his way what is said in those words, finds Being to be saying of itself. Cf. HW 302f, EGT 18f.

already a going away, as each gives place to each.[90] And yet, as whiling, those particulars also tarry (*verweilen*) in the present.[91] In their very moving forward they linger, each proffering itself as itself in the on-going openness that permits self-disclosure to take place; and precisely as thus tarrying they presence. Each has its "while (*Weile*)," the staying as which Being—the presencing that is enduring-unto—happens it by way of a particular present.[92]

But even as the particulars of what-is while unitedly in this way, they are ever surmounting a mode of self-carrying-out even more fundamental to them, namely, a happening in deviance from such whiling. The self-maintaining self-concealing that in the self-revealing of Being into what-is happens as self-withdrawing self-withholding happens in every particular entity of what-is as well. Like Being as which it happens, the single manifold of what-is happens as self-sundering. Its particulars, themselves centerings of shaped happening, do not simply pursue their way in compliant, orderly interrelation. Rather, the tarrying of each partakes inherently of hesitation. The particulars restrain themselves, "keep to themselves (*halten sich an*)";[93] they seek, each on its own behalf, to maintain themselves inviolate in the open region, the present, by way of which unconcealing brings itself to pass. In so doing they "hold out" against whiling, seeking insistently to persist in isolation, each asserting itself *as* itself and *for* itself.[94] Precisely as they presence on-goingly together, the particulars endeavor "to remain more present

---

[90]"The Anaximander Fragment," HW 322f, 337; EGT 37, 41f.

[91]HW 323; EGT 37. *Verweilen* might also be translated as "to delay."

[92]HW 328; EGT 42. Heidegger uses for this particularity of time as the present that is ever peculiar to certain particulars in their whiling the word *jeweils*, "at any given time," "from time to time." The immediate connection between *das Jeweilige* (see this chapter, p. 49, n. 89) and *jeweils* should be noted.

[93]"The Anaximander Fragment," HW 331, EGT 45. On keeping-to-self (*An-sich-halten*) as a self-concealing as which presencing manifests itself, cf. this chapter, p. 24, and n. 10.

[94]HW 331; EGT 45. In this context Heidegger uses verbs formed from the verb *harren*, to wait expectantly, to await, to tarry. "Hold out" translates *verharren*: to remain, to hold out, to persevere. Behind "persist" lies *beharren*, to persist in, to be steadfast; behind "insistently," *beharren auf*, to persist in, to insist upon.

(*anwesender*)."[95] They thrust themselves forward, attempting to intensify their presencing and therewith eclipsing one another. So vehement is this divisive striving that Heidegger can speak of the "mania for persisting"[96] that possesses the particulars as at all costs they ruthlessly exalt themselves over against one another. In this, they strive to achieve "stable continuance (*Beständigkeit*)."[97] In intense self-assertiveness they cling stubbornly to a continuance that merely concerns each individually, as though such continuance were proper whiling.[98] In so striving, however, the particulars do but remove themselves from their *while*— from their passing over together in the open region of the present in an orderliness of compliant, ever-limited tarrying—through rebelliously disrupting the interrelating order belonging to them. Therewith introducing disorder into their whiling, the particulars accomplish in themselves not genuine enduring, which is enduring-unto—even among themselves—but mere constant *lasting on*.[99]

The realm as which what-is happens as historical is, then, one where disruptive self-assertion, carrying with it contention, deception, and confusion, holds sway even to the least particular that happens. The "mania for persisting" that drives the particulars of what-is apart, threatening the manifold of what-is itself with fragmentation and chaotic instability, itself originates in the very happening of Being as presencing. For, as presencing, Being, powerfully happening what-is whilingly as what presences in particularity, so delays it in letting it tarry in the present as to subvert its whiling.[100] "In presencing itself," Heidegger says, "rendering constant rises up (*steht die Beständigung auf*)."[101] The

---

[95]HW 328; EGT 42.
[96]*Sucht des Beharrens*. HW 331; EGT 46.
[97]HW 328; EGT 43.
[98]HW 328; EGT 43.
[99]das *beständige Andauern*. HW 331; EGT 45. One might well consider the contrast between this *Andauern* and *Anwähren*, proper "enduring-unto." See this chapter, pp. 35f.
[100]HW 328; EGT 43. Heidegger's allusion to a "mania" for persisting graphically suggests the urgent forcefulness that he takes to be intrinsic to the happening of Being. Cf. EM 54, IM 59, and Ch. IX, p. 334. Cf. also Ch. II, p. 69, n. 50.
[101]*Im Anwesen selbst . . . steht die Beständigung auf*. HW 328; EGT 43.

very *while* of what-is is possessed of "uprisingness."[102] And it is through this uprisingness that belongs intrinsically to its whiling that what-is, in all its particularity, "insists on mere permanence (*blossen Beständigkeit*)."[103] The happening of presencing as the whiling of the particulars of what-is is, then, ever a surmounting of the deteriorative divisiveness belonging inherently and primally to presencing itself.[104]

In Being's happening as self-unconcealing, its "need" to maintain itself enduringly as self-concealing rules as primary, governing what-is— as a whole and in all its particulars—as the latter presences. And that "need" happens as self-sundering. Being maintains itself—through coming to appearance as the Twofold—in happening itself away from itself, as what-is. Happening as what-is, Being happens as that which is other than and yet the Same as itself and which therefore, as a whole and in every *centering* as which it happens, likewise happens so as to maintain itself as itself. This maintaining-as-itself is the divisive, disruptive drawing-apart—the self-assertive withholding-to-self—that rules so powerfully at once in the unfolding of the Twofold viewed as history and in every centering as which the latter in its unfolding displays itself in particularity.

The divisiveness that has, in the unfolding of the Twofold, the "priority" belonging to Being's happening as self-concealing inherently endangers Being as the Being of what-is. Through it Being threatens itself with disintegration, with disruption, with the decisive accomplishing of self-withholding. For the self-withdrawing as which the Being of what-is happens is tangential. Isolatedness and disjunction and disparity, affecting the relating of Being and what-is as such, grow more and more

---

[102]*das Aufständische*. HW 328; EGT 43.

[103]*besteht . . . auf der blossen Beständigkeit*. HW 328; EGT 43.

[104]Here presencing shows us the fundamental structure of the happening of Being—a differentiating that, via a separating away, happens as a gathering-together (see this chapter, pp. 29-33, and n. 36)—as wrought upon by the primal depletive deviance that follows from the priority of concealing within that happening itself. Cf. this chapter, pp. 44-46. On such divergence as pertaining to the structure in question, see this chapter, p. 46, n. 77. What-is, as what presences, in accordance with its happening in a manner the same as that of Being, shows precisely this same structuring in its happening as a manifold. Cf. this chapter, p. 55, and n. 112.

evident as the Twofold, happening historically, unfolds itself by way of time.

And yet Being as *Difference*, happening as self-sundering, is, precisely as such, the Same. Being's happening-away of what-is is always at the same time a coming-together of the differentiated, revealingly, enduringly, as the on-going unitary Happening that is the Twofold.[105] Self-sundering is transcended in restorative reciprocity of interrelation. Being delivers itself over to what-is; it allows what-is to happen of itself. Yet Being holds what-is toward itself. Happening itself as self-relating, in its self-maintaining as the Twofold, the Being of what-is, as the presencing of what presences, happens after the manner of a handing-over (*Aushändigung*)[106] of what-is that allows what-is to happen, whiling, on its own and as itself, while yet keeping what-is preservingly in hand.

Heidegger uses for this, Being's unique manner of self-relating in its self-unconcealing, the word *Brauch*, which commonly means usage or custom and comes from the verb *brauchen*, ordinarily meaning to need, to want, to utilize. *Brauch* gains the unique meaning that Heidegger finds resident in it through serving to translate the Greek present participle *chrēon* from the verb *chraomai*, related to *cheir*, hand. Used of the Being of what-is, *chrēon* does not speak of a need that is a necessity in the sense of a compulsion, as would the verb "must." It speaks, rather, precisely as a participle with duality of connotation, of the spontaneous, uncompelled self-maintaining that is the handing-over-keeping-in-hand according to which what-is presences as what is in Being.[107]

---

[105]Cf. this chapter, pp. 29-33.
[106]"The Anaximander Fragment," HW 340, EGT 55. And see following note.
[107]*In* chrēon *liegt* chrao, chraomai. *Daraus spricht* he cheir, die Hand. *So bedeutet* chrao *zugleich: in die Hand geben, einhändigen und so aushändigen, überlassen einem Gehören. Solches Aushändigen aber ist von der Art, dass es das Überlassen in der Hand behält und mit ihm das Überlassene.* "The Anaximander Fragment," HW 337, EGT 51f. It is in the light of this, Heidegger's usage, that it must be remembered that when the happening of Being is in question the word "need" and related words that seem to speak of requirement always speak, in our discussion, not of any compelling necessity but simply of the manner of happening intrinsic to Being as such. In the handing-over-keeping-in-hand that is now in question, we glimpse very clearly the

Happening thus in its self-differentiating, Being does not sever what-is to happen in isolation and so to perish. It governs it still. And what-is, happening of itself and yet *in Being*, allows to Being that preserving governance.[108] Happening as the self-sundering Same, Being happens as a "gathering that disposes and freeingly safeguards (*verfügend wahrende Versammlung*)."[109] Happening unconcealingly—happening indeed as truth (*Wahrheit*)—Being frees what-is to happen on its own and in so doing safeguards it, allowing it genuinely to maintain itself in unconcealing; and the freeing and safeguarding are one. In manifesting itself and maintaining itself in the unconcealing as which Being happens as and by way of the open clearing that is time, what-is, ruled by Being happening as unitingness—as a disposing gathering—transcends the sundering thrust of differentiating. It surmounts the powerful tendency toward rebellious self-assertion that comes upon it from out of Being. What-is, happening as particulars, does not disintegrate. Beset by divisiveness, it nevertheless so undergoes its hazardous happening by way of time as to withstand it in genuine stability.[110] What-is whiles. In traversing the open region belonging to unconcealing, it maintains itself as

---

structure—a self-differentiating that is, as such, a gathering-toward—which belongs, as its constitutive manner of self-carrying-out, to the Happening that is named as Being. On that structure, cf. this chapter, p. 33, n. 36.

[108]HW 329f; EGT 43f.

[109]HW 340; EGT 54. The verb *wahren*, whose present participle in adjectival use is here translated with "that . . . freeingly safeguards" ordinarily means to watch over, to preserve, to keep safe. For a discussion in which Heidegger shows that connotations of freeing also belong to *wahren*, see "The Turning," TK 41, QT 42, where *wahren* and its companion noun *Wahrnis* are paralleled with *retten*, to save, defined as a freeing preserving. The noun *Wahrheit*, truth, which Heidegger takes to speak fundamentally of unconcealing (see this chapter, p. 38, and n. 49, and see p. 24, n. 12) stands within the same circle of meaning. Throughout our study *Wahrheit* is ordinarily translated as freeing-safeguarding, and *Wahrnis* as freeing-safekeeping.

[110]HW 329; EGT 44. Behind the phrase "to withstand in genuine stability" is Heidegger's use of *bestehen*, which means, when used transitively as here, to undergo, to get through, to withstand, to overcome. As intransitive, *bestehen* means to be, to exist, to continue, to stand steadfast. Since in this context Heidegger connects *bestehen* with *Beständigkeit* (continuance, permanence, steadfastness) (see following note), we may assume that the verb's full range of connotation is in play in this context. See Ch. VI, pp. 244-247 & 259-264, for discussion of the related *Bestand*.

itself, in unitedness, i.e., it maintains itself as a single manifold of particulars.

In contrast to the *mere continuance* that the particulars self-assertively seek each for itself, this "withstanding (*Bestehen*)" of passing over, which ever maintains itself as whiling, is, Heidegger says, "the fitting stable-continuance (*fügliche Beständigkeit*) of what-is as what presences."[111] It is its orderly continuance as a single interrelated manifold wherein divisiveness is surmounted and the particulars, going beyond their inclination to stand fast alone in the unconcealing granted them, presence, endure toward one another on-goingly, together.[112]

Hence the unfolding Twofold, however divisiveness may rule it as it happens historically, is ruled simultaneously by orderly gathering-together. Whatever disruptive divergent withdrawing, whatever deception or confusion may characterize the happening of the Twofold as history, some unitedness, some unconcealing of the unfolding Twofold always also happens. The Being of what-is meets us always as the presencing of what presences. In one manner or another, always it unconceals itself by way of an orderly, particularized manifold of what-is, as the latter stands forth on-goingly—presencing in the present—in traversing the ever-opening clearing as which Being unconceals itself maintainingly by way of time.

It is this portrayal of Being as the Being of what-is, i.e., as a self-differentiating, single *Happening* that, in its maintaining of itself as itself, through happening as *time*, opens itself disclosively and unfolds itself via ever-changeful self-particularization, which stands central to all Heidegger's thinking. If we grasp that portrayal, we lay hold upon the

---

[111]HW 329; EGT 44. *Dieses je weilende Bestehen des Übergangs ist die fügliche Beständigkeit des Anwesenden.*
[112]Here the happening as a manifold of what-is as what presences evinces—in accordance with the happening of what-is in a manner precisely the same as that of Being, in the unitariness of the Twofold—the same carrying into play of the fundamental structure of Happening, in a sundering in depleting deviance from full-transpiring-as-itself that is ever transcended in a gathering-together, as that evinced in Being in its happening as presencing. Cf. on the structure as such, this chapter, p. 33, n. 36, and on its evincing in presencing, p. 52, and n. 104.

key that can unlock many mysteries of his thought. Failing that, the varied aspects and unique insights of his work must threaten—like the particulars of which he writes—to confront us in isolation, while the unity that pervades them remains unglimpsed, ever evasively concealing itself from view.

CHAPTER II

MAN

The Being of what-is, happening, by way of time, precisely as a self-unconcealing self-concealing, meets us as the presencing of what presences.[1] Thus far we have tacitly assumed the comprehensibility of this word "us." We characteristically think of ourselves as discrete, individual human beings who confront a world teaming with particular entities—other humans included—that constantly impinge on our awareness. And it might readily seem as though we could quite properly incorporate into this way of thinking Heidegger's notion of presencing. But so long as we retain this our customary understanding of who we are, we necessarily find ourselves still very far from what Heidegger means when he speaks of us humans as those toward whom presencing is reached out by way of the ever-opening clearing of time.

Our thinking assumes that we, each of us in isolation, confront a reality that is radically other than ourselves. And if we use the generic term "man," each human individual represents for us a separate instance of its application. For Heidegger, on the other hand, the word "man" (*Mensch*), which he uses frequently, speaks of human beings from the perspective of the oneness that prevails among them. It alludes to individuals only implicitly, as they are understood to be intrinsically united in a corporate oneness. And "man" has no standing ground in isolation from the rest of reality. Man belongs within the great manifold of what-is. He subsists among the entities as which the Twofold unfolds by way of time, and every aspect of his interrelating with them is governed by Being's governing of that unfolding.[2]

---

[1] Cf. Ch. I, pp. 24-26, 35f, & 39f.
[2] Cf. Ch. I, pp. 27-29.

1

At the beginning of his work Heidegger spoke of man with the word *Dasein*. *Dasein* is in German both a verb, meaning to be there, to be present, to exist, and a noun meaning existence. Preempting *Dasein* to speak solely of human existing, and drawing primarily on its verbal force, Heidegger finds in the word a complex meaning. For him "Dasein" does not in any way denote an "I-thing"[3] that stands in external relation to its environing world. This *Dasein* is, Heidegger can say, that which "we always and in particular are," i.e., it is the specific instancing of the happening of the Being of what-is that we humans live out, each in the specificity of time and circumstance peculiar to himself or herself.[4] Hence the word "Dasein" speaks qualitatively of the manner of existing[5] that would be called human, which happens precisely via the commonly governed existing of particular persons. Here our discussion of the Being of what-is, the unfolding Twofold, as a single Happening that both

---

[3]SZ 119; BT 155.

[4]*Dieses Seiende das wir je selbst sind . . . fassen wir terminologisch als Dasein.* SZ 7; BT 27. Cf. also SZ 9, BT 29. For other uses of *je*, "at all times," "at any given time," "each," to bespeak this particularization specified via time, cf. Ch. I, pp. 49f and n. 89, and Ch. V, p. 183. For a parallel thought, see the discussion of Being's happening as *Wesen*, Ch. VI, pp. 253f.

[5]Heidegger uses the verb *ek-sistiern* and the noun *Ek-sistenz*, which he hyphenates to emphasize their literal import, to speak of man's living-out of his existing as a "standing-out" that is a setting-exposedly-out (*Aus-setzung*) into the very disclosedness vouchsafed, from out of Being, to what-is in its happening in particularity. (*Die in der Wahrheit als Freiheit gewurzelte Ek-sistenz ist die Aus-setzung in die Entborgenheit des Seienden als eines solchen*). "On the Essence of Truth," W 85, BW 128. This standing out exposedly to meet what-is arises out of and carries into play the openness for the happening of Being of which the word *Dasein* speaks. (*Die Entborgenheit selbst wird verwahrt in dem ek-sistenten Sicheinlassen, durch das die Offenheit des Offenen, d.h., das "Da" ist, was es ist.* [paragraph] *Im Da-sein wird dem Menschen der langehin ungegrundete Wesensgrund aufbehalten, aus dem er zu ek-sistieren vermag.*) W 84; BW 128. Our employment of "existence" in speaking of *Dasein* certainly endangers the clarity of Heidegger's distinction in word usage. We can only urge the reader to keep that distinction in mind. Since Heidegger's portrayal of *Dasein* fundamentally includes the exposed standing-out of man that *ek-sistieren* names, our use of "existence," "exist," and "existing"—*unhyphenated*—does in fact seem to us to be adequately justified, particularly given the fact that the foremost meaning of *Dasein* in its ordinary uses is "existence."

happens itself away from itself as multitudinously centered shaped happening and governs unitarily in the centerings thus provided must be kept closely in mind.[6]

Da-sein, literally *"Being*-there,"[7] is a manner of being that is fundamentally and pervasively one of relational[8] involvement. The "there" in the word is not locative. *Dasein* is not already a locus of existence, already a *there*, to which movement and accomplishment are secondary. It is, rather, a bringing into play of Being, which so carries itself out as, we might say, to let "thereness" happen. As such, *Dasein* is a manner of existing that, in immediate involvement, takes place simultaneously as a directing of itself outward in attentive, active interest and a drawing to itself of that which is coming into the sphere of its concern.[9]

Despite Heidegger's stated intention in his introduction to *Being and Time*,[10] and despite his subsequent attempts to guard against misinterpretations, the *Da* of *Dasein* has often been taken to refer to a human existence isolatable and explicable solely from within itself, which, in view of its description in terms of elements such as interest and receptiveness in relation to an ever individualized world, is to be identified with the consciousness belonging to a particular human self. In reaction against this widespread tendency, Heidegger later rejected the simple meaning "there" for the *Da* of *Dasein*. Underlining *Sein* so as to stress the primacy of Being in the constituting of *Dasein* and hence for

---

[6]Cf. Ch. I, pp. 28f.

[7]Being is underlined in this translation of *Dasein* so as to stress the verbal force belonging to *Being*-there.

[8]Here and elsewhere in this study the reader should be careful rigorously to separate "relation" from its usual meaning. Here "relate," in contexts where the Being of what-is is in any way under consideration, does not speak at all of a connecting of inherently discrete elements, one of which is to be subordinately referred to the other. The relating that is now in question is a *holding-toward-one-another* of participants that, while they are distinct from one another, are ultimately one. For we are using "relate" to speak of the happening of the single Twofold. See further, Ch. IV, p. 145, n. 10.

[9]For detailed discussion of the constitutive aspects of this mode of transpiring cf. this chapter, pp. 61-75.

[10]SZ 2-40; BT 21-64.

any proper understanding of it, and directly contrasting the word as so written with *Bewusstsein* (consciousness), he insisted that the *Da* should be understood as "openness." *Dasein* becomes, then, "openness-for-Being."[11] Man's existing, his being present as relating-fraught *there*, is itself the openness "needed" and provided for itself by Being in its accomplishing of its self-maintaining.[12]

The self-maintaining self-unconcealing of Being takes place by way of man. Without man no such unconcealing could happen.[13] For in order to bring itself to pass unconcealingly as the Being of what-is and so to maintain itself as itself, Being "needs" an openness for possibility, wherein its self-concealing self-disclosure via the manifold of what-is, happening in particularity, can partake of selection and exclusion, of supersession and change. It needs, that is, the open clearing that is time.[14] And it is precisely, indeed exclusively, in man's ever-relational existing that the on-going interplay of the dimensions of genuine time takes place.[15] It is therefore solely by way of that manner of existing that Being can reach itself forth intricately as presencing, as enduring-unto, via those dimensions and hence gather itself on-goingly to manifestation as the Being of what-is.[16] *Dasein*, understood as *Being*-there-as-openness, is a word that speaks descriptively of what we might call the structure of the unique way in which man exists as an included yet definitively distinguished participant in the on-happening manifold of what-is.

Seen from one perspective, human persons belong just as integrally within that manifold as do all the other particular entities found therein. Like the latter, they are, as belonging to that manifold, "in Being."[17] And all that pertains to the particulars of what-is as such, as subsisting in

---

[11]See the "Introduction" to QT, p. xxxv, n. 2. See also "The Way Back into the Ground of Metaphysics," W 201-203, K 270-272. In "On the Essence of Truth," W 84, BW 128, Heidegger identifies the *Da* of *Dasein* as "the openness of the open (*die Offenheit des Offenen*)."
[12]On the meaning of "need" as applied to Being, cf. Ch. I, p. 53, and n. 107.
[13]See "Time and Being," SD 12, TB 12.
[14]Cf. Ch. I, pp. 40f & 37f.
[15]See this chapter, pp. 69-72.
[16]Cf. Ch. I, pp. 35f & 39f.
[17]Cf. "The Turning," TK 45, QT 47.

that relation, pertains to them. But the humanness to which Heidegger points with the word *Dasein* opens out on another perspective, in which man is seen as possessed of a role and a relation to Being that belong to him alone. That relation to Being, which is his as the being who discloses Being as the Being of whatever is, makes of his relation to all that is and is met by him one of intention that gathers anything whatever toward just that disclosure.

In keeping with its transpiring as the medium of the on-happpening of time, the thereness as which *Dasein* takes place involves a relating-toward that is a relating-back.[18] As such, it transpires via a "surpassing (*Überstieg*)" which is a going-beyond that brings-toward in a disclosing. Such relating, such accomplishing of *Being*-there, which Heidegger characterizes as "transcendence (*Transcendenz*)," is in no sense an adjunct to a previously existing human self. It is the constitutive manner of existing of every human person as such. Only via its transpiring, indeed, does selfhood ever come into play.[19]

In this, its relational happening, *Being*-there happens as Being-in-the-world (*In-der-Welt-Sein*).[20] That is, it takes place inherently as a reciprocal relating wherein a human center has and concernedly interacts with a meaningful grouping of environing elements.[21] That grouping, however, that world, is not to be thought of after the manner habitual to us, namely, as though it were existant beyond, i.e., independently of, the human existing from out of which it is apprehended. The "world" here in question is, indeed, not met at all as something that subsists, vis a vis *Being*-there itself, prior to the relating of the two. Rather, *Being*-there

---

[18]On the dual directionality in play in the happening of time, cf. Ch. I, pp. 39f.

[19]WG 36-38; ER 37-39. In this transcendence, and in the carrying out of *Being*-there as Being-in-the-world as which it transpires, the manner of happening of Being, a self-happening-away that is, precisely as such, a gathering-toward, is directly evinced. In our ensuing discussion aspect after aspect of Being-in-the-world will readily be seen to instance it. On the structure in question, and on Heidegger's portrayal of it as having been given first in just that context with which we are now chiefly concerned, namely, his explication of human openness-for-Being (*Dasein*) in *Being and Time*, see Ch. I, p. 33, and n. 36, and this chapter, pp. 72f, and n. 66.

[20]See SZ 54, BT 80.

[21]See SZ 119-121, BT 155-157; SZ 351, BT 402.

finds itself always, *as Being*-there, already in, i.e., in relation with, a
world that is simultaneously and equally constituted as meaningful by it
and constitutive of it as the meaning-oriented *Being*-there that it is.
"Whatever is in such a way as to be in the world at all," Heidegger says,
"is projected toward world," in an incorporating referral. That is, it is
projected "toward a whole of meaningfulness, within whose referential
relationships intending concern, as Being-in-the-world, has already made
itself secure."[22]

"World," so understood, far from denoting what-is as something
independently encountered by human existing, names a mode of
happening according to which human *Being*-there so grasps as an entirety
whatever for it is, including itself, as to comport itself toward it—and
indeed toward its particular elements—in keeping with the what and how
pertaining to it, i.e., in keeping with the manifesting of its Being, which
the unifying awareness intrinsic to *Being*-there as Being-in-the-world has
therewith disclosed.[23]   Reaching forth comprehendingly even as it is

---

[22]*Das innerweltlich Seiende überhaupt ist auf Welt hin entworfen, das heisst auf ein
Ganzes von Bedeutsamkeit, in deren Verweisungsbezügen das Besorgen als In-der-
Welt-sein sich im vorhinein festgemacht hat.* SZ 151; BT 192. On concern, see this
chapter, pp. 64-68. It is frequently suggested that "world" in *Being and Time* can be
taken as synonymous with Being. It is rather the case that "world" may best be
understood (to speak in terms of the later Heideggerian formulation) as the happening
Twofold to which man belongs, yet to which he relates in a peculiar manner as the one
via whom it is disclosed. For a like portrayal of world as the happening Twofold, see
Ch. V, pp. 173-179.

[23]WG 82-88; ER 83-89. In *On The Essence of Reasons* (*Vom Wesen des Grundes*),
which was written in 1928, a year after the publication of *Being and Time*, Heidegger
treats of and clarifies a number of concepts fundamental in the latter. In so doing, in
the course of his discussion he names and concerns himself overtly, as in *Being and
Time* he did not, with the Being of what-is. In *On The Essence of Reasons*, Heidegger
is concerned to portray what-is as manifest only by reason of the manifestness that
comes upon it by way of the transpiring of human existing. He is concerned with the
Ontological Difference from out of which what-is is brought to disclosure by way of its
Being, which is always other than it and yet always intrinsically disclosing itself via it.
(WG 26-28; ER 27-29.) Cf. Ch I, pp. 27 & 30-33. This dual disclosing is underway
in the providing of world. "World," then, shows us the Twofold before the word
Twofold had been brought by Heidegger into use. Cf. Ch I, pp. 27f, and n. 21. Later
Heidegger came to see disclosing of the "what and how" of anything as a manifesting
of Being only in its withdrawal from full self-disclosure. Cf. Ch. X, pp. 370-372.

situated, as immediately participant, in what-is, *Being*-there transpires as "surpassing to the world," in a launching-forth that is, as Being-in-the-world, an investing of what-is—seen in its entirety though not necessarily with explicitness—with significance.[24] Accordingly, Heidegger can identify Being-in-the-world as "the happening of projecting 'throwing world over what-is.'"[25] Transpiring thus, *Being*-there "lets world happen."[26] And "world," as a manner of happening of the Being of what-is, both permits whatever is encompassed in the entirety as which it happens to manifest itself and, in so doing and in therewith allowing to human *Being*-there the disclosure to itself of itself and its milieu, happens determiningly for the latter in its self-accomplishing. Brought into play via that accomplishing, world "worlds" for *Being*-there and, in its happening as meaningfully disclosive of what-is, therewith "rules" for it as well.[27]

As Being-in-the-world, *Being*-there is always individualized. It is intrinsically possessed of the particularity of "mineness (*Jemeinigkeit*)";[28] for it pertains always to persons each of whom is relationally "in" a world that is in some sense peculiar to himself or herself. Indeed, Heidegger can speak of "'individual' *Being*-there."[29] And he can describe man's peculiar mode of existing which, as ever-transcending Being-in-the-world, has awareness both of itself and of other beings that it itself is not, as having inherently the character of a self,[30] where "self" can be understood to speak of a "who" that sustains its identity throughout the play of changing experience and changing comportment, so relating itself to this change-fraught multiplicity as to happen always as "the selfsame in manifold otherness."[31] Yet persons do not live in isolation. Rather, their

---

[24]WG 82-84; ER 83-85.
[25]WG 88; ER 89.
[26]WG 90; ER 91.
[27]WG 90 & 102; ER 91 & 103.
[28]SZ 42; BT 68.
[29]SZ 371; BT 422.
[30]WG 38; ER 39.
[31]. . . *als Selbigen in der vielfältigen Andersheit.* SZ 114; BT 150. Cf. "The Way Back into the Ground of Metaphysics," W 204, K pp. 272ff. For a passage in which the thought of *Da-sein* as a centered "self" is of key importance see, e.g., SZ 274, BT

individualized existing itself takes place from out of a prior corporateness. Their very Being-in-the-world is itself a "Being-with (*Mitsein*)" one another.[32] It is, indeed, a being-together so characterized by a mutuality of participation as to transpire as a having-a-care-for (*Fürsorge*) directed includingly one toward another, as each centering of *Being*-there, i.e., each person, primordially acknowledges a centering like to itself in each of its fellows and, whatever may be its specific treatment of them, in carrying itself forward on its own behalf, carries itself forward on behalf of those others as well.[33] Always persons exist, in that mutuality, in a meaningful world that is theirs in common, although the world of one historical people will differ from that of another.[34]

The reflexive relational "movement" as which the *Being*-there-as-openness of man, as Being-in-the-world, carries itself forward is a self-directing-toward-so-as-to-meet and a drawing-toward of that which comes to meet. Just as that movement manifests itself in the having-a-care-for via which particular centerings of *Being*-there accomplish their participant relating in their Being-with one another, so it shows itself also—and

---

319; and cf. this chapter, pp. 94f.

[32]SZ 121; BT 157.

[33]SZ 120-124; BT 156-161. The central emphasis in *Fürsorge* (solicitude, care,) as used by Heidegger to speak of the mode of relating as which human Being-with is carried out, is on the care (*Sorge*) that is *Dasein*'s apprehensive concern for its on-going sustaining of itself and hence on the "for-the-sake-of," related always to itself, that informs that concern. SZ 121, 123; BT 157, 160. See also this chapter, pp. 75f & 74. Since *Fürsorge* is intended to speak primarily of the manner of interrelating obtaining for centerings of *Dasein*, as that interrelating is considered thus structurally, it can be in play as that relating no matter in what manner human existing may happen as, manifoldly centered, in its fundamental unitariness it carries itself forward. That happening can indeed take place diversely, for it can be both inauthentic and authentic. (Cf. this chapter, pp. 85f.) *Fürsorge* therefore speaks of a having-a-care-for that, far from being simply a kindly solicitude, as both the German word itself and our translation might suggest, as a concern-with-respect-to can manifest itself in comportment that ranges from self-interested coerciveness to self-forgetful generosity toward the other, depending on the way in which human existing is being carried out in particular instancings of relationship and on how the treatment of the "other" is perceived as furthering the self-sustaining of the existing that is, after one manner or another, via some particular congeries of circumstances underway. (SZ 121f; BT 158.) Cf. also this chapter, pp. 86f & 99f.

[34]Cf. Ch. III, pp. 136f.

indeed concomitantly, in the complexness wherein individualized *Being*-there, as Being-in-the-world, has inherently to do with whatever belongs to its own and its fellows' world[35]—in a counterpart of that mode of *Being*-there's transpiring. For there is intrinsic to human Being-there as Being-in-the-world an ever-particularized, intent-fraught concern (*Besorgen*)[36] wherein *Being*-there reaches toward and draws toward itself non-human components of its "world." These latter lack the capacity answeringly to evoke from *Being*-there the self-mirroring response of having-a-care-for that it accords to other instantiations of itself. But they are possessed of their own sort of evocation of concerned response. They are met as offering themselves, in the meaningfully met configuration of world, as ready-to-hand (*zuhanden*), i.e., as intending themselves for some particular use.

Here "use" should not be taken simply in a narrow practical sense, even though Heidegger's discussion of the ready-to-hand relies on an employment of the word *Zeug*, which can perhaps best be rendered as equipment, and treats frequently of spheres of quite practical activity and of various sorts of tools.[37] *Zeug* speaks of a totality of all manner of entities united in a manifold interrelatedness in referral toward one another and toward some accomplishing. As such it is characterized in

---

[35]SZ 123f; BT 160f. Cf. also this chapter, pp. 86f & 99f.

[36]The translation "concern" for *Besorgen* falls far short of true adequacy. The verb *besorgen*, from which the noun comes, is transitive and has a very wide range of meaning. It can mean to take care of, procure, see to, manage, execute. The prefix *be-* directs the action of the verb always toward the accomplishing of something specific. The English word "concern," again as both noun and verb, seems weak and non-descript beside its German counterparts, but no comparable word that has active connotations suggests itself. Wherever possible we have endeavored to suggest for both verb and noun something of the needed active force. *Besorgen* is closely akin to *Fürsorge*, our "having-a-care-for," and *Sorge*, here translated as "care" (see this chapter, pp. 75f.) We have, however, sought in vain for a form of "care," again a pallid English word and one quite incapable of direct transitive use, that would yield for *Besorgen* even the distant approximation to Heidegger's meaning that "concern" provides. *Fürsorge* and *Besorgen* could be said to speak of outworkings of *Sorge*. Hence this lack of congruence in translations is particularly regrettable, and the close kinship among the German words in question should be kept attentively in mind.

[37]Cf. e.g., SZ 70, BT 99f; SZ 157, BT 199f; SZ 360f, BT 412.

each of its instancings by an in-order-to. It is as thus understood that
"equipment" is possessed of ready-to-handness *(Zuhandenheit)*,[38] an
already intending manner of taking place that Heidegger repeatedly
contrasts with on-handness *(Vor-handenheit)*, a neutral sort of presentness
of entities in manipulatable isolation vis a vis *Being*-there in its
encountering of them, which is derivative from its intent-marked
counterpart.[39] This ready-to-handness can characterize anything whatever
that goes to comprise the meaningfully met world wherein *Being*-there,
precisely as Being-in-the-world, always takes its way.[40] Whatever is
ready-to-hand, from within the use-oriented complex to which it belongs
and from out of a keeping-to-self *(An-sich-halten)* that lends to it its own
integrity in relationship,[41] proffers itself as handleable *(handlich)*, i.e., as
disposable to that use toward which, from out of the serviceability and
usefulness that belong inherently to it, it is already tending.[42] This, its
self-proffering, is answered by a purposeful transpiring of human *Being*-
there in an intending concern that accommodates itself to the ready-to-
handness of that with which it has to do. That concern, in its meeting in
thus being met, is informed by an intentional looking, a sight itself
characterized always by an in-order-to *(Umsicht)*.[43] Transpiring in
accordance with it, *Being*-there, drawing "equipment" close in the
nearness—not geographical but relational—of that which concerns, puts
into play a having-to-do-with *(Hantieren)* and a using *(Gebrauchen)* that
reckon on and take account of the ready-to-hand after the manner of just
this purposeful looking.[44]

---

[38]SZ 68f; BT 97f.
[39]SZ 71-74; BT 101-104.
[40]SZ 70f; BT 99-101.
[41]SZ 75, 69; BT 106, 99.
[42]SZ 69; BT 98. On serviceability *(Dienlichkeit)* and usability *(Verwendbarkeit)* see
SZ 83, BT 114. See also this chapter, p. 68, n. 47.
[43]SZ 69; BT 98. The noun *Umsicht* regularly means circumspection or prudence.
The prefix um- can mean around or in order to. Since in the discussion of the ready-
to-hand in which *Umsicht* is defined Heidegger stresses the happening of an intending,
with a repeated use of um- to speak of purposefulness, we have so rendered *Umsicht* as
to stress that aspect of the connotations resident in its prefix.
[44]*Diese Nähe regelt sich aus dem umsichtig "berechnenden" Hantieren und*

This handling having-to-do-with is not what we would call manipulative.[45]   Rather, it is accordant and enabling.   So too is the "using" in question. *Being*-there as openness-for-Being transpires always as Being-in the meaningfully interrelated totality that is world. Transpiring thus, it discloses that world to itself.   It answers to the tending-toward of the complex of entities there met and, in so doing, frees those entities to be as they already are.   This is a freeing to the intricate, intent-configured, ever-particularized circumstancing *(Bewandtnis)* belonging always to the ready-to-hand.[46]   It is this that the intending concern that gathers the ready-to-hand into play carries out, for it accomplishes a letting-take-its-course *(Bewenden Lassen)* that carries to fulfillment the tending-toward of the latter, as it "lets something ready-to-

---

Gebrauchen. SZ 102; BT 135.

[45]Striking evidence of this lies in the fact that the quotation cited in the preceding note, in which the verb *hantieren* is used, is followed by a discussion in which "regions," the realms of interrelating toward which—i.e., into whose compass—what is ready-to-hand belongs as something contextually "placed," (SZ 102; BT 135) are themselves identified as possessed of ready-to-handness (SZ 104; BT 137). Here the having-to-do-with that is in question can clearly not be simply a manipulative dealing with. The verb that would have that meaning would, indeed, be *hantieren mit*; but see the misleading translation in *Being and Time*. In fact, the discussion of "region" now in question includes a mention of the sun's rising and setting as "regions of life and death, which are determinative for *Being*-there itself with regard to the possibilities of Being-in-the-world that are most its own" (SZ 104; BT 137). These "regions" offer to *Being*-there as Being-in-the-world orienting realms within which and according to which whatever pertains to life and death fundamentally receives its context and is set forth therein as possessed of meaning. Here, in light of Heidegger's allusion to ready-to-handness as pertaining to regions, it is abundantly clear that ready-to-handness is, as a proffering for use, a manner of taking place whose character and meaning are, to his mind, most inclusive; and the response that can correspond to it, far from being a mere manipulative using, must be a purposeful taking-hold-of that avails itself of the "useful" that offers—whether entity or orienting "region"—in order that the self-proffering of the ready-to-hand may be carried to fruition even in the carrying out of human existing. Surely we must have in view in this context, and hence always elsewhere also, a self-presenting by that which is met with which has as its counterpart an unreserved acceding in self-committal to the "intending" in play in that which is met. The presence in the context under discussion of a reference to death confirms this. (On the extreme importance of death for Heidegger, and on the self-commitment with which death should be met, see this chapter, pp. 98-101).
[46]SZ 83-85; BT 114-117.

hand *be* thus and thus *as* it is already and *in order that* it be in that
way."[47]

The taking place of *Being*-there, in its transpiring at once
constitutingly and interactingly as Being-in-the-world,[48] as the self-
directing-outward and returning-to-self that having-a-care-for and
intending concern carry into play is enabled by reason of an already given
basis for orientation. For in its transpiring as Being-in-the-world, *Being*-
there is already possessed of a stance within some sphere of meaning,
from out of which it involves itself and copes in one way or another with
whatever encounters it in that sphere. Heidegger describes this graphically
by speaking of *Being*-there-as-openness as always finding itself in
specificity *(sich befinden)* in "thrownness *(Geworfenheit)*."[49] Whether in
its pertaining to group or to individual, human *Being*-there-as-openness
always happens in an acknowledging of given circumstancing.

---

[47]SZ 84; BT 117. The words *Bewandtnis* (state, condition of things) and *bewenden* (to
remain), together with *bewenden lassen* (ordinarily meaning to let alone) are akin to
the verb *wenden* meaning turn. Heidegger clearly avails himself of this nuance of
meaning in using these words to speak variously of the directionally interlocking
relatedness toward use of the complex of the ready-to-hand that is met as the world in
which *Being*-there always is. The translation "circumstancing" for *Bewandtnis* is
intended to suggest this ever-tending interrelatedness. In the context where *Bewandtnis*
and *Bewenden* are in frequent use, Heidegger employs for "usability" not the usual
Brauchbarkeit but *Verwendbarkeit*, thereby pointing up the sort of tending that is in
question.
[48]Cf. this chapter, pp. 61-63.
[49]SZ 135; BT 174. Heidegger uses the familiar German verb *sich befinden*, which—as
in the question: *"Wie befinden Sie sich?"* "How are you"—alludes, with the use of the
prefix *be-*, to some specific manner of being in which one might find oneself, to speak
of such an aspect intrinsic to the happening of *Being*-there as such. He also uses a
noun of his own providing, *Befindlichkeit*, to speak of the "foundness" in question.
The latter is always specific, but it is anterior to any particular state of which one
might be immediately aware. We use "foundness" for *Befindlichkeit* here with
hesitation, for it might seem to suggest a passive foundness that could be observed, and
would in fact be an appropriate rendering only for a noun such as "Befundenheit."
The foundness here intended is a self-finding, a "befindingness" that is ever underway.
As an active verb *befinden* means to think or judge. Thus both *sich befinden* and
Befindlichkeit, whose employment by Heidegger is presupposed throughout the
remainder of our paragraph, suggest immediately the meaning-finding response that
belongs to *Being*-there as Being-in-the-world.

Accordingly, it can never stand clear, so as to provide for itself either the bent of its own responding in meeting meanifully what is met, or its own determinative milieu. Always from some given basis of circumstanced, determinate transpiring, and in virtue of it, it launches itself out toward that which approaches it and, in so doing, takes the latter to itself.

This "projecting" (*Entwerfen*), wherein *Being*-there, casting itself forward, draws to itself that which meets it, involves a selective self-commitment, a centering on something to the exclusion of something else.[50] Heidegger is not here speaking of choice understood as a conscious act. He has in mind, rather, what might be described as a fundamental structural element in the happening of *Being*-there-as-openness—a self-casting ahead by which what is coming to meet is received and gathered-toward—that is always and everywhere in play as human openness-for-Being relates in any way whatever to its world.

At the crux of this venture-fraught structuring that pertains to the carrying out of man's existing as openness-for-Being lies the self-opening of Being as time.[51] In casting itself out beyond itself, as it finds itself in its thrownness in the world, in order to meet and draw to itself as it already is whatever is encountering it, *Being*-there, in what we have called its reflexive, relational movement, traverses a gap between what already is and what is coming to be. This gap is that crucial openness that we have already seen from a later Heideggerian perspective, namely, the openness for possibility that Being, happening by way of the ceaseless interplay of the three dimensions of time, provides for itself.[52]

The three dimensions of time—arriving, having-been, and

---

[50]See SZ 285, BT 331. The close kinship between the words *Entwerfen* and *Geworfenheit* should be noted. One might say that the "projecting," the self-casting-forward, here in question follows from out of the thrownness on the basis of which *Being*-there as Being-in-the-world ever carries itself out. Heidegger intends that a strong connotation of an impetus given should be heard in *Geworfenheit*. Cf. SZ 145, BT 185. Later he would characterize the happening of Being, which is implicated therein, as a powerful going-forth. EM 54; IM 59. Cf. Ch. I, pp. 50-52, and n. 100, and Ch. IX, pp. 334.
[51]On that self-opening, cf. Ch. I, pp. 37f.
[52]Cf. Ch. I, pp. 40f.

present—reach forth the presencing of what presences.[53] In the transpiring of man's *Being*-there these three dimensions, freighted with presencing, are in play. When writing specifically of man's *Being*-there as openness, Heidegger speaks of it as characterized by temporality (*Zeitlichkeit*).[54] That temporality is comprised of the aspects that he would later call the dimensions of genuine time. "Having-been" belongs to *Being*-there as possessed of thrownness in the world; "arriving" belongs to it in its outward-directed receiving to itself of that which, in its relating to its world, comes toward it to be received; and "present" belongs to it in its holding-together of those other two aspects of its manner of happening, which, ever sundered from one another though they be, are drawn together—poised yet ever changing—into the ceaselessly happening centeredness that allows *Being*-there-as-openness the on-going immediacy of its concerned involvement in its happening as Being-in-the-world.

In writing of man's *Being*-there-as-openness, Heidegger speaks not of "dimensions" but of three "ecstases," i.e., standings-away, of temporality.[55] In accordance with each, *Being*-there stands always outside itself. As thus governed via temporality, it transpires specifically as "ek-sisting (*Ek-sistieren*)," i.e., as an openness that is ever a standing-out-away that lets disclosedness happen.[56] In happening as Being-in-the-world, *Being*-there is forever slipping away from itself as having-been; it is forever overreaching itself toward itself as not yet; and it is, even in the

---

[53]Cf. Ch. I, p. 39.
[54]See SZ 351, BT 402.
[55]SZ 329; BT 377.
[56]"On the Essence of Truth," W 84, BW 128. Cf. this chapter, p. 58, n. 5. Heidegger's words *ekstasis* and *ek-sistieren* must be seen as closely akin. Both are derived, as Heidegger's use of the prefix *ek-* indicates (the usual German verb is *existieren*), from the Greek. They presuppose the verb *histanai*, meaning to stand, and contain the preposition *ek-*, out, away. In speaking of "ekstases" of temporality, then, as characterizing the *Dasein* of man who "ek-sists," Heidegger suggests the unity belonging to human existing, as a standing-out-into-the-open, and the aspects of the opening-up, time, via which, and indeed as which, man exists. On just such an identity of the "dimensions" of time and the happening of Being, the thought already suggested here, cf. Ch. I, p. 39, n. 52.

immediacy of its taking place, forever finding and orienting itself anew. It has no graspable state that it can call its own. It can never lay hold of itself, for it is always, in each of its elemental aspects, already beyond itself. These three ecstases are always happening together. Viewed in respect to each, *Being*-there-as-openness is Being-in-the-world. The world-aspect of this relational existing as openness is always present and always presenting itself anew.

It is important to emphasize that to speak of *Being*-there-as-openness in these terms is not to speak simply of subjective human "experience." Heidegger's description is intended to apply to man as man. He is presenting his understanding of the way in which, viewed most fundamentally, whole human persons live in a world. Those persons and their world belong always to the manifold by way of which Being accomplishes its self-maintaining self-unconcealing as the Being of what-is. Persons and all that environs them, via and within what we may call the structure of openness that is the meaning-accomplishing, relating-allowing transpiring of *Being*-there as Being-in-the-world that permits the disclosive happening of Being as such[57] presence together and toward one another ruled by Being as presencing. But they meet in such a way that he who, as man, both presences toward and receives whatever presences toward him[58] can lay firm and certain hold neither on himself nor on that which comes within his reach, and that which comes is possessed always of illusiveness. It is always slipping away and refusing itself to any final grasp,[59] as the keeping-to-self intrinsic to presencing here brings itself to bear pervasively as evasive withdrawal.

This manner of meeting takes place only inasmuch as *Being*-there-as-openness, which is Being-in-the-world, happens in its temporality. Here a centering transpires that yet partakes of the ever-opening disjunction among having-been, yet-to-come, and present. And solely via it does *Being*-there, as Being-in-the-world, happen as openness-for-Being. As such, it offers the very locus of that openness for possibility that Being

[57]Cf. this chapter, pp. 61-63.
[58]Cf. "The Age of the World Picture," HW 83f, QT 131.
[59]Cf. Ch. I, pp. 41f.

provides for itself so as to maintain itself in happening on-goingly as self-unconcealing self-concealing.[60] In happening, via the disjunction inhering in its temporality, as a transcending toward world that, as thrown, is already implicated in the latter, *Being*-there happens as "freedom."[61] Here "freedom" does not allude to an initiating in any ordinary sense.[62] Heidegger can speak of "will," but this "will" is in no sense a specific, decision-impelled, self-directed "wanting." It is, rather, the self-casting-forward of *Being*-there, as Being-in-the-world, as that self-casting-forward takes place as an acknowledging affirming of proffered possibilities that are discerned as confirmatory of *Being*-there's self-carrying-out as such.[63] As a bringing into play of will so understood, freedom is the transpiring of a forth-reaching openness toward the possibilities discerned in world which, as openness, is already constrained by the circumstancing that world antecedently affords it in its thrownness. Taking place after the manner of this freedom, *Being*-there, by way of the temporality that ever distances it irrecoverably from itself, simultaneously outstrips itself as it already finds itself and is deprived, by that antecedent constraining, of the possibility of fully accomplishing itself, as Being-in-the-world, in accordance with the world that it reaches forth to meet.[64] And precisely via this happening of openness structured by way of time, in which possibilities are ceaselessly provided to human *Being*-there via the world by way of which it itself constitutingly meets what-is, its already determining milieu, the Happening named as the Being of what-is, as the multitudinously varied and changeful Twofold, continually brings itself to light.

2

Here we must remember our characterization of the Being of what-is as pure Happening that, unfolding as the Twofold, happens itself away

[60]Cf. Ch. I, pp. 24-26 & 42f.
[61]WG 102, 106, & 108; ER 103, 107, & 109.
[62]WG 102; ER 103.
[63]WG 100-102; ER 101-103.
[64]WG 110; ER 111.

from itself as many-centered shaped happening while yet maintaining itself, in its self-differentiating, as the Same, and that does so by happening self-sunderingly as the ever opening clearing of time, by way of which it gathers itself forth into disclosure.[65] For it is that very structuring of the Happening named as Being—a going-forth that is a gathering-toward accomplished by way of the openness that time affords— that is shown us, at first (in *Being and Time*) without any explicit reference to Being as the Being of what-is and as it were in anticipation, in the manner of transpiring of human existing that we have just been considering.[66] It is *as* and *via* that existing that Being brings its self-disclosive happening centrally to pass in its accomplishing of self-maintaining. In happening specifically as the Being of man, of man who subsists after the manner of *Being*-there-as-openness, the Being of what-is, bringing itself to pass as self-unconcealing self-concealing, happens as that light-permitting clearing.[67] And it is by way of man in the temporality of his *Being*-there-as-openness as Being-in-the-world that, via that clearing in its ever self-differentiating unitariness, the Being of whatever is unfolds itself revealingly, enduringly, as the Twofold.

The temporality belonging to human *Being*-there-as-openness, with its three ecstases, "lightingly clears the 'there' primordially,"[68] first opening man's existing to the accomplishing of disclosure. Temporality provides the disjunction among having-been, yet-to-come, and present that allows the having-a-care-for and intent-fraught concern belonging to *Being*-there as Being-in-the-world to be in play, as, rooted in an orienting givenness, *Being*-there simultaneously reaches out receptively toward what-is-to-come and gathers itself and the environing reality to which it so intimately belongs together in an immediacy of involvement. Every instancing of care-fraught being-with-others or of accomplishing concern

---

[65]On Being as self-differentiating Same, cf. Ch. I, pp. 29-33. On time as clearing, cf. Ch. I, pp. 37f.
[66]Cf. Ch. I, p. 33, n. 36. Human existing is explicitly considered with respect to the happening of the Being of what-is in *On the Essence of Reasons*.
[67]Cf. SZ 133, BT 171.
[68]SZ 351; BT 402.

(*Besorgen*), with its ever-intending looking (*Umsicht*), presupposes an illumining and a seeing. Fundamental always to *Being*-there-as-openness is some manner of being-able-to-be (*Seinkönnen*).[69] The latter intrinsically carries with it, structurally speaking, a "toward-which" and a "for-the-sake-of-which" that pertain ultimately to itself. To "being-able-to-be" there belongs, therefore, an illumining disclosing that makes accessible, through which "for-the-sake-of" accomplishes itself.

This disclosing Heidegger calls "understanding (*Verstehen*)." The word "understanding" as here used does not denote one cognitive process among others. Understanding is *Being*-there's primordial disclosing to itself of its own manner of transpiring and, accordingly, of the possibilities for self-carrying-out belonging to it.[70] As such, it is *Being*-there's disclosing of Being-in-the-world as constitutive for it. Taking place always by way of a prior seeing and prior grasping that permit a meaning descrying looking-toward to be in play,[71] understanding so manifests the meaningfulness already found for what is met as world by *Being*-there as Being-in-the-world[72] that *Being*-there can, at a most fundamental level, be aware that its ability to be takes place only as ability to be in the world and can, through existing via intending concern and having-a-care-for, by way of the freedom—the forth-reaching openness that is its transcending toward world—accomplish precisely as this Being-in the for-the-sake-of itself that fundamentally directs and governs it.[73]

Temporality is the sine qua non of this illumining disclosing and of its concomitant "seeing." The ecstases of temporality necessitate that *Being*-there, as openness, in accordance with its on-going self-maintaining in its thrownness, so transpire as to cast itself ahead of itself and back upon itself so that it may, in the ever-given milieu of its world, orient and direct itself in some meaningful way and so preserve itself as itself. On-goingly opened up after the manner of the light-permitting clearing

---

[69]SZ 143; BT 183.
[70]SZ 143; BT 182.
[71]See SZ 150ff, BT 191ff. Cf. also Ch. III, pp. 113f.
[72]Cf. this chapter, pp. 61-63. Cf. also SZ 151, BT 192.
[73]SZ 143, BT 182; SZ 336, BT 385. WG 100-102; ER 101-103.

brought to pass as time, *Being*-there discloses itself to itself as Being-in-the-world and simultaneously discloses itself to itself—on the most fundamental level where one might begin to speak of awareness at all—as already able to be in a world but as having always, in order to maintain itself, to draw the components of that world toward itself by way of a hiatus between having-been and yet-to-come. That very gap is the juncture wherein *Being*-there finds itself to be taking place in the world in immediate centeredness, but only as precariously poised, only as ceaselessly bereft of the self-sustaining it has just sought, and as having in its self-directedness to strike out anew in order to continue as itself.

The self-sustaining via which *Being*-there carries itself out in this precariousness underlies and inheres in its every movement and mode of taking place. It is a suffering of the openness as which and via which *Being*-there itself happens. "Standing," i.e., carrying itself out on-goingly, in that openness—which is the very self-opening of Being as unconcealment by way of time—*Being*-there, itself suffering, "standing," the openness wherein it transpires, takes place as an in-standingness (*Inständigkeit*) that is a self-sustaining bearing of the openness to which it is committed and to which it commits itself. Heidegger's name for this, the specific carrying out in all its demandingness of *Being*-there's self-sustaining in openness is *Sorge*.[74] *Sorge* ordinarily means care, solicitude, uneasiness, or apprehension. Here *Sorge* does not allude to any particular preoccupation of *Being*-there with that with which, as Being-in-the-world, it has to do. Rather, it speaks of the care for its own accomplishing of itself, the intent, risk-fraught self-committing to self-continuance, that belongs to *Being*-there as such. In this "care," *Being*-there, in advance of its carrying out of itself as itself, accepts itself in the thrownness that, ever before its self-undertaking, commits it prior to its committing of itself; and in face of the ungraspability of its own self-proffering that is therewith laid upon it, it carries staunchly into play the being-able-to-be that belongs intrinsically to it and carries itself forward, as thrown Being-in-the-world that is itself Being-in-close-company-with (*sein bei*) whatever

---

[74]"The Way Back into the Ground of Metaphysics," W 203, K 271f.

entities may therewith environ it, for the sake, in the broadest sense, of its own self-maintaining.[75] Every specific reaching-responsively-toward-so-as-to-draw-toward of intending concern (*Besorgen*) undertaken by *Being*-there, and every instancing of having-a-care-for (*Fürsorge*) in acknowledged interrelation with other persons, other centerings of *Being*-there, that is put into play in the latter's transpiring, takes place as a manifesting in the on-going immediacy of Being-in-the-world of this ever-self-hazarding care for a self-accomplishing from out of the very precariousness belonging to the openness brought to pass via temporality, as which and in which *Being*-there ceaselessly takes place.[76] This "care" is nothing less than the precarious openness that is *Being*-there's manner of existing, as that openness, via the givenness in specificity that is its ever-already-provided constitutive milieu, maintains itself, holds its way, as the unique, crucially "needed" locus of unconcealing that it is.

Thus it is that *Being*-there, happening as openness by way of temporality, constitutes man as an arena of disclosure. By way of man so constituted the Being of what-is, unfolding as the Twofold, attains to self-maintaining self-unconcealing, and that in two complementary ways. First of all, man's manner of existing involves self-disclosure; it involves a glimpsing of the way in which it itself happens. But equally it involves a disclosing to itself of that amid which it finds itself, that which comes toward it, that with which it has to do. In this dual disclosing taking place in *Being*-there-as-openness-for-Being, what-is, in what we might call its double happening as man and other-than-man, comes meaningfully to light as it is laid hold of, albeit always provisionally, in primordial, self-concerned awareness. And in this way Being, which is constantly happening itself away as what-is, unconcealing itself by way of the latter, finds what it is indeed itself providing, namely, an apprehending that corresponds to its intrinsic "need," which can allow it to happen as self-manifesting.[77]

---

[75]SZ 192; BT 236f. On *bei*, cf. this chapter, p. 90, n. 118.
[76]SZ 193; BT 237.
[77]On the fundamental manner of happening of the Being of what-is, cf. Ch. I, pp. 29-33.

Man's existing, his *Being*-there-as-openness, involves a catching sight of the ceaseless holding-apart of having-been, yet-to-come, and present, as *Being*-there discovers to itself its taking-place in a precariousness wherein its self and its world are forever illusive, even as it goes beyond itself in gathering them together into a meaningful centeredness so that it may preserve itself as itself. In this, his existing by way of ecstatic temporality, man experiences a manner of happening that corresponds directly to the fundamental manner in which Being itself happens. For, happening as presencing, in the multitudinous unfolding of the Twofold, everywhere and always by way of the opening obtaining among time's three dimensions, Being simultaneously withholds itself from itself and reaches itself toward itself. In man's awareness of his manner of existing, then, Being manifests itself to a congruent apprehending.[78] Happening as the single Twofold, Being shows its manner of happening to man precisely in the manner of happening that is his in his world and that he, in his self-orienting self-directing, fundamentally knows. Here the Being of what-is, happening by way of man in his living-out of time, accomplishes with peculiar immediacy, yet always as a self-concealing—a withholding from disclosure—its self-unconcealing.

In man's *Being*-there-as-openness, self-awareness is, as such, at once awareness of Being-in-the-world. Here man encounters the Being of what-is via another perspective. In the self-projecting toward what-is as seen as "world," from a situatedness within what-is as already accepted as world, which is the transcending that distinguishes his existing as such in its meaningful meeting with whatever is met at any given time, man has awareness not only of what-is as manifest but of the unconcealment as which Being, disclosing what-is in the meaningfulness of the what and how of its appearing via the entirety that world confers, permits to it any

---

[78]Cf. "The Anaximander Fragment," HW 311, EGT 27: *Der ekstatische Charakter des Da-seins jedoch ist die für uns zunächst erfahrbare Entsprechung zum epochalen Charakter des Seins.* "The ecstatic character of *Da-sein* is, however, what is for us the most immediately experienceable correspondence to the epochal character of Being." On the meaning of epochal, see "Time and Being," SD 9, TB 9.

disclosure at all.[79] In that awareness, which alone gives a knowledge of possibilities and of bases, and which therefore alone permits the self-carrying-forward of human *Being*-there as Being-in-the-world in its transpiring as being-able-to-be,[80] the "Ontological Difference" between Being and what-is, via which pure Happening brings to pass its self-maintaining self-disclosure[81] and opens the way, therewith, for identifying distinguishing as such,[82] brings itself specifically into play. "Provided that," Heidegger says, "that which distinguishes *Being*-there lies in this, that it comports itself toward what-is in a way characterized by an understanding of Being, then the capacity to differentiate" (in this ultimate sense) "in which the Ontological Difference becomes factical"— i.e., in which it manifests itself in instancing itself in particular disclosings—"must have struck the roots of its own possibility in the ground that is the constitutive enduring (*Wesen*) of *Being*-there." That is to say, it must have taken root, after the manner of a possibility peculiar to itself, in the transcendence intrinsic to *Being*-there as Being-in-the-world.[83] Here, then, an awareness of Being vouchsafes itself via man's encountering of what-is which enables that encountering, permitting its every accomplishing.

Thus, precisely in his meeting with what-is understood as world, man from within the very constitutive transpiring of his existing once again discerns Being. Here again also, as is the case with respect to human existing considered in its aspect as *Being*-there, i.e., as openness-for-Being, this discerning of Being opens out immediately upon the ruling of Being that is directly named with respect to time in Heidegger's central

---

[79]WG 20-22; ER 21-23. Cf. this chapter, pp. 61-63.
[80]WG 112-116; ER 113-117.
[81]Cf. Ch. I, pp. 27 & 29-33.
[82]Ch. I, pp. 34f.
[83]*Wenn anders nun das Auszeichnende des Daseins darin liegt, dass es Sein-verstehend zu Seiendem sich verhält, dann muss das Unterscheidenkönnen, in dem die ontologische Differenz faktisch wird, die Wurzel seiner eigenen Möglichkeit im Grunde des Wesens des Daseins geschlagen haben. Diesen Grund der ontologischen Differenz nennen wir vorgreifend die* Transzendenz des Daseins. WG 26-28; ER 27-29. On facticity, see SZ 56, BT 82, and n. 1. On the meaning of *Wesen*, cf. Ch. I, pp. 35f, and Ch. VI, pp. 252-257.

identification of Being as presencing. Although Heidegger himself does not overtly bring these aspects of his thought together, it is readily evident that they are thoroughly in agreement. It is as Being-in-the-world that human existing meets with the Being of what-is as the presencing of what presences. For it is precisely in the disclosedness belonging to his meaningful, self-orienting self-preserving by way of his comprehending of what-is as world that man receives, via a manifesting as entirety, whatever therewith shows itself to him. And that which thus shows itself is identifiable with equal accuracy as that which presences toward him in a presencing which, as an enduring-toward that gathers together, bestows just such entirety.[84] Likewise, also, it is precisely by way of the transcending toward world through which he carries himself forward as able to be that, via that disclosedness, man presences self-directingly toward whatever is thus met.

In the relational interplay that temporality allows and necessitates for *Being*-there as Being-in-the-world, presencing—enduring unto—finds the locus of its happening. Accordingly, the manner of transpiring that temporality entails, which shows itself first in the constitutive transpiring of human *Being*-there, everywhere marks presencing's sway.

Being, itself a plenitude happening by way of the temporality that characterizes man's existing, unfolds itself in a complex manifold whose particulars, human beings included, sundered though they be from one another by their very particularity and by the impulse toward self-preserving that governs each, yet constantly find themselves in indissoluble and determinative relation.[85] From within that manifold, human beings, ever particularized in their *Being*-there-as-openness and ever wrought upon individually by the pull of self-preserving, nevertheless ever subsist as individuals integrally belonging to groups. Ceaselessly they presence, each as a center of mutually undertaken disclosure. Ever living out an inherently unitary existing and ever gathered together, as that existing transpires in specificity, via the

---

[84]Cf. Ch. I, pp. 55 & 49f, and this chapter, p. 71.
[85]Cf. Ch. I, pp. 50-52 & 52-55.

bringing into play of world, as individuals they endure by way of the on-going opening-up of time. Continually they withstand time's disjuncture, maintaining themselves against that ephemerality of every self-preserving already achieved that time entails. Thus do human beings endure, and their enduring is an enduring-unto, i.e., unto that reality which—always as interrelated particulars, including other persons—meets them even as they, in their complex care-fraught reaching forth, draw it to themselves for the sake of their self-continuance.

Concomitantly, the particulars gathered to light after the manner of world presence toward those human beings, who thus make their way in individualized corporateness, through entering on-goingly the sphere of intent-fraught taking-heed-to via which, as openness, disclosive *Being*-there as Being-in-the-world, in its temporality, is taking place. In coming toward and being drawn toward man via his existing those particulars endure. Herewith they traverse and surmount that same disjunction among the dimensions of time that man himself, in his peculiar existing as openness-for-Being in individualized corporateness, must overleap. For each happens on-goingly in an arriving that is already a passing-away.[86] In the continuity of its self-sameness, each presents itself at once as arriving, as transiently present, and as having-been. None can escape these disjunctive dimensions of its happening, so as to arrest itself and be only present; rather, each and all of the particulars present themselves only provisionally. They linger in the open region of unconcealing, as the three modes of their happening by way of temporality are centered into the illumined arena of disclosure that allows *Being*-there, as Being-in-the-world, to happen self-maintainingly. But they yield to one another and depart, as the precarious self-preserving that belongs to *Being*-there-as-openness ceaselessly demands a new reaching forth beyond a new thrownness, so that grasped as world, what-is, in this, its particularity, may be newly disclosed, met, and brought near for the sake of that preserving.

The particulars of what-is presence together toward man both

---

[86]Cf. Ch. I, pp. 49f.

meaningfully and changefully. By way of his existing they are brought to light as an ever fresh and transitory "world." The fundamental awareness that, in its openness, characterizes man's existing as *Being*-there that is Being-in-the-world involves a disclosing apprehending of a unitedness that transcends discontinuity and diverseness as, in the intending reaching-forth intrinsic to it in its self-sustaining, that awareness brings itself to bear for the accomplishing of the toward-one-another of man and what-is—itself both met and presupposed as world—that allows them, precisely in their interrelation, to endure in the milieu of the illumined openness that the temporality of *Being*-there provides. And via that awareness Being, as presencing, i.e., as enduring-unto which conjoins the sundered that together they may remain, offers itself, by way of what-is as what presences, seen as a meaningful world, to the human apprehending that, itself structured by way of temporality congruently to Being's rule as presencing, can alone disclose and catch sight of its, Being's, manifesting of itself.

3

In happening as presencing, Being happens as unconcealing that is at once and even more fundamentally concealing.[87] As such, it provides for itself, in man's *Being*-there as Being-in-the-world, interrelated centers of meaningful disclosing. What lies outside these centers—even, to choose an obvious example, the reality that subsisted before man as man was present—Heidegger does not say. Indeed he cannot. For to him all knowing and naming take place solely via man's existing by way of the open clearing of time that he lives out. Man has no access whatever to an "outside of" beyond that ever-opening clearing by way of which his disclosive existing happens.[88] Nevertheless, certain indications point to such a "beyond." Whatever presences by way of the open clearing that is

---

[87]Cf. Ch. I, pp. 24-26.
[88]Here one is reminded of the answer that St. Augustine maintains a certain person gave when asked what God was doing before he created the world. He replied, "Preparing hell for those who pry into mysteries." See St. Augustine, *Confessions*, Bk XI, Ch. XII.

time arrives and departs. Thus it *is*. For Heidegger nothing that is—and this means no particularized centering of happening—is without some antecedence.[89] Heidegger's use of words like "opening" and "closing" in speaking of Being alludes by implication to Being as "beyond" its happening as the light-permitting clearing of time. Indeed, he can say that were man not constantly to receive presencing as he does, then Being would remain shut away and concealed.[90] In such a thought we can glimpse the implied "elsewhere" or perhaps, more accurately, the implied "how-else" that must be inferred as the antecedence "out of" which what-is comes. That "how-else" is *Being*—the Being of what-is—solely as concealedness. This is to say, it is Being as Nothing.

The word "Nothing" (*Nichts*) does not for Heidegger have a privative meaning, as though it spoke of a lack. "Nothing" is, in fact, Being itself[91]—inherently a plenitude; but it is Being precisely and totally as self-concealing. Using the word with this meaning, we may say that "beyond" the centers of disclosing that man's existing by way of temporality provides—where even the words Being and Nothing cannot sound—there "is" Nothing.[92]

In a relatively early work, Heidegger speaks of "earth"—which is what-is, happening concretely as rock and storm, animal and plant, tone and color—as the self-closing; earth's every component is as such intrinsically impenetrable to elucidation.[93] When we raise our question regarding what "was" before man's advent, this thought of "impenetrable" reality hints at an answer. And yet, from Heidegger's point of view, it can once again only tell us that "beyond" man's existing via temporality, by way of which the manifold of what-is can find disclosure, there is Nothing. It is only with Being as concealing

---

[89]Cf. "The Origin of the Work of Art," UK 87, PLT 76.
[90]"Time and Being," SD 12, TB 12. Cf. "The Thing," VA 178f, PLT 179f.
[91]"The Age of the World Picture," HW 104, QT 154. Cf. "The Thing," VA 177, PLT 178f. See also "What is Metaphysics?" W 17, BW 110.
[92]It is certainly improper, according to Heidegger's way of thinking, to use the word "is" in this connection; but its provisional use offers the only viable alternative to silence at this point.
[93]"The Origin of the Work of Art," UK 41-47, PLT 42-47.

happening as unconcealing—i.e., with the Being of what-is—that we have to do.

Being happens as unconcealing by way of the interrelated centers of disclosure as which man's existing takes place. But those centers are by no means unreservedly bright. For Being's happening as unconcealing is simultaneously a happening as concealing;[94] indeed it is ever a transcending of the latter.[95] And the tangential pull of self-maintaining as self-withholding keeping-to-self is, as it were, ever the shadowing "not"— indeed, the Nothing—that every unconcealing must surmount. Man in his existing is constantly required to suffer its impact. In the distancing of itself from itself that ceaselessly befalls his existing in its transpiring by way of temporality; in the precariousness of its being-able-to-be that afflicts that existing from out of this, its constituting by way of time; in the discrepant correlation between encountering possibility and already prevailing actuality that ever pertains to it in its happening as Being-in-the-world as, carrying itself forward on the basis of thrownness, it does endeavor to "be"—in all this, at the crux of Being's holding sway via man's *Being*-there as openness-for-Being so as to accomplish its self-disclosure, Being's withholding of itself in its on-happening-forth rules as that "not," i.e., as the "Nothing" as which it happens as concealing. Both what-is, as world, and the human existing that, inhering in what-is and yet so transcending the latter as to disclose it as a world that both allows that existing itself to accomplish itself on-goingly and, as prior, restricts the possibilities open to it in its so doing, are everywhere denied Being's unreserved ruling. Precariousness pervades every time-structured disclosing as which Being brings itself into play. And man, existing via that precariousness, beset by discrepency, by change, by the need to distinguish and understand, to comport-toward and to compose-with, in his unceasing confronting of that provision of the possible that brings itself to pass by way of what is the finitude of his own self-reaching-forward from out of the constraint of thrownness, can but comport

---

[94]Cf. Ch. I, pp. 43f.
[95]Cf. Ch. I, pp. 24-26.

himself by way of an existing that is never without question and can but achieve a disclosing that is never without the obscuring that concealing, ever bringing itself to bear, continually puts into play.[96]

By way of that very precariousness, Being's ruling as concealment-fraught keeping-to-self so centers its pervasive dominion in the mode of happening—a carrying of itself forward toward what-is from the midst of what-is—via which human *Being*-there as Being-in-the-world from out of Being pursues its self-maintaining course as to set human existing as it were against itself and to affect most profoundly the disclosing that that existing alone permits.

Here we must bring to mind, from a context chronologically later than the one where that existing is overtly under discussion, the manner of happening of what-is that we have already found Heidegger to portray. What-is, in its happening in particularity as what presences, is so happened by Being as presencing as ever to transpire in a surmounting of a disruptedness occasioned by an urgent tendency toward a self-concerned achieving of permanence, in its accomplishing of any proper whiling, that is imparted to it from out of presencing as such. In their passion for self-maintaining, the particulars of what-is continually obscure and eclipse one another to man's self-orienting seeing. And again, happening self-concealingly, those particulars refuse themselves to disclosure, offering not genuine appearing, but mere semblance.[97] From out of such happening of what-is, confusion assails man.[98] Moreover, that which draws near and is drawn near in the intending, accomplishing, "care"-imbued heedfulness intrinsic to his *Being*-there remains present to him only transitorily. Always, soon or late, it falls away, withdrawn via forgottenness (*Vergessenheit*).[99]

---

[96]Cf. WG 110-116, ER 111-117.

[97]Cf. "The Anaximander Fragment," HW 322f, EGT 37, and Ch. I, pp. 50-52. Cf. also "The Origin of the Work of Art," UK 57, PLT 53f.

[98]Cf. "The Age of the World Picture," HW 84, QT 131. See also "On the Essence of Truth," W 92, BW 136, and this chapter, p. 89.

[99]Forgetting, oblivion, is never for Heidegger human forgetting. It is always, first of all, referable to the happening of Being itself, which, as concealing, rules as and by way of forgottnness. See ZS 90, QB 91; "*Alḗtheia*," VA 265, EGT 109.

This manner of being met must, however, not be thought of simply as something that befalls man from beyond himself. Rather, as first found in its taking place, his very existing brings just such a mode of meeting into play.[100] The latter is depicted as arising, indeed, from the very fact that, in a way peculiar to himself as the one who lives out openness-for-Being by way of time, man himself ever strives self-assertively to persist as secure.[101] The precariousness intrinsic to his existing daunts him, and he falls into an inauthentic (*uneigentliches*)[102] *Being*-there, one not

---

[100]SZ 370f; BT 422.

[101]It should here be remembered that, although his existing has no parallel among the modes of transpiring belonging to other particulars that go to make up the manifold of what-is, man himself, viewed from one perspective, belongs within that manifold. Cf. Ch. I, p. 28f. Even as whatever presences toward him is ruled by presencing, so too is he. Any impulse that pertains to presencing, here the urgent, self-assertive impulse toward persisting that so directly manifests the powerful thrust of Being's ruling (cf. Ch. I, p. 51, n. 100), pertains alike to what presences to him and to himself.

[102]Cf. SZ 175ff, BT 219ff. In our present discussion, in accordance with the now accepted adequate English usage, "inauthentic" translates *uneigentlich*; "inauthenticity," *Uneigentlichkeit*. "Authentic (*eigentlich*)" and "authenticity (*Eigentlichkeit*)" will also be found here. Elsewhere in our study, save in discussions where human existing is in question, *eigentlich* is regularly rendered with "genuine," and "genuine" is used in formulations of our own in which Heidegger's use of *eigentlich* is presupposed. This practice is intended to prevent those other characterizations made with or presupposing *eigentlich* from suggesting first of all a transpiring within the context of human existing rather than within that constituted via the provisive Happening from which such existing itself arises, since the latter context is that to which *eigentlich* (and hence *uneigentlich*) always ultimately points. In fact, neither authentic nor genuine ever provides truly adequate translations. Both *eigentlich* and *uneigentlich* should, if it were possible, always be translated with words that would overtly express the meaning resident in our word "own." We have, however, sought in vain for any such equivalents, and the reader must simply be urged to keep in mind this vitally important aspect of meaning (cf. Ch. I, p. 37, n. 46.) Words built on *eigen* (own) always carry connotations of ownness. They must, in a Heideggerian context, always be heard as closely akin to one another and as having that import as primary. (Cf. Ch. IV, p. 149, n. 28.) The meaning that Heidegger intends for "authentic" and hence, in contrast, that intended for "inauthentic" can be seen strikingly in a statement made in the course of his discussion of inauthenticity, namely, that the "they-self," as "the self of everyday" as which Being-there transpires inauthentically (see our ensuing discussion), is to be distinguished from the "authentic (*eigentliche*)" self, the self that has been expressly (*eigens*) taken hold of," i.e., that has been taken hold of self-accomplishingly in the way that is properly its own. (SZ 128; BT 166) For a discussion of *eigen* in conjunction with the elucidation of the

genuinely his own.

In that inauthentic mode of happening human existing, in its on-going taking place in a diversity of centerings, protects itself by abandoning its individualized centeredness and surrendering itself to a corporate manner of its transpiring. An individual undertaking of openness in an ever given context of world met in relationally determined specificity is lacking to it. It is, indeed, this failure of *Being*-there to transpire as an openness wherein an individually centered self-carrying forward takes place that makes this mode of happening inauthentic. For here *Being*-there lacks what we may perhaps call the own-ness, the intensely individualized self-accomplishing, that is its "authentic" way of carrying itself out.[103]

Happening inauthentically, human *Being*-there-as-openness-for-Being, as Being-in-the-world, does not simply accept its thrownness, reach forth answeringly in self-accomplishment toward what approaches it meaningfully as world, and gather itself together ever afresh in the ever happening present. Instead, that *Being*-there so transpires as Being-in-the-world as at once to find that world readily comprehensible, to conceal the relational complexity that is intrinsic to it, and therewith to mask the true nature of its own proper transpiring together with it *as* world.

The transpiring of *Being*-there in inauthenticity is its transpiring as what Heidegger calls "the they (*das Man*)"; therein, the persons in whom the happening of *Being*-there, is instanced, in the having-a-care-for (*Fürsorge*) in play among them meet one another each from out of the for-the-sake-of-itself that intrinsically belongs to *Being*-there as an accomplishing of being-able-to-be.[104] Those persons meet always, in one way or another, concernedly via their undertakings. And as such meeting takes place, in the intending, accomplishing "concern (*Besorgen*)" of each with that which he—from the perspective of his own self-carrying-out—has taken hold of, there is constantly in play apprehensive, risk-sensitive "care (*Sorge*)" regarding whatever distinction in accomplishment or

---

words *Ereignis* and *ereignen*, see Ch. IV, pp. 149-151.
[103]SZ 128; BT 166.
[104]Cf. this chapter, pp. 64 & 73f. On *das Man*, cf. SZ 126ff, BT 163ff.

relative power subsists between himself and the others with whom he finds himself together in a world. The Being-with-one-another that is thus in play is, though hiddenly, disturbed (*beunruhigt*) by this its care about disparity (*Abstand*); and in transpiring thus, it is, as a mode of Being-with, characterized by deterioratedness (*Abständigkeit*).[105] For *Being*-there, in its happening in individualized centerings as Being-with, fails to maintain itself steadily in face of the precariousness that inheres in the awareness of the sort of disparity in the self-carrying-out of those centerings that is here in question. In each of its instancings, there prevails an insecureness vis a vis the others in their interrelating with it—an interrelating everywhere marked by self-interest like its own. And with this, in each case the dominion of the others, to whom determining reference is made, is tacitly acceded to.[106] It is in just such transpiring, in a falling away from individualized being-able-to-be as Being-in-the-world—in all the specificity of particular circumstancing and particular self-carrying-forward belonging thereto—that the happening of *Being*-there as the "they" consists.

In happening thus as the "they," *Being*-there could be said to forfeit particularized centering as such. The "they" is no one in particular; yet it is everyone, including the person who, cloaking his subservient submission to its dominion, invokes it as though it were quite other than himself.[107] As this "they," *Being*-there, having obviated the burdensomeness of its transpiring as Being-with, likewise relieves itself in each of its centerings of a concomitant burden, namely the demand for self-direction intrinsic to its manner of happening as a being-able-to-be that must in each case, from out of a thrownness ever particular to itself,

---

[105]SZ 126; BT 164. *Abstand* can mean either distance or disparity. Heidegger has formed the noun *Abständigkeit*, which is not in ordinary use, from the adjective *abständig*: deteriorated, stale, flat, dessicated. It is therefore clear that the distance here in question is a disparity and that the preoccupation with differences among persons of which Heidegger is here speaking is itself of a severely wasting character. This is an interpretation that is amply borne out by the context in which these words appear. Unfortunately the translation of *Abstand* and *Abständigkeit* in *Being and Time* does not display these important nuances of meaning. On Being-with, see this chapter, pp. 63f, and n. 33.
[106]SZ 126; BT 163f.
[107]SZ 126f; BT 163f.

meet, understand as meaningful, and draw affirmingly toward itself for its own self-accomplishing whatever, in the broadest sense of the word, proffers itself in its own particular relational context for its *use*.[108]

This mode of human existing that manifests itself as the "they" transpires as a Being-with-one-another that, in its deterioratedness, puts into play as determinative an undistinguishedness that obviates the uneasiness that afflicts *Being*-there in the very happening of Being-with. In carrying itself out via an accomplishing concern, it makes averageness regnant.   Assiduously suppressing the exceptional, it levels down everything with which it has to do. In accordance with this, it meets and understands   everything   via   what   Heidegger   calls   "publicness (*Offenbarkeit*)." Everyone is taken to have access to everything. Correspondingly, everything, including human *Being*-there itself, is conceived as something on-hand, i.e., as something discrete, something detachedly   present,   something   unencumbered   by   specificities   of circumstancing and relatedness that would demand a genuinely centered, individual commitment of openness as a mode of human existing adequately accordant with it, and hence as something equally accessible to all who might wish to have to do with it in any way.[109] In all this, readiness-to-hand remains in play; Heidegger can even apply the word to publicness itself.[110] Circumstancings in relatedness-toward pertain to the entities met meaningfully from out of the world in which human existing as the "they" transpires. But the unexpectedness belonging to that which proffers itself via the ungraspable givenness of thrownness is avoided. The circumstancings, bespeaking interrelatednesses, that belong to entities encountered by the "they" are such as are already known to the latter— even as everything is known to it as everyone and no one—and they are governed by the limitations dictated by the averageness in whose accomplishing the "they" takes its way.[111]

---

[108]On this disburdening see SZ 128, BT 165f. On the self-proffering for use of that which is met see this chapter, pp. 64-67.
[109]SZ 129; BT 167.
[110]SZ 127; BT 165.
[111]SZ 129; BT 167.

This manner of human existing that is its transpiring as the "they" is called by Heidegger "everydayness *(Alltäglichkeit)*," for it is precisely *"the* mode of existing in which *Being*-there sustains itself every day." As such it must be understood as the very "how" of on-going existing that pervasively governs *Being*-there "lifelong."[112]

Thus ubiquitous is the everydayness in which, in inauthentic transpiring, *Being*-there as Being-in-the-world contrives to escape the vulnerability that its happening by way of temporality thrusts upon it. The wasting falling-away *(Verfallen)*[113] from its authentic mode of happening that therewith marks it carries it, in virtue of the prescriptive Being-with-one-another of the "they" to which it submits itself, into a complacent tranquility *(Beruhigung)* from which disquieting precariousness is hidden.[114] This tranquility is, however, by no means an inert calm. Even when its thrownness is hidden by *Being*-there from itself, the on-going movement provisive of that thrownness is ever in play. Hence the very falling away that lets *Being*-there find itself at ease carries it into a busy preoccupation with all manner of matters and with itself, a preoccupation in which fullness of understanding is assumed and boldly exercised, even where deceptive ambiguity is, quite unnoticed, constantly in play.[115]

Happening thus self-vitiatingly as falling, *"Being*-there has fallen sunderingly away *(abgefallen)* from itself as an authentic potentiality for being itself," i.e., it has drawn rebelliously away from itself[116] as an individually centered potentiality for Being that, acknowledging its particularity, owns to the precarious transpiring that is, as so centered, its own. As such it "has fallen toward and into *(an)* the world."[117] As lost

---

[112]SZ 370; BT 422.
[113]*Verfallen* means to decay, to decline, to lapse.
[114]SZ 177; BT 222.
[115]SZ 178; BT 222f. On ambiguity *(Zweideutigkeit)*, see SZ 173ff, BT 217ff.
[116]The verb *abfallen* means to fall off, to decrease, to degenerate, to desert, to revolt, or to secede. Its use must suggest to us the notion of self-assertion with whose mention our discussion of the inauthentic transpiring of human existing was begun. Cf. this chapter, pp. 84-86. See also this chapter, pp. 90-93.
[117]SZ 175; BT 220. On *an-* as connoting simultaneously for Heidegger both "toward" and "into" see Ch. I, p. 36, n. 43.

in the deceptively transparent publicness that characterizes the "they," *Being*-there is so absorbed in the Being-with-one-another that the latter bespeaks as to trust to the idle talk (*Gerede*), with its ever-informative curiosity and its seemingly authoritative yet ambiguous utterance, that habitually guides that sort of Being-with. In transpiring thus, *Being*-there is in its everydayness "*in close company with* its concern-met world." And just this "being depleted in intimate company with" is manifest in the erosive publicness to which, as the "they," it abandons itself.[118]

Heidegger insists that this self-wasting fallen-awayness (*Verfallenheit*) of human *Being*-there is not a secondary, defective state to which, from an already full carrying out of itself authentically, it can be said to come. Such fallen-awayness is, rather, a manner of transpiring that is intrinsic to *Being*-there as such, and one that is found always to be in play in the on-going course of the latter's inherently precarious self-carrying-forward. "As something decayingly falling (*als Verfallendes*)," Heidegger says, "*Being*-there, as factically Being-in-the-world, has already fallen away, drawn asunder (*abgefallen*), from itself."[119] As immediately found as *Being*-in-the-world—i.e., as found in the specificity of its taking place as itself—precisely in its self-accomplishing from out of thrownness, *Being*-there displays itself as already having struck self-vitiatingly out into its inauthentic manner of happening. It is and can be known only in a duality of transpiring.

Thus can the inauthenticity of human *Being*-there-as-openness-for-

---

[118]SZ 175; BT 220. Heidegger writes: *Der Titel* ["*Verfallen*"], *der keine negative Bewertung ausdrückt, soll bedeuten: das Dasein ist zunächst und zumeist bei der besorgten "Welt." Dieses Aufgehen bei . . . hat meist den Charackter des Verlorenseins in die Öffentlichkeit des Man.* The preposition *bei* means near, with, alongside. Connotations of closeness are strong in the word. Like the French *chez* it means at home with. Heidegger's repetition of *bei* here shows the importance of the preposition in the context and demands careful attention. The use of *bei* with *aufgehen* (rise, come loose, be spent) is unusual. Almost immediately Heidegger uses the ordinary phrase "*aufgehen in*," meaning to be absorbed in, but with Being-with-one-another as object of the preposition, rather than world. Our somewhat difficult translation attempts to display the distinction that he clearly intends.

[119]SZ 176; BT 220. On facticity, see SZ 56, BT 82, and n. 1. For the instancing, in a later context, of this same wasting diverging-from-self as intrinsic throughout to the happening of the Twofold, the Being of what-is, cf. Ch. I, pp. 45f, and n. 77.

Being be seen to evince the impulse toward self-maintaining—here manifesting itself as the impulse toward being able to be securely—that we have previously found, in considering a chronologically later context, to characterize whatever is. That impulse arises out of the very Happening that is named as Being.[120] In the case of man, it brings itself to bear in a way accordant with his unique manner of carrying himself forward within the manifold of what-is. In the unfailing sensitivity of his existing, thanks to the latter's taking place in awareness, to the precariousness that besets it in its on-going happening by way of time, the fundamental manner of Being's happening can be seen to be determiningly in play.

To speak thus of Being's fundamental manner of happening must be, when this phrase is accorded its most basic meaning, to speak of Being's happening as self-unconcealing self-concealing. And indeed, as Heidegger shows in a discussion in "On the Essence of Truth,"[121] published only a few years after that from *Being and Time* which has just occupied us, at the root of the inauthentic, ever-falling transpiring in everydayness in which, in a way at once self-assertive and self-vitiating, man's existing carries itself out in face of precariousness, there lies the happening of Being as concealing that maintains itself through happening as unconcealing.

That concealing, the unplumbable mystery of keeping-to-self that happens ever transitorily by way of ever-opening time,[122] at once rules what-is in its entirety and pervades man's existing as both instancing the latter and allowing its happening to come to consummation in the manifesting of all that is.[123] The revealing harboring-forth of the manifold of what-is into the particularity proper to it, which is accomplished via man, maintainingly carries into play primal concealing.[124] At the same time, in happening thus revealingly, Being conceals its happening as pervasive

---

[120]Cf. HW 328, EGT 43, and Ch I, pp. 44f & 50-52.
[121]W 73ff; BW 117ff.
[122]Cf. Ch I, pp. 24-26 & 41f.
[123]Cf. "On the Essence of Truth," W 89, BW 132f.
[124]W 93; BW 137. *Die Entbergung des Seienden als eines solchen ist in sich zugleich die Verbergung des Seienden im Ganzen.*

concealing. What-is, happened and happening disclosively in specificity—both as a single manifold and as the particulars of that manifold that manifest themselves at any given time[125]—stands forth from its inherent concealedness, thereby eclipsing its Being. Likewise, in this concealing of concealing, the particulars of what-is themselves obtrude, thereby eclipsing their Being.[126] For his part, in his carrying out of his existing as openness-for-Being, man has, in his awareness of the mystery-governed happening in ungraspable precariousness of all that meets him and, indeed, of himself, an awareness of that very concealing that is concealed. For, as man, he "stands out (ek-sists)" in just such precariousness, in the very grip of concealing, to meet what, from out of its keeping of itself concealingly to itself, presents itself to him.[127] But just in so doing, in that awareness he, like all other particular beings, steadies himself in a self-preserving that masks in its fullness the ruling Happening in accordance with which he exists. He turns from the mystery of invincible, all-ruling concealing that withdraws everything, including himself, from his grasp, and turns toward what-is in its seeming standing-alone.[128]

So turning, man "stands in" (*in-sistiert*) toward what-is, as the latter proffers itself—in the self-assertiveness intrinsic to it—as isolated, reliable particulars.[129] In so doing he turns to this and that as current and

---

[125]For Heidegger's presenting of this dual character of what-is, see, e.g., "On the Essence of Truth," W 85, BW 129: . . . *dass einzig sie (die Freiheit) einem Menschentum den alle Geschichte erst begründenden und auszeichnenden Bezug zu einem Seienden im Ganzen als einem solchen gewährt.*" In discussing the concealing of concealing in this essay, Heidegger repeatedly uses together the phrases *Seiendes im Ganzen* and *Seiendes als eines solchen*. The latter phrase is unusual. It is clearly not the same as *Seiendes als solches* (what-is as such), for in the same discussion Heidegger also uses *als solches* with its ordinary meaning (W 93f, BW 137). The presence of the indefinite article makes it necessary to understand *Seiendes als eines solchen* to refer to what-is as something particular in contrast to what-is as a whole, in its entirety.
[126]Cf. Ch I, p. 40.
[127]"On the Essence of Truth," W 84f, BW 128f.
[128]W 89; BW 132.
[129]W 91; BW 135. *In-sistieren* (standing-in-toward) is counterpart to *ex-istieren* (standing-out exposedly) (cf. this chapter, p. 58, n. 5). The manner of transpiring that

available, as practicable and manageable (*das Gangbare*). Therewith, he goes mistakingly astray (*irrt*) in the deviant disruptedness (*Irre*) that arises precisely from Being's happening revealingly as *concealing*.[130] The pervasive concealing that holds sway in what-is as a whole is forgotten, although, as forgotten, it haunts man still.[131] Man mistakes what meets him and mistakes himself.[132] Disturbed by withheldness, by precariousness, yet relying on himself and on what meets him as accessible, in what is, ultimately, his abandonment by Being in its ruling happening as self-concealing, man trusts to his assessing of the familiar and to his own capacities to assess and control. Confidently he undertakes collectively to make himself secure through purposeful intentions and multifaceted planning with reference to currently perceived needs and commonly held outlooks.[133] Thus, thanks simultaneously to the withdrawal of primal concealing in virtue of the revealing of what-is into its manifestness in particularity and to the implicating of man, answeringly, in that deceptive revealing, that very inauthenticity is brought into play in which, in the transpiring of his existing in everydayness, man masks from himself his intrinsic awareness that insecureness and transitoriness, ungraspability and sheer mystery, belong decisively to all that meets him in his world and to himself.

In the inauthentic transpiring of human *Being*-there that awareness is masked, but it is not quelled. Sometimes it obtrudes, indeed, in what Heidegger calls anxiety (*Angst*), an oddly estranged calmness wherein, for an individual, the components of a world so slip away from his identifying apprehending as to show themselves as radically other, in disclosing their utterly unreachable contingency, which is, as it were, overspread by the appalling Nothing that such sheerly given thatness bespeaks.[134] Here Being itself, as the radically other than whatever is,

---

it names is precisely an inauthentic happening of human existing as such. For a parallel portrayal of the mode of existing to which it points, cf. this chapter, pp. 89f.
[130]On the inception of the rule of such disruptive divergence, cf. Ch. I, pp. 45f.
[131]W 91; BW 134f.
[132]W 92f; BW 136.
[133]W 90f; BW 134f.
[134]*In der hellen Nacht des Nichts der Angst ersteht erst die ursprüngliche Offenheit des*

i.e., as "Nothing," vouchsafes itself in direct confrontation.[135]

Such a confronting implies the correspondingly radical question that may or may not be overtly asked then, or in other all-enveloping moods, but that reaches to the very relation of Being and what-is: "Why is there what-is"—the single, particularized given of that which confronts—"rather than Nothing?"[136] And the anxiety that thus confronts and may thus question cannot but concomitantly expose to him whom it befalls the security-undercutting Nothing belonging to himself as well.

Whether or not, however, such an experience be overtly known, always and for everyone there is in play a fundamental anxiety, born of the sheer "thatness" belonging to *Being*-there-as-openness itself. In it *Being*-there, manifesting itself as intently self-accomplishing care (*Sorge*), in its very carrying forward of itself discovers itself always as already given in specificity, i.e., it discovers itself as transpiring, as Being-in-the-world, in the particularity of thrownness; and it finds itself inescapably engaged, from out of that already-givenness and by way of the precariousness of temporality, in accomplishing itself in putting into play its own ability to be.[137]

In this anxiety, at a level where it comes thus face to face with itself in its primordial givenness, *Being*-there is aware of its own "uncanniness (*Unheimlichkeit*)." This is to say that, in an acknowledgement that is the countering counterpart of the lulling assumption of a subsisting in confident familiarity among all that comprises the world, which belongs to the inauthentic mode of its transpiring, *Being*-there is aware of itself as "Being-in-the-world as without abode (*nicht-zu-Hause*)." For it discloses itself to itself directly as "the bare 'that-it-is [*Dass*]' in the 'Nothing' of the world."[138] It

---

Seienden als eines solchen: dass es Seiendes ist—und nicht Nichts. "What is Metaphysics?" W 11, BW 105. Cf. W 8ff, BW 102ff.

[135]"Postscript to 'What is Metaphysics?'" W 101f, EB 384.

[136]*Warum ist überhaupt Seiendes und nicht vielmehr Nichts?* EM 1; IM 1. See also "What Is Metaphysics?" W 11, BW 105.

[137]SZ 191f; BT 235-237.

[138]. . . *das nackte "Dass" im Nichts der Welt.* SZ 276f; BT 321. Cf. SZ 134ff, BT 173ff.

discloses itself to itself, that is, as ever provided to itself, although never by itself, and discloses itself as ever in a world that it, in the disclosedness of its understanding, structures meaningfully in gathering it toward itself in its own casting forward of itself, but that is as world likewise ever given and is never graspable, never possessed but is, in that very nullity, forever being given anew.

Here *Being*-there knows and acknowledges itself in precisely that demanding centeredness that is the authentic carrying forward of itself from which, as Being-in-the-world, it falls away in its inauthentic, everyday transpiring in vitiating itself as the "they." For this acknowledging of itself in its "uncanniness" is wholly individual. Reaching to the root of the understanding of its capability-for-Being that belongs to *Being*-there as such, it requires the acknowledgement of a potentiality-for-Being that, from out of the thrownness of *Being*-there as Being-in-the-world—in all the specificity of the latter—is wholly limited to whatever possibilities are, if on-goingly, therein ever already given to *Being*-there as this particular centering of disclosure, this particular "self." In any authentic transpiring of human existing, as the latter, casting itself ahead of itself toward the accomplishing of particular possibilities, carries itself forward in the way that belongs to it as peculiarly its own, an individual "I" must make this acknowledgement and, in so doing, must acknowledge that to it, in its inescapable thrownness, certain possibilities only, namely those inhering in the specific configuring into which it has been and is ever being thrown, lie open to it.[139] Here "nullity (*Nichtigkeit*)" is disclosively acknowledged— not nullity as a lack, but the nullity of restrictedness, a not-other-than-this that is yet the positive, truly accomplishing mode wherein *Being*-there as Being-in-the-world alone fully happens as itself.[140]

The disclosing of this nullity to itself is never absent for *Being*-there. In the uncanniness that declares it, *Being*-there, precisely in the midst of its self-deceptive falling away into the "they," ever calls

---

[139]SZ 276, 279, 285; BT 321, 324f, 331.
[140]SZ 283-285; BT 329-331.

summoningly to itself to return to itself." Calling thus, it calls as
"conscience (*Gewissen*)."[141] And in that very calling, it proclaims itself
to be guilty; that is, it proclaims itself to be "the null basis of a nullity,"
with respect to whatever, from out of its own limited givenness as Being-
in-the-world, it carries out in unavoidable restrictedness in its transpiring
as Being-with-others.[142] This being-guilty (*Schuldigsein*) is intrinsic to
Being-there. Every ordinary understanding of conscience and of being
guilty, every establishing of norms, and every attribution of value or
assumption of a reality of good or evil, far from possessing any ultimacy
in its own right, is in fact a manifesting in the ubiquitous inauthentic
sphere of the everyday of this authentic summons of *Being*-there to self-
acknowledgement and acknowledgement of guilt, which the inauthentic
manner of existing of the "they" sedulously conceals from view.[143] A
genuine harkening to conscience accepts as pertaining properly to itself
the implied "Guilty!" that lies behind all such conceiving, and,
committing itself to those possibilities that alone lie open to it in its
thrownness, carries itself forward self-accomplishingly via a having-to-do-
with (*Handeln*) that is intrinsic solely to itself and bears no reference to
any externally imposed "ought."[144] This is a "wanting to have a
conscience" that is as such "conscienceless." Since in its acknowledgement
of its inherent nullity it already acknowledges its guilt toward others, it
pursues affirmingly and without question the specific possibilities—and
hence takes the specific actions—into which it has been thrown. It is
indeed, Heidegger says, solely in this mode of existing that the authentic
human "possibility of being good" actually consists.[145]

  Being-there summons itself back to itself from its falling away, in a
summons that lays bare to it the wholly individualized, restricted
potentiality for Being, inherently imbued with nullity, that can alone be
the authentic accomplishment of itself that is accordant with its own

---

[141]SZ 276f; BT 321f.
[142]SZ 283-285; BT 329-331.
[143]SZ 281-283; BT 326-328. Cf. also SZ 286, BT 332.
[144]Cf. SZ 287f, BT 333f.
[145]SZ 288; BT 334.

transpiring as the intent, self-sustentative "care (*Sorge*)" that enables it to pursue its way in the precariousness of a self-carrying-forward that is a standing out into the openness via which on-goingly, ungraspably, it as Being-in-the-world takes place. Via each of its centerings human existing demands an acknowledging of the radical already-givenness, utterly beyond its own bestowing, that pertains to whatever it encounters and that, precisely in the pursuing of its own intent self-accomplishing, pertains through and through to itself. In the discussion in *Being and Time* where this portrayal is given, Heidegger does not inquire into the whence of this givenness, the well-spring alike of thrownness and of outreaching encountering. In it we can readily see a clear anticipatory instancing of the all-provisive, enabling happening of Being that he later portrayed. We have seen that Being is, as ruling self-concealing, the Happening that comes revealingly upon what-is—thought as an ever-particularized whole—in such a way as to allow what-is to arrive into unconcealment.[146] And just such a structuring of disclosive Happening can be seen to be in play in the imperious providing from beyond itself of the human existing that, precisely as so provided in its thrownness, as Being-in-the-world allows what-is to arrive into disclosure. Indeed in the disclosing and acknowledgement of nullity, concealing's self-evidencing, to which Heidegger accords such a central role in the authentic transpiring of human *Being*-there, the happening of Being must be seen to be bringing itself into play. For man's existing, as a standing-out into the time-structured open, is itself held into the Nothing, the nullity, that is the "completely other of what-is," in order that therewith meeting and disclosure may take place. Man, Heidegger can say, in the on-going openness of his existing ever holds open the very locale for that "complete other," i.e., for the not-Being of whatever meets him and of himself, so that in the openness thus lived out, and indeed solely via it, "there can be such a thing as presencing (Being)."[147]

In the taking place of *Being*-there in the anxious individualization

[146]Cf. Ch. I, pp. 30-32 & 27-29.
[147]ZS 96; QB 97.

that manifests to it the "Nothing" belonging to it in every aspect of its
transpiring, *Being*-there is not only aware of that nullity as attesting its
thrownness as Being-in-the-world that is Being-with-others.   It is aware
also of the selfsame nullity as it pertains directly to itself, i.e., to its own
continuance in the self-maintaining that is wrought out via its happening
as self-sustaining care.   Specifically, it is aware that its precarious,
Nothing-fraught being-able-to-be includes as its extreme possibility its
being-able-not-to-be; and it is aware, from out of the very circumstancing
givenness according to which—in the latter's ceaseless renewedness—it
carries itself forward, that that ultimate possibility is a given that is
vouchsafed to it as the possibility that belongs to it most surely, i.e., that
is "most its own (*eigenste*)."[148]

   In its falling away from itself in inauthenticity, man's *Being*-there-
as-openness flees this awareness also.[149]   On the other hand, when in
happening authentically human *Being*-there-as-openness bears its anxiety
precisely as an affirming acceptance of its own nullity, that very
acceptance of nullity in this, its most intimate impinging as the manner of
happening most intrinsic yet most opposed to *Being*-there's carrying
forward of itself,[150] is such as to carry to the most intense pitch possible
the latter's demanding maintaining of itself as an openness to the radical
accomplishing of that Nothing from out of which it happens as it does.
Here we catch sight in an ultimate way of an accepting of that negating
which, via the coming to pass of thrownness, encounters man's existing
from beyond itself, for we find *Being*-there making affirmative individual
answer to the radical happening of Being as concealing.   Specifically, we
have here to do with the way in which *Being*-there determines itself
positively in accordance with the ineluctable fact that in its intently self-
sustaining happening as Being-in-the-world, it is always accomplishing
Being-toward-death (*Sein zum Tode*).[151]

---

[148]SZ 250; BT 294.
[149]SZ 254; BT 298.
[150]SZ 307, BT 354; SZ 266, BT 310.
[151]Cf. SZ 260f, 265ff; BT 304f, 310ff.

Man alone among all living beings dies,[152] and every individual must constantly withstand the encroaching nearness of death. Death is not an entity of some sort. Rather, Being itself happens as death. In so doing, it happens, in most immediate relation to man, as Nothing. In death, Being utterly hides itself away,[153] and man as man finds himself always vis a vis this death.

Heidegger came to use as a central designation for human beings "those who die," "mortals *(die Sterblichen)*."[154] Mortals live out their relation to Being *in extremis.* Ceaselessly and in varied ways they withstand concealing. Whereas an inauthentic existing obscures to itself the unconcealing of Being's concealed, negating manner of happening that ever brings itself to bear upon it, the existing that is authentic expressly offers to Being as unconcealing that is yet concealing a center of disclosure wherein it can directly rule. Always, in the pervasive nullity known to their self-preserving awareness, mortals are beset by that provisional concealing, itself harbinger of the ultimate threat, individually encountered death itself. Death overshadows them and one day will lay claim to each. Those who succumb seek to escape this confrontation. But those who sustain an authentic existing accept and withstand it. For them, death becomes not a grim necessity from which to flee, but the most extreme possibility of which they are capable. It determines the perspective in which they view all else.[155] And they pursue their way in readiness to yield to this final claim of Being.

A *Being*-there-as-openness that accomplishes itself authentically, as in disclosingly orienting and directing itself it surmounts the disjunctions that its temporality entails and so transpires in centered individuality as to maintain itself properly as itself, transpires after the manner of a

---

[152]"The Thing," VA 177, PLT 178. For Heidegger's contrast between human life and the life of other living beings *(Lebewesen)*, see the "Letter on Humanism," W 157, BW 206.
[153]Cf. "The Thing," VA 177f, PLT 178f.
[154]Cf., e.g.,"The Thing," VA 172ff, PLT 173ff; " . . . Poetically Man Dwells . . . ," VA 196, PLT 222.
[155]Cf. SZ 264, 383; BT 308, 434.

perseverant self-opening-up (*Entschlossenheit*)[156] that, as a restrictive closedness that permits forward movement, is nothing less than its urgent self-committing to whatever possibilities, from out of its unflinching acknowledgement of the nullity that belongs to it as ever thrown into existing, it finds to pertain to itself. This self-opening-up is indeed, as a fully accomplished disclosedness, the authentic happening of *Being*-there-as-openness-for-Being in its happening as Being-in-the-world. As such, at any particular time it both brings that individualizedly centered self as which *Being*-there authentically exists "precisely into its concernful Being-in-close-company-with (*Sein bei*) that which is ready-to-hand"—that which proffers itself for rightly discerned use—and "thrusts" that self into a "Being-with others" that genuinely has a care for those met and that, accordingly, so happens as to assist them to come, from within themselves, to an authentic existing.[157] In like manner, the resolute self-

---

[156]*Entschlossenheit* is ordinarily translated "resoluteness" or "resolve," in keeping with its usual meaning. Behind the noun lies the verb *entschliessen*, comprised of *schliessen* (to close) and the prefix *ent-*, which signifies a movement away from one state to another. (Cf. Ch. I, p. 24, n. 12.) In his use of *Ent-schlossenheit*, Heidegger intends to bring forward this, its fundamental meaning. The closing in question opens forth toward the authentic possibilities for Being belonging, always in particularity, to *Being*-there. (SZ 300; BT 347.) In "The Origin of the Work of Art," (HW 76, PLT 67) Heidegger writes, hyphenating the noun so as to display its content: *Die in Sein und Zeit gedachte Ent-schlossenheit ist nicht die decidierte Aktion eines Subjekts, sondern die Eröffnung des Daseins aus der Befangenheit im Seienden zur Offenheit des Seins.* ("The *Ent-schlossenheit* thought in *Being and Time* is not the decision-determined action of a subject, but the opening of *Being*-there-as-openness out of its caughtness in what-is toward the openness of Being.") Cf. also "Conversation on a Country Path," G 61, DT 81.

[157]*Die Entschlossenheit löst als eigentliches Selbstsein das Dasein nicht von seiner Welt ab, isoliert es nicht auf ein freischwebendes Ich. Wie sollte sie das auch—wo sie doch als eigentliche Erschlossenheit nichts anderes als das* In-der-Welt-sein *eigentlich ist. Die Entschlossenheit bringt das Selbst gerade in das jeweilige besorgende Sein bei Zuhandenem und stösst es in das fürsorgende Mitsein mit den Anderen.* SZ 298; BT 344. Heidegger's use here of *stösst* (thrusts) shows us that the powerful impetus that he sees as in play in the thrownness of human *Being*-there-as-openness-for-Being and in the self-casting-ahead of the latter that follows from it must be seen to pertain also to the resolute commitment in openness that *Ent-schlossenheit* names. Cf. this chapter, pp. 68f, and n. 50. Indeed, the connotations of *Ent-werfen* and of *Ent-schlossenheit* are closely akin. On the authentic manner in which having-a-care-for (*Fürsorge*) takes place, see SZ 122, BT 159.

opening in question happens as the authentic self-commitment in awareness of *Being*-there to that ultimate possibility that pertains so intimately to it as itself, the ever-given possibility of death. Transpiring authentically, *Being*-there is possessed of a disclosive understanding, at once persevering and forth-goingly open, of the "nullity (*Nichtigkeit*)" that belongs to it precisely in its carrying forward of itself as itself via temporality. In accordance with its pursuing of its way in acknowledgement of the nullity that pervades it in its transpiring as Being-in-the-world, it acknowledges that nullity to the uttermost, in the latter's happening as death. It acceptingly affirms that for it on-going temporality will end and that, whereas it itself is, there will "then" be Nothing.[158]

Like conscience, which is indeed in play in its transpiring, this resolute openness belongs to *Being*-there as accomplished in centered awareness of individuality.[159] At the same time, however, it so takes place as to bring the individual, in his very awareness of himself *as* an individual, decisively into the community of which the word *Da-sein* equally speaks. For the resolute openness in which *Being*-there so launches itself forth as unreservedly to encounter even the utter negating of itself in death is, precisely as such, a turning back in awareness toward the thrownness out of which such self-projecting springs. And, caught sight of thus in extremis, the individual's thrownness shows itself as a Being-in-the-world that is a Being-with-others which is, as known thus as antecedent, a having-been-there in-a-world-that-has-been, even as it is, simultaneously, directly and presently in play for him who apprehends it, just in being determiningly antecedent for him.[160] Resolute-openness, as an authentic coming to pass of the being-able-to-be that accomplishes itself as care, is able to discern in the disclosure of having-been-there that it thus vouchsafes to itself any like authentic accomplishing of being-able-to-be. Glimpsing such accomplishing—together with the carrying into play of specific possibilities that arise from it—in the having-been of the world that presently determines it itself, it likewise glimpses the presently

102   *Chapter II*

transpiring world, wherein it itself under the clarifying impending of death is as existing, in the light of the insight into the meaning of things and events thus gained.[161] Its "moment-of-insight (*Augenblick*)" is an accomplishing of authentic presence in an openness that is a present immediately determined by coming-toward and having-been.[162] Via that moment, resolute-openness so responds to the having-been-there that it discerns, as to bring again into play, in a fresh achieving that is in a manner a "repeating (*Wiederholung*)" the authentic possibility, the self-carrying-out as a resolutely accomplished being-able-to-be, that it sees. Catching sight of such a possibility in the having-been-there as having-been-in-the-world on which it itself rests back, it so grasps it as to hand it down determiningly to itself.[163] It makes rejoinder to that possibility precisely in carrying it into play in the moment-of-insight that is its own accomplishing of the present via which its discerning of specific possibilities proper to itself in the world is taking place.[164] Thus it brings itself into one via its affirming "choice" with the already accomplished capability-for-Being, specifically wrought out, that is presently determining it as, from out of the already accomplished communal context—where that possibility has been fulfilling itself—and in accordance with it, it exists as thrown.

　　In the transpiring of such resolute-openness conscience, as the summons to the acknowledgement of the nullity belonging intrinsically to *Being*-there and to the acceptance of the Being-guilty that is affirming self-restriction to whatever particular possibilities such nullity entails, comes fully into play.[165] Here it is that such affirming acceptance of the possibilities already belonging to itself that authentically transpiring *Being*-there properly undertakes in accordance with that "conscience" is carried out. The undertaking is that of an individual, but the discovering of the possibilities in question as belonging to itself as Being-in-the-world

---

161SZ 382-385; BT 435-437.
162SZ 328; BT 376.
163SZ 437; BT 385.
164SZ 386, 390; BT 438, 442.
165SZ 382f, BT 434; SZ 384f, BT 436f.

via a particular present so opens back into the having-been in-a-world-that-has-been that the individual's relying upon his own discerning of what properly belongs to himself becomes, when thoroughly considered, by no means an individual accomplishment of openness for possibilities somehow intuitively perceived, as might seem to be the case. Rather, there is here decisively in play an acceding to commitments made and possibilities provided via a tradition wherein the already determined happening of the community as having-been-able-to-be as having-been-in-a-world, and as constituting, as such, the thrown basis of the individual's nexus of taking-place, is finally in question.[166] This it is that constitutes the determinative wellspring of the choice, the self-committing acceding to the already given, that is individually undertaken in resolute-openness. And likewise from out of this arises, ultimately, any individually centered bringing into play of the "conscienceless" conscience that enjoins guilt and summons to the wholly unjudgeable, self-affirming accomplishings in which alone any possibility of authentic human goodness lies.

The transpiring of *Being*-there in a manner that thus brings authentically into play the embeddedness of the individual in the extended existing of a community is its transpiring as historical.[167] Historical-happening (*Geschehen*) is, in this context from *Being and Time*, *Being*-there's "movement (*Bewegtheit*)," wholly peculiar to itself, "of extended self-extending."[168] It evinces a self-constancy (*Selbstständigkeit*) that is intrinsic to *Being*-there from out of the very temporalness that structures its carrying forward of itself.[169] Accordingly, an extendedness in a constancy of self-carrying-out belongs to *Being*-there in its individualized transpiring as seen in itself in that very resolute-openness that, in the self-casting-ahead that is a turning back to initiatory thrownness, sustains itself against descent into inauthenticity. It is yet more readily manifest when

---

[166]SZ 383f; BT 435.
[167]For the meaning of "history" as Heidegger understands it primarily from the perspective of the happening of Being, cf. Ch. I, pp. 47-49, and Ch. XIII, pp. 493-495.
[168]*Die specifische Bewegtheit des* erstreckten Sicherstreckens *nennen wir das* Geschehen *des Daseins.* SZ 375f; BT 427.
[169]SZ 375; BT 427.

such turning toward thrownness is seen fully as the bringing on-goingly into play of possibilities from the having-been-there-in-a-world that is therewith disclosed as having-been in continuity with the presently happening *Being*-there that, in insight into a world—itself "historical" as the world wherein *Being*-there has-been—that is also marked by the same continuity, is in such resolute-openness presently catching sight of that on-going world and opening itself in an authentic way toward the possibilities proffered therein.[170]

In so transpiring as to carry authentically into play in resolute-openness, from out of an appropriating of its own thrownness, this historical-happening that is constitutive for *Being*-there as Being-in-the-world as such, *Being*-there accomplishes itself as fate (*Schicksal*), i.e., it accomplishes itself as the intent, individualized happening in which it "hands itself down to itself free for death, in a possibility that it has inherited and yet has chosen."[171] Precisely in so happening, *Being*-there accomplishes itself from out of "destiny (*Geschick*)," that is, from out of the historical-happening-together (*Mitgeschehen*) that belongs to a "people (*Volk*)," the human grouping that is the specific form of community as which, as Being-in-the-world, *Being*-there takes place as Being-with-others. Our individual fates, Heidegger says, "have already been guided in advance in our Being-with-one-another in the same world and in our resolute-openness for definite possibilities."[172] Here, indeed, the individuality in corporateness that belongs to *Being*-there as such stands strikingly manifested. In the "repetition" in resolute-openness of possibilities that have-been, an individual "choice," a self-affirming commitment, lifts into the immediacy of present undertaking the like undertaking of an exceptional individual, a "hero," whose self-carrying out has brought the capability-for-Being of the people decisively into play.[173] It therewith carries forward that capability from out of and on

[170]SZ 390f, 391f; BT 442f, 444.
[171]SZ 383f; BT 435.
[172]SZ 384; BT 436. For Heidegger's use of *Geschick* with reference to the ruling happening of Being, cf. Ch. I, pp. 47f.
[173]SZ 385; BT 437.

behalf of the community as a whole. And the individual's thus achieving, in the acknowledged face of death and in a communicative unity with others, of the affirming, through renewing "repetition," of the possibilities that carry into play against all countering tendencies the being-able-to-be that has triumphed in the on-going existing of his people, instances, in a particular accomplishing of that people's destiny, "the full and authentic historical-happening of *Being*-there."[174]

Individually undertaken resolute-openness, as properly understood in the full context of *Being*-there-as-openness-for-Being as an individualized interplay within on-going corporateness, constitutes, Heidegger says, "the loyalty (*Treue*) of existence to its own self." This is the "loyalty" of a "free existing." In it as such, an individualized existing steadfastly open in its transpiring can, in acceptance of anxiety and in a gathering to itself of the having-been-there-in-a-world antecedent, in a continuity of extendedness, to itself, in a fitting response properly acknowledge the "sole authority" that it itself may have, namely, "the repeatable possibilities of existing."[175]

The freedom in which *Being*-there discovers itself to exist in recovering itself from inauthenticity and accomplishing its authentic self-opening in the individuality of historically transpiring resoluteness is, clearly, as far as possible removed from a volitional capacity that chooses among alternative entities or possibilities that it ranges before itself.[176] Indeed freedom, *Freiheit*, ultimately considered, is nothing less than the very self-opening, as instanced via man, as which Being as unconcealing

---

[174]SZ 384; BT 436.
[175]SZ 391; BT 443.
[176]Cf. "On the Essence of Truth," W 86, BW 129. In the transpiring of the human manner of existing as inauthentic and authentic we should see again, in a particular guise, what we have called the fundamental structure of the happening of the Being of whatever is, namely, an on-going going-forth-from-itself that is yet a gathering-back-toward-itself (cf. Ch. I, p. 33, n. 36). That existing is ever falling away from itself and is yet calling itself, gathering itself, back toward itself. When genuine freedom takes place, this structure is transcended. But it must be remembered that such transcending is never final and that the structure in question as evidenced in the transpiring of *Dasein* in everydayness is for everyone always continually in play. It is always there to be surmounted. It can never be left behind.

happens what-is revealingly into the open (*das Offene*) as that which is met and disclosed.[177] It happens as "the letting-be (*Sein-lassen*) of what-is"[178]—i.e, of the single manifold as a whole, which, met with as world, *Being*-there as Being-in-the-world brings to light, in the latter's time-borne proffering of itself as the particulars of that manifold—that alone allows any disclosure whatever to take place. This letting-be, freedom, as lived by man, receives its structuring by way of ever-opening time— that time as which Being, happening as unconcealing, opens itself for the happening into manifestation of what-is—immediately from out of Being.[179] In freedom, we catch sight of Being in its direct governing, in a specificity of its happening as self-presenting self-withholding, of the self-presenting of man. For as letting-be, freedom transpires as a setting-exposedly-out (*Aus-setzung*), indeed, as the very standing-out into the open wherein what-is presents itself, as which man exists.[180] Transpiring

---

[177]Heidegger alludes to "truth (*Wahrheit*)" as freedom ("On the Essence of Truth," W 85, BW 128) and to freedom as the *Wesen* of truth (W 87, BW 130), where "truth" is the unconcealment in question (W 84, BW 127). He writes: *Die so verstandene Freiheit als das Sein-lassen des Seienden erfüllt und vollzieht das Wesen der Wahrheit im Sinne der Entbergung von Seiendem.* W 86; BW 129. On *Entbergung*, regularly rendered by revealing, as bespeaking unconcealing, see Ch. I, p. 24, n. 12. On truth as freeing unconcealing, cf. Ch. I, p. 38, and n. 49, and p. 54, n. 109.

[178]*Freiheit enthüllt sich jetzt als das Sein-lassen von Seiendem.* "On the Essence of Truth," W 83, BW 127.

[179]Heidegger, in speaking of Being as presencing, could identify it, with respect to its happening of what-is, as letting-presence (*Anwesenlassen*) ("Time and Being," SD 12, TB 12). He does not similarly speak of Being's happening as letting-be (*Seinlassen*). Nevertheless, it is unquestionable that Being lets what-is be. Cf. "The Onto-Theo-Logical Constitution of Metaphysics," ID 132, 64. It is evident that by his use of *Seinlassen* to speak of freedom he suggests the way in which Being, once again in self-differentiation, happens rulingly specifically as the Being of man in relation to what we may call Being's inclusive happening as the Being of whatever is. For a like structuring, cf. the discussion of *logos*, Ch. IX, p. 338, and of *Gemüt*, Ch XIV, pp. 554-556. This letting-be, as the openness via which man comports himself as man, shows us one aspect, i.e., the manward aspect, of the happening of Being for which man, in his existing, proffers and lives out openness as such. It is, as according openness to man, the particular mode of Being's happening that permits the vastly varied happening of Being as openness for whatever discloses itself to reach consummation.

[180] . . . *die Freiheit ist in sich aus-setzend, ek-sistent. Das auf das Wesen der Wahrheit hin erblickte Wesen der Freiheit zeigt sich als die Aussetzung in die Entborgenheit des*

thus, it engages itself immediately with the particulars of what-is in their manifestness. But this engaging (*Sich-einlassen*) is at once a self-withholding. It is a "stepping-back" that allows what-is governingly to proffer itself and that receives its every configuring directing for specific disclosure from the latter.[181]

This freedom, happening as restrained engagement with what-is, as "admittedness (*Eingelassenheit*) into the revealing, the harboring-forth, of what-is as a particular something,"[182] carries into consummation the intricate self-differentiating Happening named as the Being of what-is.[183] As an openness that goes forth into the openness that what-is may be said to provide for itself, it lays itself open (*setzt . . . sich . . . aus*) to what-is in its taking place in particularity. As such, it is antecedent to all specific human comportment vis a vis the particulars of the manifold of what-is, and, indeed, as a setting-out-exposedly-into-the-open it carries away (*versetzt*) all comportment whatever into the "open" of time-accomplished unconcealing, wherein every such meeting takes place.[184] For in happening participantly as admittedness into Being's revealing of the single manifold of what-is in the unitariness of its entirety, "freedom has already attuned (*abgestimmt*) all comportment to what-is as a whole,"[185] toward which it itself is disclosively transpiring in self-restrained response.

Happening ultimately in this way, freedom is, when considered specifically with reference to the structuring of man's existing (*Dasein*), in the latter's transpiring in self-awareness via temporality, that very self-orienting self projecting wherein that existing, *Being*-there as Being-in-the-world, on the basis of its given thrownness, so casts itself forward as

Seienden. ("On the Essence of Truth," W 84, BW 128.) See also this chapter, p. 58, n. 5. In quite another context Heidegger speaks of dwelling, in a way reminiscent of his portrayal of freedom as letting-be. Cf. Ch. V, pp. 187f, and n. 69.
[181]"On the Essence of Truth," W 84, BW 128.
[182] *. . . die Eingelassenheit in die Entbergung des Seienden als eines solchen.* W 84, BW 128.
[183]Cf. Ch. I, pp. 29-33.
[184]"On the Essence of Truth," W 84, BW 128.
[185]W 87; BW 130f.

to meet that which is meeting it precisely after the manner that that thrownness has provided. From out of the restrictedness of possibility that has already been accorded to it, *Being*-there singles out that possibility to which it has already been confined and, in so doing, unflinchingly suffers the brute circumstance that naught else was or could be chosen.[186] Such freedom comes to the fore when, in the authentic existing that surmounts inauthenticity, self-transparency is achieved and there is, at an advanced level of self-awareness, a freedom for the full bringing into play of the self-maintaining self-accomplishing "care (*Sorge*)" as which and via which, as on-going openness, *Being*-there as Being-in-the-world takes place.[187]

When *Being*-there encounters acceptingly the nullity belonging to it as thrown and as directing itself so as to select and exclude, it discovers "its Being-toward the capability-for-Being that is most its own." That is, it discovers a "Being-free-for" that is an inclination toward the authenticity of the Being that belongs to it and that does so both "as a possibility that it always already is" and as that to which, as Being-in-the-world, it "has been consigned."[188]

What-is, in all its particulars, is possessed of a certain autonomy in its relation to Being in the unfolding of the Twofold. It happens as unconcealed of itself; although such happening is initiated from out of Being—pure Happening—and so unfolds the latter in specificity that Being and what-is, while sundered, are yet the Same. Being and what-is are intrinsically one in a relating that transcends Being's self-differentiating. Being hands what-is over to its own happening but yet keeps it in hand, ruling always within it.[189] Man, as a being in the manifold of what-is, partakes of the same manner of relating to Being that obtains for what-is as such.[190] But the provisional autonomy belonging to man is unique,

---

[186]SZ 285; BT 331. On the importance elsewhere in Heidegger's thinking of the acceptance of restriction, see Ch. V, p. 188, n. 69.
[187]SZ 122; BT 159. On "care," see this chapter, pp. 75f. On the meaning of transparency (*Durchsichtigkeit*), cf. SZ 146, BT 186f.
[188]SZ 188; BT 232f. Cf. SZ 285, BT 331.
[189]Cf. Ch. I, pp. 53-55.
[190]For a statement that affirms just such relating with respect to man see Heidegger's

precisely because it is man and only man who, in his unique manner of existing, accomplishes disclosure. Man alone so happens as to unconceal himself to himself, just as he alone also unconceals to himself in his receptiveness whatever unconceals itself to him. In his everyday manner of existing, such disclosure may be all but hidden from him. Even if but hiddenly, however, man, in his *Being*-there-as-openness, in accomplishing his self-maintaining, always goes out beyond himself only to discover himself as already having been in a world and always reaches to meet a world that is already coming to be met. Being's happening happens him wholly accordantly with itself—even to the details of his self-carrying-out—as a center of openness vis a vis a world, and yet the self-orienting self-directing that informs his self-maintaining, although it is decisively governed by the inclusive happening of the manifold Twofold of the Being of what-is, is also intrinsically his own. It is here that, when in genuine awareness he is able to be as authentically himself, his freedom is found. It is not a freedom that is an initiating; it is a freedom that is an according-with. It happens as a response to the happening of the Being of what-is, a response that indeed, in the last analysis, instances that happening. As such, it offers to the latter, via man's capability for disclosure in his existing by way of temporality, the arena for unconcealing that Being as concealing "needs" and provides for itself. "Freedom," Heidegger says, "administers the free-open (*verwaltet das Freie*) in the sense of the cleared and lighted up": it administers "revealedness."[191]

The Being of what-is, in accomplishing its self-opening by way of temporality through happening as the Being of man, provides for itself an arena of unconcealing by happening away from itself a being who subsists in peculiar juxtaposition to it. That being, man, the accomplisher of disclosure, does not, like all other entities, simply accord in immediacy— albeit in a complexity of happening—with the manner of Being's happening and with its governance. Man, in the openness-for-Being that

---

words that define the meaning of the word Greek, Ch. IX, p. 332.
[191] "The Question Concerning Technology," VA 32f, QT 25.

is his manner of existing, whether clearly or deceptively, lets the manifold of the Twofold come to light. He sees, and he reaches forth to gather to himself, directed by that seeing. His manner of existing—the *Being*-there-as-openness that is Being-in-the-world—is, whether hiddenly or overtly, fraught with awareness. Its already-givenness evokes and underlies a self-orienting accomplishing of self-maintaining that, as the locus of disclosure, is itself of fundamental and vital significance. Man too accords with Being's manner of happening and its governance, but he does so as the one who discloses, as the one who is aware. Being, in its governance of man, takes him into dialogue with itself. His according with Being is a corresponding, an answering to it (*Entsprechen*).[192] It may be unacknowledged; it may be openly, self-dedicatedly, carried out. Freedom inheres intrinsically in his very self-accomplishing, as the unconcealing openness that lets anything whatever be. It comes fully into play when, as a "hearer" and not merely as one enslaved in a response not self-affirmed,[193] he, adequate to Being's ruling claim, responsively accords with Being by on-goingly, heedfully maintaining himself as the arena of disclosure via which Being as self-concealing accomplishes its own self-maintaining self-unconcealing.

The decisive way in which this "dialogue" of complex accordance between Being and man takes place is through language. Indeed language, as Heidegger understands it, is the most fundamental mode of Being's achieving of revealing by way of the existence of man.

---

[192]Cf. "The Turning," TK 39, QT 39. On *Entsprechen*, see further Ch. III, p. 118, n. 36.
[193]"The Question Concerning Technology," VA 32, QT 25.

CHAPTER III

LANGUAGE

If asked, we might well define language first of all as a means of communication consisting primarily of spoken words but composed also of looks and gestures, of a variety of actions that "speak" to others. We ourselves think of language as belonging centrally to human beings, although we quite readily extend the concept to the communicative actions of animals as well. Spoken language commonly appears to us to be a complex system of signs, i.e., words that exist in expressive relation to actions or objects, thoughts or feelings, indeed to whatever we experience. Such words refer to realities other than themselves. They "mean" this or that. Thus they permit us, we would say, to communicate in a multitude of ways regarding all manner of things, from the most abstruse ideas to the most familiar articles of everyday use. Ordinarily we take this language to be a tool of man's devising. We see it as something brought into existence and subsequently changed and ever changed again—albeit almost always through the unself-conscious working of the common mind—to meet the needs and exigencies of new conditions and situations so that the members of human groups might successfully pursue their life in community.

For Heidegger, in contrast, language is in no sense an instrument that, once forged, subsists in itself as a kind of fund transmitted from generation to generation, to be drawn upon separately by countless individuals and put to use by them in their dealings with one another. Rather, language is wholly integral to man's manner of existing in individualized corporateness. It takes place, indeed, as constitutive for it. Only man is possessed of language. For it belongs exclusively and determinatively to *Being*-there as Being-in-the-world, i.e., to man as the locus of the illumined open clearing via which Being unconceals itself.[1]

---

[1]Cf. Ch. II, pp. 72f.

1

Early in his work, in analyzing human existing as *Being*-there-as-openness-for-Being *(Dasein)*,[2] Heidegger gives to discourse a central place. This "discourse *(Rede)*"[3] is, however, not in the first instance an utterance in words. In the self-orienting, self-directing fundamental to its self-maintaining, *Being*-there-as-openness, transpiring as Being-in-the-world, discloses itself and its world meaningfully to itself.[4] Such disclosing requires a separating out of elements that can therewith be seen in their interrelation.[5] That which *Being*-there as Being-in-the-world sees and discloses so as to bring it to manifestation is met and drawn near in a clarifying, literally an understandingness *(Verständlichkeit)*, a receiving into intelligibility, that *Being*-there-as-openness in its taking place self-maintainingly as understanding *(Verstehen)*, brings to bear.[6] This clarifying puts into play meaning, the orienting "toward-which" in accordance with which man's existing, as Being-in-the-world, on the basis of its thrownness casts itself concernedly ahead of itself so as to meet and gather to itself, in order to preserve itself as itself.[7] The disclosive articulation of such clarifying that in accomplishing distinction and relation allows meaning *(Sinn)* to be displayed is what Heidegger means by discourse.[8]

Discourse, then, belongs, first of all, among the constitutive ways in which *Being*-there-as-openness carries itself out as Being-in-the-world and takes place decisively at the primary level of the latter's determinative awareness.[9] In it a network of intertwining significations *(Bedeutungen)* is wrought out, in the elucidating that permits self-orienting and self-

---

[2]On *Dasein* as openness-for-Being, cf. Ch. II, pp. 59f & 76f.
[3]In ordinary usage, *Rede* means talk, conversation, address, rumor.
[4]Cf. Ch. II, pp. 73f.
[5]Cf. SZ 150, BT 191.
[6]SZ 151; BT 192f. *Verständlichkeit* ordinarily means "intelligibility." Heidegger is using it to speak of the understandingness wherein that which is met with appears as possessed of intelligibility. On understanding *(Verstehen)*, see Ch. II, p. 74.
[7]On this manner of transpiring as belonging to human existing, cf. Ch. II, pp. 69f.
[8]See SZ 162, BT 203f.
[9]On Being-in-the-world, cf. Ch. II, pp. 61-63.

directing as such. Yet utterance in words also belongs integrally to this discourse. Words are not discrete and isolated "word-things," external to discourse and merely supplied with meaning secondarily as they are manipulated in its service. Rather, the totality of significations articulated in discourse of itself "comes to word." Spontaneously "words overgrow significations." Speech (*Sprache*), which like the fabric of meaning presented in it is always possessed of a wholeness, is itself the very "spokenforthness (*Hinausgesprochenheit*) of discourse."[10]

This is not to say that discourse always eventuates in spoken words. The disclosure that it provides may also take place by way of a refraining from speaking, for there may be a silence that, juxtaposed to speech, can itself reverberate with meaning.[11] But for the most part discourse comes to spoken utterance,[12] and it does so for reasons intrinsic to man's existing as such. Since the latter, as *Being*-there-as-openness, is inherently Being-in-the-world, it necessarily bodies itself forth in words. The "worldly" character of those words, their availability among the entities constituting a world, does not betoken their subsistence in isolation from some sphere of human experience out of which meaning is assigned to them. It arises, rather, from the inherent character of man's manner of existing as always and immediately in the world.[13]

Happening as Being-in-the-world, the *Being*-there-as-openness as which man exists clarifies via discourse that with which it has to do. Even as it does so, it finds itself, in its thrownness, always in some given relation[14] to the latter. From out of this relation it interprets to itself its world, presenting that world's components to itself in meaningful

---

[10]SZ 161; BT 204.
[11]SZ 164f; BT 208. Cf. "The Way to Language," US 252, WL 122.
[12]SZ 277; BT 317.
[13]SZ 162; BT 205.
[14]Here and elsewhere in this study the reader should be careful rigorously to separate "relation" from its usual meaning. Here "relate," in contexts where the Being of what-is is in any way under consideration, does not speak at all of a connecting of inherently discrete elements, one of which is to be subordinately referred to the other. The relating that is now in question is a *holding-toward-one-another* of participants that, while they are distinct from one another, are ultimately one. For we are using "relate" to speak of the happening of the single Twofold. See further, Ch. IV, p. 145, n. 10.

interrelation and doing so in accordance with the self-directing self-orienting that allows it to maintain itself as itself even as it casts itself ahead of itself so as to accomplish the precarious on-goingness belonging to it.[15] Thanks to such interpreting, literally a laying-out (*Auslegung*), entities met with are never first encountered via some sort of pure apprehending, only then to be fixed and affirmed into a meaningful pattern. Rather, always on the basis of some outlook that has already been assumed, each entity is viewed *as* something. As such it takes its place in a totality that has already been seen together from a particular point of view.[16] For in every interpreting whatever a fore-having (*Vorhaben*)—an understanding-fraught appropriating of the interpreted by way of a prior awareness of its totality of circumstancings—which is characterized both by an already-having-seen (*Vorsicht*) and a prior conceiving (*Vorgriff*), is, as what Heidegger can call the "hermeneutical situation (*hermemeutische Situation*)," always determiningly in play.[17]

Such elucidative interpreting eventuates and displays itself in various language structures fraught with meaning, notably in declarative assertions that offer a judgment regarding that which has come under interpretation.[18] These assertions spring directly out of that intending-fraught encountering of whatever comes-toward that *Being*-there, as Being-in-the-world, continually accomplishes.[19] Assertion (*Aussage*) always immediately exhibits something that is. That is, it lets something that is be seen from out of itself. In addition, it predicates something regarding that of which it speaks, thus defining and delimiting the latter with specificity.[20]

---

[15]On that precariousness, cf. Ch. II, pp. 74f. On the structure of Happening evident in this interpreting and in the fundamental "seeing" via which it carries itself out, which is a going-forth-from-out-of-itself that is inherently a gathering toward and which, as such, evidences via the transpiring of human existing as Being-in-the-world the intrinsic structure after whose manner the Happening named as Being brings itself to pass, cf. Ch. II, pp. 72f and p. 61, n. 19, and Ch. I, pp. 29-33, and n. 36.

[16]See SZ 150ff, BT 191ff.

[17]SZ 150, 152f; BT 191, 194f.

[18]See SZ 153ff, BT 195ff.

[19]Cf. Ch. II, pp. 64f.

[20]See SZ 154, BT 196.

Here a manifesting takes place. But here too resides the possibility of obscuring. In predicating something of a subject, assertion embodies a looking-forward that does not take place purely, but that is always governed by a harking-back to something already in view. However, the disjunction obtaining between what is to come and what has been permits a lack of congruence between what is viewed and the orientation from out of which it is viewed. In seeing on the basis of its having seen, human *Being*-there, as the accomplishing of openness, must surmount the gap opened by a temporality that prevents it, in its on-goingness from out of thrownness, from ever fully laying hold of itself or its world.[21] And in relying decisively upon the orientation wherein it already finds itself, it risks perpetrating a rigidity of outlook that can allow it to mis-see and misconstrue what meets it. This possibility is inherent in interpreting as such.[22] In the synthetic character of predicative assertions it comes overtly into play.[23] Thus the elucidation that discourse as such achieves can be, if hiddenly, an obscuring.

The taking place of this elucidating that may disclose but may also becloud is, in accordance with the individualized corporate character of man's existing, intricate and far-ranging. Via discourse in the full sense of the word, the individual articulates to himself on his own behalf the meaning of his world. But his doing so always remains incidental to the primary function of discourse. It is incidental, that is, to the latter's taking place fundamentally as a communicating transpiring among individuals who are always intrinsically together with one another.[24]

In this context, communication is in no sense to be thought of as a conveying of experiential elements that first exist as internal to discrete human subjects and that must then be externalized, whether through expressive actions or through words that stand for what is conveyed, if they are to be imparted at all by one such subject to another. For to

---

[21]Cf. Ch. II, pp. 69-71.
[22]See SZ 150, BT 191.
[23]SZ 34; BT 57.
[24]On Heidegger's portrayal of man's existing as carried out in individualized corporateness, cf. Ch. II, pp. 57 & 63f.

Heidegger there is no isolated or purely "inner" experience. *Being*-there-as-openness transpiring as Being-in-the-world is always "outside." As such, it is always inherently Being-with (*Mitsein*), i.e., with others.[25] As *Being*-there happens as Being-in-the-world, a taking place in common characterizes its individualized centerings. Hence the Being-toward that, in variously transpiring taking-heed-of, orients those individuals to a world and governs their encountering of entities and persons that there meet them[26] can be immediately shared.

Since *Being*-there-as-openness as Being-in-the-world is always intrinsically Being-with, the orientation in which the individual finds himself in relation to his world is always one with that of his fellows. Such an orientation, in which the primordial "foundness (*Befindlichkeit*)" that ever pertains to thrown Being-there is in play, involves, as a fundamental characteristic of Being-in-the-world as such, some specific mood (*Stimmung*), some specific way of being attuned (*Gestimmtheit*) to the world.[27] This attunement obtains in common for individuals. In their communicating it is expressly shared and so becomes manifest. Yet it is, more importantly, antecedent to all communication, for it underlies it as a concomitant condition of its taking place. Co-attuned in, and indeed from out of, their very Being-in-the-world, individuals can directly accomplish a sharing-with—the literal meaning of the German verb *mitteilen*, to communicate. Out of their shared perspective, shared discourse comes to utterance. Meaning, disclosed in common as individuals find themselves and seek to maintain themselves together in a world, is articulated in common.[28]

Those individuals can interpret their world to one another. They can, through words, show one another all manner of entities discovered there. For whatever is in view is seen from a decisive prior point of view that they already share. Thus via discourse, using centrally the words that

---

[25]Cf. SZ 162, BT 205. On Being-with, cf. Ch. II, pp. 63f.
[26]On such heedfulness, see the discussions of *Fürsorge* (Ch. II, p. 64) and *Besorgen* (Ch. II, pp. 64-68).
[27]SZ 134ff; BT 172ff. Cf. also Ch. II, pp. 68f, and n. 49.
[28]SZ 162; BT 205.

body it forth, individuals mutually display to themselves and to one another the components of their world, continually and in a multiplicity of ways elucidating their meaning. The disclosure therewith accomplished may be genuine. More often it may have the character of mis-presenting that characterizes human *Being*-there as, in happening in the everydayness that belongs to its carrying out of itself in inauthenticity, it speaks out of an assured adequacy of understanding, which is in fact only a self-content misconstruing whose utterance is mere idle talk (*Gerede*) that shows nothing aright.[29]   But every such elucidation, however it may take place and to whatever extent it may genuinely clarify or concealingly obscure, forms, as it were, one strand in a vast network of disclosure constantly wrought out by whatever community is engaged in exhibiting to and within itself a given world.[30]

2

Having first focused his inquiry on the existing peculiar to man, in an endeavor to discover the structures of Being manifest in the unique openness that is man's "thereness," Heidegger subsequently shifted and widened his perspective.   His questioning came to be oriented more directly toward the Being of what-is, as such, and, with this, the unique relating of Being and man was in various ways brought immediately into view.   With this change of perspective, many of the themes important for Heidegger's description of human existing fell into the background of his thinking.   The theme of language, on the other hand, came to the fore. The effort to think comprehendingly of language is, Heidegger insists, fraught with the profoundest difficulty.[31] Yet its role is crucial.   For it is precisely by way of the clarifyingly discursive apprehending that carries itself out in a speaking that is constitutive for man's manner of existing that the self-maintaining self-unconcealing that accomplishes itself in the

---

[29]SZ 167ff; BT 211ff. Cf. Ch. II, pp. 87-90.
[30]For Heidegger's view of communication among persons, see SZ 162, BT 205, & SZ 155, BT 197f.
[31]"The Nature of Language," US 160, 185; WL 58, 80f.

unfolding of the Twofold brings itself to pass.[32]   For to Heidegger to
speak of language is not merely to speak of a fundamentally constitutive
element in man's manner of existing.   It is to speak, even more
fundamentally, of the bringing to pass of that very unconcealing.
"Language," Heidegger can say, "is the illumining-concealing arrival of
Being itself."[33]   Hence any genuine engagement with language is a
crucial submitting to the ruling sway of originative Happening that must
"touch the innermost nexus of our existence."[34]

   The interpretive clarifying discourse of man, in all its aspects,
answers to and fulfills that self-unconcealing.   In the disclosive unfolding
of the Twofold there rules an utterance, albeit unspoken.   It governs the
complex disclosing wrought out in man, and it eventuates accordingly in
his own uttered word.   So thorough-going is Heidegger's attribution of
speech (*Sprache*), in its original provisive transpiring, to this happening of
the Being of what-is that he can assert that "actually (*eigentlich*)[35] it is
language that speaks (*spricht*) and not man.   Man speaks only insofar as
he, at any given time, answers to and accords with (*ent-spricht*)
language."[36]

   Through the language that we speak we continually *say* to one

[32]On Being as a self-concealing that, as self-unconcealing, maintains itself in
happening as the Twofold, cf. Ch. I, pp. 24-29.
[33]*Sprache ist lichtend-verbergende Ankunft des Seins selbst.* "Letter on Humanism,"
PL 70, BW 206.
[34]"The Nature of Language," US 160, WL 58.
[35]On *eigentlich*, which, where feasible, is here ordinarily translated with "genuine,"
see Ch. I, p. 37, n. 46, and Ch. II, p. 85, n. 102. Especially in later Heideggerian
contexts, *eigentlich* must suggest to us the meaning that Heidegger finds for *Ereignis*.
(Cf. Ch. IV, p. 149, n. 28. On the meaning in question cf. Ch. IV, pp. 149-151.) In
the present instance, where the evidently adequate translation "actually" has been
chosen, we should be aware also that in the happening of language as such—i.e., its
"genuine," its "actual" happening—it is precisely *Ereignis* that is in play. Cf. Ch. IV,
pp. 154f.
[36]*Eigentlich spricht die Sprache, nicht der Mensch. Der Mensch spricht erst insofern
er jeweils der Sprache ent-spricht.* HHF 34; HFH 99. On the prefix *ent-* as indicating
a movement to a new state, see Ch. I, pp. 24, n. 12. Almost certainly for Heidegger
entsprechen always carries connotations not only of an according-with as an answering-
to, but of a speaking disclosingly forth that is the culmination of the happening of
language (*Sprache*), which, ultimately, speaks.

another. That is, we continually show one another that which in our common orientation toward a world met with in common by all who constitute our particular community is showing itself to us. Heidegger traces the German verb *sagen*, to say, back to the verb *zeigen*, to show. The early *sagan*, he says, means "to show, to let appear, to let be seen and heard."[37] This coalescence of the meanings "say" and "show" is of crucial importance for him. It must always be heard when Heidegger speaks of saying, of speaking, or of language.

In our language, even as in everything that constitutes the reality that we know, the Happening that is named as the Being of what-is is bringing itself to pass. Indeed, it is in the taking place of language that the self-presenting of whatever is finds its consummation. For according as a word proffers itself or is lacking, the status of an entity disclosed as itself, which is spoken forth in the "is," pertains or does not pertain to that which the word names. The word (*Wort*) takes place as a specific instancing of Being's granting of itself. Only the word "provides (*verschaft*) Being" to anything whatever; it alone brings the entity in question "into the is" via which its self-disclosure can happen.[38] The word itself, thus properly understood, does not take place as an entity like all the entities that *are*, which it may call forth. Rather, a word is an instancing of the provisive Happening named as Being that, even in its physicality, takes place peculiarly and decisively as a bringing into play of that originative Happening itself.[39]

The showing-saying that is accomplished in our spoken words is, then, a bringing into play of Being itself. It is governed by the Being of what-is happening as showing, that is, as *saying*. Heidegger uses for this primordial Saying as which the Happening that Being names holds sway the word *Sage*.[40] *Sage* speaks of a showing-Saying that, after the manner

---

[37]"The Way To Language," US 252, WL 122. A like dual meaning is found for Greek and Latin verbs meaning "to say." Cf. "The Nature of Language," US 167, WL 65.
[38]"The Nature of Language," US 162-164 & 168f; WL 60-62 & 65f. See also US 194, WL 89.
[39]US 193f; WL 87f. Cf. Ch. IV, pp. 154f, and n. 46.
[40]"The Way to Language," US 253, WL 123. Cf. "The Nature of Language," US

of the self-unconcealing self-concealing as which Being happens, accomplishes "the lighting-concealing-freeing proffering of world," of reality as meaningfully gathered forth.[41] This primal Saying at once displays by delineating and conjoins in a patterned whole, intricately structuring and uniting the multitudinous disclosures taking place.[42] This is the bringing to appearance that rules in the appearing of every particular of the manifold of what-is, on-goingly happening both the self-disclosive particulars themselves and the human apprehending that launches itself forth receptively to meet and disclose to itself and others whatever offers itself to it. Hence, man's apprehending and that which it apprehends, in their self-disclosure, are both equally "something said," something "brought to appear."[43] As the Being of what-is unfolds itself as the Twofold, each of them is disclosively happened away to happen of itself, while yet it is kept in hand and governed from out of Being.[44]

In the case of man, who so belongs to the Twofold as to offer its arena of disclosure, primal showing-Saying (*Sage*) simultaneously brings to pass and keeps underway a dialogical relation between itself and man's own disclosive apprehending. In so doing it happens as an ever self-specifying saying (*Sagen*) that impinges immediately upon that apprehending. As such, it provides for itself the avenue whereby it, as the disposing Happening ruling in language (*Sprachwesen*),[45] brings itself to consummation. Primal Saying ceaselessly so brings itself to bear as a proffered saying, a proffered language, as to accomplish itself by way of the concrete happening that is man's saying through the spoken word.[46]

199f, WL 93.
[41]"The Nature of Language," US 214, WL 107. On the meaning of "world" that is in question in this passage, cf. Ch V, pp. 173-179.
[42]Cf. US 252-255, WL 121-124.
[43]*Das* noein, *das in-die-Acht-nehmen und das, was es vernommen, sind ein Gesagtes, zum Vorschein Gebrachtes.* "*Moira,*" VA 244, EGT 90.
[44]Cf Ch. I, pp. 53f. There man is not specifically mentioned, but man, with all his uniqueness, belongs with all else within the manifold of what-is, which is in question. Cf. Ch. II, p. 108, and n. 190.
[45]"The Way to Language," US 253, WL 123. On the meaning of *Wesen* see Ch. VI, pp. 252-257.
[46]US 253; WL 122f.

Hence the word language (*Sprache*) bespeaks at once for Heidegger the Being of what-is, happening as primal showing-Saying, the coming to pass of the latter as the saying vouchsafed to man, and the uttered speech of man over which and from out of which primal Saying holds sway. And he can, accordingly, speak of the "bringing of language as language to language."[47]

All our saying is, accordingly, a saying-after (*Nachsagen*).[48] We are enabled to speak only through listening to primal Saying which summons us into its employ. As that Saying, the very happening of Being as self-maintaining self-unconcealing, welling forth from the happening as self-concealing from out of which and indeed as which that unconcealing brings itself to pass,[49] as it rules continually structures into play both the disposing saying—the proffered language—that reaches us and our speaking of language that answers to and bodies that saying forth. Both the on-coming language and our ever-renewed speaking rest back upon primal showing-Saying as the ever on-happening provisive flow, one with puissant Stillness,[50] that ceaselessly builds them both onward like banks that, as such, are inherently united with that flow and carry into accomplishment the specificity with which it brings itself to pass.[51]

Indeed, even our keeping silent answers directly to the governing happening of primal Saying. That Saying, showing whatever is, "speaks" unspokenly. It is in silence that we listen to the unspoken that it says to us. And primal Saying closely governs such silence, even as it does the speaking that arises from it. Silence, indeed, points to the depths of primal Saying. For it "accords with the soundless sounding of Stillness," the initiatory self-containedness, a happening of concealing "intrinsic to primal Saying which, in showing, brings anything into its own."[52]

---

[47]*Die Sprache als die Sprache zur Sprache bringen.* (Italics Heidegger's.) US 242; WL 112.
[48]See US 255, WL 124.
[49]Cf. Ch. I, pp. 24-26.
[50]Cf. Ch. I, p. 26. On *Sage* as a manifesting happening of Stillness, see Ch. V, pp. 197-199.
[51]"The Way to Language," US 255, WL 124.
[52]*Das Schweigen entspricht dem lautlosen Geläut der Stille des ereignend-zeigenden*

Our "listening" to primal Saying is, then, in no sense an act of sensory perception. It is the responsiveness to the governance of the Being of what-is that intrinsically characterizes man as maintained and maintaining himself within the Twofold. We hear (*hören*) primal Saying, Heidegger says, precisely "because we belong (*gehören*) into it."[53] We are attuned to the revelatory happening of Being, which is of such a character that it reaches us through the language that we know and eventuates in our own speaking.

The Being of what-is, as an unconcealing, a harboring-forth that disposes, happens as a providing of meaning.[54] And our language, as vouchsafed from out of the ruling of that unconcealing as primal showing-Saying (*Sage*), displays meaningfully.[55] In the interest of our orientation within and toward our world, language allows for the assigning of specificity, for interrelating, and hence for elucidation.

Language simultaneously particularizes and draws together that of which it speaks. The very structure of our sentences—and sentences (*Sätze*) are the foundational units of our speaking[56]—shows this. Subject and predicate and all their syntactical adjuncts speak each with its own specificity while yet functioning expressively together as an organic whole.[57]

These characteristics of language display the manner of happening of the Being of what-is that Heidegger intends to present in naming as showing-Saying the self-disclosing that characterizes the unfolding of the Twofold. This Saying that brings to appearance is a showing that so rules as to let everything that is appear or disappear precisely in its

---

Sage. US 262; WL 131. Cf. Ch. V, pp. 197-199. On the meaning of *ereignen*, see Ch. IV, pp. 149-151.
[53]US 262; WL 131.
[54]"*Alētheia*," VA 276, EGT 118.
[55]Cf. "*Alētheia*," VA 275f, EGT 117f, where, in elucidating a Heraclitean fragment, Heidegger, speaking of the manner of happening of the Being of what-is in its provisive ruling, couples *sinnen* (to intend, to meditate), which is immediately related to *Sinn* (sense or meaning), and the verb *weisen* (to direct), thus giving to *sinnen* an unusual transitive force. Cf. Ch. XIV, pp. 569f.
[56]Cf. SZ 62, BT 89.
[57]See SZ 154, BT 196.

particularity,[58] its specificity of time and circumstance. And just in so doing it gathers those particulars together into an on-goingly interrelated whole, the manifold of what-is as what presences. It is this meaning-according Saying that "frees whatever presences into its presencing at any given time,"[59] for it is this that rules and structures the light-bringing clearing—time as lived out by man—by way of which all appearing and disappearing takes place.[60]

The enduring (*Wesen*) as which the Happening named as Being brings itself to pass[61] is, when ultimately considered, an initiatory happening which opens a way for everything that comes to disclosure and which therewith disposes everything into the way that permits its arrival as disclosed, through letting it presence, i.e., endure unto us human beings so as to concern us, to summon us and bring us upon our way to arrive as ourselves.[62] And language, the mode as which pure Happening, ruling as primal showing-Saying (*Sage*), consummates its self-maintaining self-unconcealing as the unfolding Twofold via human speaking, is precisely that as which the enduring in question as such be-waying (*Bewegung*)[63] carries itself into play. "Language"—viewed in its full complexity—"belongs into" that initiatory enduring. It "is suited (*eignet*)" to the latter as that which prepares the way for and disposes everything toward disclosure, "as that which is most its own (*als dessen eigenste*)." The Happening that thus disposes everything "be-ways (*be-*

---

[58]"The Way to Language," US 255, WL 124. "The Origin of the Work of Art," UK 83, PLT 72, & UK 84, PLT 73. "*Alētheia*," VA 276, EGT 118.

[59]US 257; WL 126.

[60]UK 84; PLT 73. On man's living-out of the "clearing" of time, cf. Ch. II, pp. 72f.

[61]Cf. Ch. I, pp. 35f. For an elucidation of the meaning of *Wesen* see Ch. VI, pp. 252-257.

[62]"The Nature of Language," US 201, WL 94f. Cf. Ch. I, pp. 36f.

[63]On the meaning of *be-wegen* and the meaning of "way" intrinsic to it, see Ch. IV, pp. 146-148, and "The Nature of Language," US 197f, WL 91f. Heidegger's hyphenating of the noun *Bewegung*, movement, and the verb *bewegen*, to move, to alter, strongly stresses their central component, *Weg*, way. He accords to both the double meaning just given, namely, to prepare and, we might say, to consummate a way. See further, Ch. IV, pp. 146f. Our translation with "be-waying" and "be-way" is intended to mirror the import intended by his orthography and also to suggest in one the double meaning that the words bear. On the "movement (Bewegung)" in question, cf. also Ch. IX, p. 353.

wegt)," Heidegger says, "in that it speaks."[64]

The Greek word *logos*, which *Sage* parallels, both bespeaks the initiatory way-opening disposing as which Heidegger would have originative Happening thus ultimately understood and evinces with peculiar force the unique integral relation obtaining for that Happening as a Saying that lets appear and man's saying couched in elucidative speech. *Logos* bespeaks the all-provisive "Way" that is brought into play as the disposing enduring as which the Being of what-is rules as the presencing of what presences.[65] Yet in thus bespeaking originative Happening, it simultaneously names the human speaking that fulfills it. Heidegger elucidates the import of *logos*, which he can characterize as "the name for Being and for saying"[66] through an explication of the purport of Heraclitus' Fragment 2, in which he brings out in the verb *legein*, in a manner that readily suggests his portrayal of "be-waying," the meaning "to let lie forth together."[67] It is as such a letting lie forth that *logos*, as the happening of Being, holds sway. *Logos* is the Being of what-is ruling disposingly as a gathering that precisely as such lets happen in particularity. It is "the laying-forth-together that selectingly gathers (*die lesende Lege*)," which, in a manner that carries-toward via on-going time, "lets lie forth before." It "restricts everything that presences and absents itself to its place and course," thus shelteringly gathering everything into the single manifold of what-is.[68] And this ruling accomplishes itself through happening as a saying that shows. For just this letting lie forth together is carried out in the human speaking that the Greek word *logos* ordinarily named.[69] Hence the word *logos*, viewed simultaneously from the perspective of ordinary Greek usage and from that of Heraclitus'

---

[64]"The Nature of Language," US 198, WL 92.
[65]US 198; WL 92.
[66]"The Nature of Language," US 185, WL 80.
[67]*beisammen-vorliegen-Lassen*. See the discussion in *"Logos,"* VA 209-211, EGT 60-62.
[68]*Die lesende Lege versammelt alles Schicken bei sich, insofern es zubringend vorliegen lässt, jegliches An- und Abwesende auf seinen Ort und seine Bahn zuhält und alles ins All versammelnd birgt.* *"Logos,"* VA 222, EGT 72. *Zubringen* means to spend time.
[69]*"Logos,"* VA 212, EGT 63. Cf. VA 208f, EGT 60.

insight, serves strikingly to show man's speaking, ruled by the Being of what-is, as Heidegger sees it, i.e., as "the gathering letting lie forth before of what presences in its presencing," as which that originative Happening, through ruling after the manner of a provisive showing-Saying, fulfills its self-maintaining in accomplishing itself as self-unconcealing.[70] Heidegger, indeed, identifies *logos*, which bears always intrinsically this dual meaning, as unconcealment (*alētheia*) itself.[71]

Speaking illumines what is spoken about. What-is, powerfully ruled by Being's happening as self-concealing, ceaselessly crowds in upon us confusedly.[72] Our speaking, as saying, strikes out into that obscureness. It *names*, not merely in single words as nouns, but in groupings of words of diverse form and import. And this naming lets unconcealing happen. Ruled by Being happening as showing-Saying, our naming-saying, which always exhibits something in specificity, although only in the complex milieu of some wider relational context, brings that something to light, letting it appear. Naming allows what-is to display itself in the particularity belonging to it as a manifold. It reaches forth toward what-is in its approaching, drawing it into the arena of disclosure ever opening with the self-maintaining self-surpassing achieved via the interpretive understanding that naming is itself bringing to fruition. Such naming-saying is, Heidegger says, "a projecting of light-bringing opening-up wherein is announced *as what* what-is is coming into the open."[73]

In this unitary occurrence wherein naming shows what-is as what it shows itself to be, Being as the unitary Twofold, happening as the unconcealment of that which, from out of that very happening, unconceals

---

[70]"*Logos*," VA 228, EGT 77. See also "The Nature of Language," US 185, WL 80, and EM 98, IM 108f.

[71]VA 220; EGT 70. Cf. VA 212, EGT 63.

[72]"The Origin of the Work of Art," UK 84, PLT 73f. Cf. Ch. I, pp. 45f & 50-52.

[73]Italics ours. . . . *ein Entwerfen des Lichten, darin angesagt wird, als was das Seiende ins Offene kommt.* "The Origin of the Work of Art," UK 84, PLT 73. Cf. "On the Essence of Truth," W 80, BW 124, where reference is made not specifically to "naming," but to asseverative statement (*Aussage*).

itself, brings itself to pass as a concrete happening of what-is.[74] Naming,
then, as a putting into play of showing-Saying, brings to completion
Being's happening as unconcealing. A word that names forth, precisely
in what we may call its physicality, is that wherein primal Saying so
happens in self-disclosive specificity as therewith to gather and let
appear.[75] Our naming, identifying what-is, names it forth "*to* its Being
out of the latter."[76] That is, it lets what-is appear in the unconcealment
vouchsafed to it at any given time, and it does so precisely out of the
immediate holding sway of Being happening as showing-Saying in naming
and named alike. Thus, in our speaking of language, which always
centers in naming, we constantly show what-is to ourselves and to one
another in elucidative flashes that constitute nothing less than the pivotal
central core of that intricate self-opening via which the Being of what-is,
as self-concealing that maintains itself as self-unconcealing, happens self-
manifestingly by way of time as the unfolding Twofold.[77]

     All language and all man's speaking of it partake of this disclosive
character. Yet the unconcealing as which Being rules is precisely self-
unconcealing self-concealing. Unconcealing never happens solely and
unreservedly. Always in its ruling as self-disclosive the Being of what-is
happens as evasive self-withdrawing, and in its every happening as
unconcealing, concealing is underway.[78] Hence concealing is ever in
play in the showing-Saying that rules our speaking. In keeping with what
Heidegger sees as the constitutive manner of Being's happening according
to which it conceals itself behind the particularized happening of what-is,
that Saying, as which Being brings itself rulingly to bear as the gathering,
disposing enduring (*Wesen*) holding sway in language, in the course of its
very bestowing of spoken language continually withholds itself, rarely
letting itself be glimpsed within and beyond the words it permits to be

---

[74]Cf. Ch. I, pp. 30-32, and n. 32.
[75]"The Nature of Language," US 207f, WL 100f.
[76] . . . *zu seinem Sein aus diesem.* (Italics ours.) UK 84; PLT 73.
[77]On the Twofold, cf. Ch. I, pp. 27-33.
[78]Cf. Ch. I, pp. 24-26 & 43f.

voiced.[79] More than this, the very manner of ruling of primal Saying is a "lighting-hiding, veiling reaching forth of world."[80] Accordingly, whenever in the multitudinous occurrings of showing-saying that take place by way of spoken language disclosure happens, its pattern is complex, and the illumining of what-is that is achieved is ever shadowed with concealing.

Taking place thus, language is profoundly deceptive in its happening.[81] In keeping with the ceaselessly shifting intricacy of Being's ruling, now and again unconcealment flashes forth brightly, but more often it takes place as variously veiled. Accordingly, the naming-saying of man may be a showing that brings what-is to genuine appearing; however, most often it is a designating that in some manner indeed discloses, but that does so only in such a way as actually to obscure.

Genuine naming is a gathering of "words (*Worte*)" in a single, organic utterance, or a speaking of a single naming word which gathers forth of itself, that lets what-is appear as at once particularized and subsisting unitedly in interrelation. Mere designating, on the other hand, is a putting together of "terms (*Wörter*)."[82] That is, it is an assembling of words considered primarily in their sounded form and hence viewed and treated as isolated, manipulatable units. As such, designating fragments what-is, letting it appear only as a congeries of obtrusive particulars whose fundamental interrelation is hidden from view.[83]

Even the specific words that thus disclose or mask what-is in its appearing take place under the immediate governance of Being's happening as primordial Saying. For the showing-Saying that, by way of our speaking, directs each particular of the manifold of what-is toward the

[79]"The Nature of Language," US 186, WL 81. On the withdrawing of Being behind what-is, cf. Ch. I, pp. 44f.
[80] . . . *lichtend verhüllende, schleiernde Reichen von Welt* . . . . US 200; WL 93. Cf. also US 211, WL 104.
[81]"Hölderlin and the Essence of Poetry," EHD 35, EB 275f.
[82]*Wörter* as opposed to *Worte*. "*Moira*," VA 254, EGT 99.
[83]"*Moira*," VA 253-255, EGT 98-100. Here, surely, the thought of that inauthentic "everydayness" of human existing in which everything, as something on-hand (*Vorhandenes*), is met and viewed isolatedly and hence as readily manipulatable must be seen to be unspokenly in play. Cf. Ch. II, pp. 88-90.

appearing that it achieves, itself so happens the on-going, delicately shifting fabric of language[84] as ever to provide it immediately to us. The Being of what-is, happening as showing-Saying, reaches toward us by way of language. In the hearing of that Saying, which is our self-submission to its governance, we accept the detailedly transpiring language it provides. To accomplish our saying as a saying-after, "we demand the sound already stored up for us, reach out toward it and summon it."[85] Sometimes the word we seek is denied to us, and we are touched subtly by language as something whose provision is not simply within our human control.[86] But most often we make our way by way of diverse "modes of saying," unquestioningly living out of their kinship,[87] as words in their variousness and multiplicity are unobtrusively provided to us.

Again and again Heidegger himself works from out of this understanding of language. For him, in language at a particular time the Being of what-is is bringing itself to utterance. In its historically structured unfolding as the Twofold,[88] Being is so ruling human speaking and what is spoken about as to manifest its own manner of happening at any given time in the import of words and in the manner of appearing that they vouchsafe to the particulars of the manifold of what-is. Accordingly, Heidegger often looks back to the original meanings of words,[89] traces changes in connotation that words have undergone,[90] or alludes to radical changes of meaning occasioned when, as in the shift from Greek to Latin,

---

[84]"The Principle of Identity," ID 102, 38.
[85]"The Way to Language," US 255, WL 124.
[86]"The Nature of Language," US 161, WL 59.
[87]US 208, 161; WL 101, 59.
[88]Cf. Ch. I, pp. 47-49.
[89]See, e.g., the lengthy explication of the meaning of *bauen*, to build, in "Building, Dwelling, Thinking," VA 146-149, PLT 146-149; the discussion of the meaning of the German word *Ding* (thing) and its Latin parallel *res* in "The Thing," VA 172-177, PLT 174-176; and, perhaps most importantly, the complex elucidation of the meaning of the verb *sein* (to be) in terms of the archaic verbs that have coalesced into it and that lend it their diverse ranges of connotation, in EM 54f, IM 58f.
[90]See, e.g., the discussion of *wirken* (to work) in "Science and Reflection," VA 48f, QT 160f, and of *betrachten* (to contemplate, observe) in the same essay, VA 56ff, QT 166ff.

the words of one language are used to translate those of another.[91] This he does because he finds the manner of happening of the Being of what-is immediately disclosed in the meanings that words bear. For the same reason his own utterances are wrought out with great care. For his words are consciously intended to offer to Being an adequate avenue of self-manifestation.[92]

3

Human language, as the elucidative speech via which at any given time the Being of what-is happens in manifoldness as the unfolding Twofold, provides openness—the openness of interpretive disclosure—wherein, in the coming together of what-is in its appearing and of the apprehending that lets appear, possibilities of human orientation and on-going accomplishment can be lived out. All human doing whatever takes place in this arena.[93]

Foremost here among human accomplishings is *thinking*. Thinking has its inception and is everywhere underway in man as man at the primary level of awareness where human apprehending is first in play.[94] Out of that origin, thinking is by certain individuals entered into as a particular undertaking[95] and is accomplished, via considered utterance, as an elucidation of the Twofold, as, in one way or another, the latter vouchsafes itself.

Thinking (*Denken*) is a human response to the governance of Being that takes place with peculiar directness. In making clear the meaning of thinking, Heidegger repeatedly expounds the saying of Parmenides, "For the same is thinking and Being."[96] In this statement the word "same," he

[91]See, e.g., the discussion of the disparity between *aition* and *causa* in "The Question Concerning Technology," VA 16ff, QT 6ff, and between *thēoria* and *contemplatio* in "Science and Reflection," VA 54f, QT 164-166.
[92]Cf. Ch. XV, pp. 602-614.
[93]"The Turning," TK 40, QT 40f. Cf. "On the Nature of Language," US 208, WL 101.
[94]Cf. Ch. XIV, pp. 552f.
[95]Cf. Ch. XIV, pp. 567f & 571-573.
[96]*To gar auto noein estin te kai einai.* (*Denn dasselbe ist Denken und Sein.*) "*Moira*," VA 231ff, EGT 79ff. Cf. also, e.g., WHD 147ff, WCT 240ff; "The

says, connotes the belonging-together of thinking and Being.[97] It speaks of the unique belonging of thinking within the single Twofold as which the Being of what-is happens. Happened away by Being, thinking, out of its provisional autonomy, directs itself attentively back toward Being. Thus it heeds the Twofold as such. In so doing, it answers a summons laid upon it, in that it corresponds to and accords with the Being of what-is. As such answering to (*Entsprechen*), thinking attends upon the Being of what-is in its disclosive happening as showing-Saying. In the most fundamental manner of disclosure possible to man, it responsively and immediately says, i.e., shows, the Being of what-is in the way in which the latter is manifesting itself. And it does so by way of language as the spoken word.[98]

Thinking obeys a summons. In the sphere of history, those who are called to dedicate themselves specifically to its accomplishment fulfill a crucial role. Every thinker follows an assigned path.[99] It is granted to him to glimpse what-is according to the manner of Being's happening ruling it at a given time and therewith to shape an utterance that incisively says what he sees. Not only can a thinker like Anaximander or Parmenides be said to utter a decisive saying (*Spruch*),[100] but Heidegger can speak of the "word" of Descartes or of Nietzsche in alluding to seminal statements from which their thinking opens out.[101] Again, a single word—Plato's *idea*, Aristotle's *energeia*, Kant's *Position*—shows the way in which at a particular time Being is happening in governing what-is.[102]

A thinker need not himself know the true import of his word. Indeed, Heidegger believes that even those who at the beginning spoke seminal words did not fully *think* Being.[103] And to his mind, from Plato

---

Principle of Identity," ID 90ff, 27ff.
[97]"The Principle of Identity," ID 102, 38.
[98]Cf. "The Turning," TK 40, QT 40. Cf. Ch. XIV, pp. 573f. On *Entsprechen*, cf. this chapter, p. 118, n. 36.
[99]See "The Word of Nietzsche, 'God Is Dead,'" HW 194f, QT 55.
[100]"The Anaximander Fragment," HW 296, EGT 13.
[101]See HW 215, QT 76 & HW 220, QT 82.
[102]"Time and Being," SD 9, TB 9.
[103]"The Word of Nietzsche, 'God Is Dead,'" HW 243, QT 108f. Cf. also *"Logos,"* VA 228, EGT 77.

onward thinkers were able to consider only what-is; for although the Being of what-is called forth and informed their thinking, Being, as the self-differentiating Happening that comes upon and shows itself via what-is but remains rulingly other than the latter,[104] was concealing itself. It was disguising itself variously as the underlying ground of the totality of what-is, a ground that could be grasped as belonging to a single structuring in company with that which it grounded, and hence was so withholding itself as to go undetected as the wholly sovereign self-unconcealing self-concealing that is intrinsically beyond that structuring as such and yet holds sway within it in happening self-consummatingly as the unfolding Twofold.[105]

In its self-withholding, Being progressively masked its fundamental manner of happening as the presencing of what presences. Happening on-goingly in a sequence of self-withholdings that were its adaptive self-sendings-forth as what Heidegger calls the "history of Being (*Seinsgeschichte*)," Being successively displayed itself according to various stampings (*Prägungen*)—various instantiations of itself that, while bearing its likeness and partaking of its sway, concealed the Happening itself that ruled—which were caught sight of in the taking place of what-is and were spoken forth manifestingly by a series of thinkers in word after word.[106] Heidegger's own thinking took place, he believed, from a fresh vantage ground. Unlike most of his predecessors, he was permitted so to catch sight of reality as properly to discern that originative Happening, bringing itself to pass as the Being of what-is, in play at once within *and*

---

[104]Cf. Ch. I, pp. 29-33.
[105]See "The End of Philosophy and the Task of Thinking," SD 62, TB 56; "The Onto-Theo-Logical Constitution of Metaphysics," ID 139f, 71f; and "Overcoming Metaphysics," VA 77f, EP 90. Cf. also "The Word of Nietzsche, 'God Is Dead,'" HW 243, QT 109, and see "A Heidegger Seminar on Hegel's *Differenzschrift*," translated by William Lovitt, *The Southwestern Journal of Philosophy*, Vol. XI, No. 3, Fall, 980, pp. 27ff. For a complete German rendering of the latter passage, which is a translation from the reworked French original (*Questions IV*, Paris, Gallimard, 1976, pp. 240ff), see VS 47ff.
[106]"Time and Being," SD 8f, TB 8f, and "The Onto-Theo-Logical Constitution of Metaphysics," ID 134f, 66f. Cf. ZS 100, QB 101.

beyond it. And the word that he speaks is a word that bodies forth that Happening.[107]

When a "thinking of Being" takes place, in it preeminently and with determining priority "language," primordial showing-Saying as which Being happens, "comes to language," comes to disclosive utterance in words.[108] The thinking of such thinkers, then, taking place by way of language, is peculiarly endowed with the power to accomplish the elucidative illumining of what-is in its appearing that language as such achieves. Time and time again, the pronouncements of the thinkers of the West, wittingly or not, have shown to the ages wherein they spoke the manner of Being's happening there holding sway, in trenchant words specifically suited to whatever revealing was taking place. Even as they brought to light the manner of happening implicit in what was transpiring in a given time, those pronouncements and whatever expository thinking followed from them provided a stated basis that, in one way or another, underlay and conditioned the outlook and cultural endeavors of each age.[109]

So direct is the ruling of Being over any thinking that gives it utterance that Heidegger can characterize such thinking as "the dictation (*Diktat*) of the truth of Being."[110] In it Being, happening as self-maintaining self-unconcealing,[111] transcribes itself into human speech. Such thinking, in whatever context and whatever guise it takes place, is always bound by the dictates of Being's happening as unconcealing.

To speak of this decisive immediacy of the rule of unconcealing Heidegger uses the verb *dichten* and the noun *Dichtung*. *Dichten* ordinarily means to compose as would an author or poet, while *Dichtung* means poem or poetical work. Heidegger extends the meanings of these words, taking them to speak of the light-bringing opening-up as which

---

[107]Cf. Ch. XV, pp. 602-614.
[108]"The Anaximander Fragment," HW 303, EGT 19.   On the uttering of the happening of Being, cf. Ch. XV, pp. 592f & 599-614.
[109]Cf. "The Age of the World Picture," HW 69, QT 115.
[110]"The Anaximander Fragment," HW 303, EGT 19.
[111]Cf. Ch. I, pp. 24-26. On truth as unconcealment, see Ch. I, p. 38, and n. 49.

Being as unconcealing brings itself to pass in the midst of what-is in a flashing forth, into the prevailing obscurity that so persistently shrouds the latter, which allows what-is, in its particularity, genuinely to disclose itself.[112] With special stress on the imperative resident in Being's ruling as a happening that makes manifest, *Dichten* and *Dichtung* name that primordial showing-Saying as which Being, unfolding as the Twofold, pivotally accomplishes itself as such unconcealing. Hence spoken language, as that wherein primordial showing-Saying, as such manifesting, holds sway, can be said to be "*Dichtung* in the essential sense,"[113] i.e., in the sense that, as most immediately disclosive of Being's ruling, governs every other usage of the word. And since in its taking place by way of language thinking, in responding to primordial Saying, most fundamentally and most directly lets unconcealing happen, Heidegger, indicating this primacy with the prefix *Ur-* (original, preeminent), can speak of thinking as *Urdichtung*.[114]

This characterization of thinking in terms of its primacy in according with and bringing to fruition the self-maintaining self-unconcealing happening of the Being of what-is, via Being's disposing ruling as showing-Saying, places it, as Heidegger's central use of *Dichtung* suggests, in immediate relation with art, which the poetic, indeed, instances in a preeminent way.[115] Like thinking, art too, in accordance with the governance of that ruling, is itself a decisive avenue for Being's accomplishing of its self-unconcealing by way of man. In art as such also the primordially dictative unconcealing as which Being brings itself to pass as primordial showing-Saying, as language, holds direct

---

[112]"The Origin of the Work of Art," UK 82f, PLT 72f. Cf. the description of poetic disclosure in " . . . Poetically Man Dwells . . . ," VA 196, PLT 221. Cf. Ch. V, pp. 199-201.

[113]. . . *Dichtung im wesentlichen Sinne.* See "The Origin of the Work of Art," UK 84-86, PLT 74f. On the meaning of *Wesen*, see Ch. VI, pp. 252-257.

[114]"The Anaximander Fragment," HW 303, EGT 19. On thinking as possessed of a primary role, cf. Ch. XIV, pp. 552-571.

[115]"The Origin of the Work of Art," UK 83, PLT 73. On the close relation obtaining between thinking and poetic composing (*Dichten*), cf. "The Nature of Language," US 188f, 195f; WL 83f, 89f. Cf. also Ch. XVI, pp. 627f & 661. On the broader question of the relation of thinking and art, cf. Ch. XVI, pp. 639-645.

determining sway. Accordingly, not only is poetry *Dichtung*, in the narrower sense of the word, but for Heidegger the word *Dichtung* can be applied more broadly to the way in which every art takes place.[116] Like thinking, art is immediately ruled by the imperative disclosive decreeing as which the Being of what-is happens unconcealingly as showing-Saying.

Here poetry (*Poesie*) is foremost, precisely because it brings itself to pass immediately in language.[117] Thinking and poetry take their way in but one "element of saying," and together they are superior among the many modes in which, from out of its origin in Being's ruling, via spoken words that saying brings itself into play.[118] The word of a poet can be as directly revelatory of the manner of happening of the Being of what-is as can that of any thinker, a fact that Heidegger's own use of the words of poets as immediate bases for his own thinking clearly evinces.[119] On the other hand, the utterance that poetry accords to Being is, characteristically, so presented in words dense with meaning and/or so couched in image upon image[120] that, while its disclosing of the manner of unfolding of the Twofold is no less direct, a probing of the words or image in question will be needed if that disclosure as such is to be brought fully to light.[121]

All the other arts also, from painting to architecture, although they are themselves wrought out in media other than words, always take place within and from out of the elucidative disclosing of what-is in its historically structured appearing that primordial showing-Saying is

---

[116]See "The Origin of the Work of Art," UK 83, PLT 73.
[117]UK 83; PLT 74.
[118]"The Nature of Language," US 208, 188; WL 101, 83.
[119]See. e.g., "' . . . Poetically Man Dwells . . . ,'" VA 187ff, PLT 213ff; "The Question Concerning Technology," VA 36ff, QT 28ff; and "The Nature of Language," US 162ff, WL 60ff. Cf. Ch. V, pp. 195f & 220; Ch. XII, pp. 486-488; and Ch. XIII, pp. 494f & 536.
[120]On poetic imaging, see Ch. V, pp. 208f.
[121]For such a disclosure of meaning, cf. "The Thinker as Poet," AED 5ff, PLT 3ff. Cf. also Ch. XV, pp. 606-609. See also, e.g., "Words," US 219ff, WL 139ff. And for a presentation in which, as such probing is undertaken, the question of the relation between thinking and poetry is one of the chief matters under consideration, see "The Nature of Language," US 162-216, WL 60-108.

forever accomplishing by way of spoken language. They are determinatively rooted in the illumining of what-is that language ceaselessly but unobtrusively provides. Thus they take place under the pervasive governance of primordial showing-Saying that is there in play. But more than this, these arts, each in its own way, are themselves channels for the accomplishing of such illumining, i.e., for the immediate taking place of that ruling Saying that is a showing. In the works that they bring forth they too, in accordance with its dictates, let Being, happening as unconcealing, achieve a manifesting as direct and as specific as that achieved by way of a revelatory spoken word.[122]

In a genuine work of art of whatever sort truth as unconcealing powerfully resides. It accommodates itself immediately into the shaped, specifically delineated happening as which the art work takes place.[123] Hence Being, happening as unconcealment, manifests itself directly in the work. It therewith lets the latter stand forth in astonishing uniqueness as a particular entity.[124] But it also displays the latter's relatedness within the manifold of what-is. Seen from one perspective, the art work so immediately manifests the happening of Being as to make surrounding entities viewed simply in themselves seem bereft of Being.[125] But from another perspective, when all is seen in relation, the illumining vouchsafed in the particularity of the given work gathers surrounding entities to itself so that, viewed in relation to it, they too can appear in the particularity, even the uniqueness, peculiar to each.[126]

Here the Being of what-is, happening as the unfolding Twofold, takes place as the work in its milieu. In its so doing, the responsive attentiveness of man is again centrally in play. As artists and as those who, accomplishing "preserving (*Bewahren*)," genuinely behold their works in standing in openness toward the work set forth, men receive the

[122]"The Origin of the Work of Art," UK 82ff, PLT 71ff.
[123]Cf. HW 81, PLT 71: *Im Werk ist das Geschehnis der Wahrheit und zwar nach der Weise eines Werkes am Werk.*
[124]See UK 73, PLT 65.
[125]UK 82; PLT 72.
[126]See UK 50f, PLT 41f.

illumining thrust of Being's happening as unconcealing, shape for it a dwelling place in the midst of what-is, and hold open the avenue of revealing that it has wrought out.[127]

It is because art in all its varieties can, from out of its basis in language, which names and disposes into view, fulfill with obedient directness the happening of Being as primordial showing-Saying—i.e., as the letting appear that, decisively ruling, displays the manifold of what-is in particularity—that Heidegger can extend the meaning of the word *Dichtung* to include, with poetry, every branch of art. This extension displays for him the close conjunction of thinking and art as twin executers of the dictates of that primal Saying. Indeed, because there is in man's accomplishing of art from out of his carrying forward of his existing in the domain of language a submissive looking toward Being that receives and specifically bodies forth the latter in its happening as revealing, and because this responding is ever underlain by the primary apprehending by which, as that apprehending transpires as a thinking, language first brings itself disclosingly into play, Heidegger can say that all art is "a thinking."[128]

Like thinking, too, art performs a foundational role in the on-going, ever particularized life of man. Even as the revelatory word of a thinker is constitutive for an age, so the revelatory work of an artist is constitutive for a given people, a coherent group possessed of its own language. In a genuine work of art, Being as directly ruling all that is comes to manifestation. Every such work is brought forth always in the midst of a given people and can be seen to spring, indeed, specifically out of the concrete milieu that the people calls its homeland.[129] In gathering everything to itself, in its illumining of the particulars of the manifold of

---

[127]See UK 81 & 74-77, PLT 71 & 66-68.

[128]"The Anaximander Fragment," HW 303, EGT 19. Heidegger writes in "The Way to Language," (US 267, WL 135): *Alles sinnende Denken ist ein Dichten, alle Dichtung aber ein Denken. Beide gehören zueinander aus jenem Sagen, das sich schon dem Ungesagten zugesagt hat, weil es der Gedanke ist als der Dank.* For further consideration of the interrelation of thinking and art cf. Ch. XVI, pp. 627f & 639-645.

[129]"Memorial Address," G 16, DT 47.

what-is wherein it itself is taking place, the work accomplishes an opening-up, a providing of meaning, centered in the life of that people. In such a work, in all its concreteness—whether the work be an ancient Greek temple or a Van Gogh painting showing a peasant woman's shoes— the directive meaning informing the life of some particular community is embodied and immediately in play.[130] That meaning can be discerned in the way in which the work is taking place, and at the same time it sheds light on all the doings of the community within which the work belongs. Just as the art work illumines and gathers all entities to itself, it gathers and illumines those doings as well. Paths to be trodden, events to be lived out, decisions to be made by the individual or by the group in their corporate life are thrown into relief and seen interrelatedly in an on-going context of significance, i.e., as a world, because the meaning governing and disposing that life is decisively manifest in the work of art.[131]

A destining rules in the life of any people, for every people belongs to the manifold of what-is that is ruled by Being's self-sending-forth.[132] A genuine work of art lets that destining be glimpsed as, in disclosing meaning in the events and relationships whose totality constitutes the people's "world," it discloses the direction that the people's life is taking. Undergirded by the taking place of such orienting disclosure, the people lives as historical, i.e., as a community destined, sent on its appointed way. Catching sight in the openness of disclosure provided in the art work of the meaning meted out to it, it has the possibility of moving forward in awareness toward the fulfilling of the prior determination that Being, as destining, has laid upon it.[133]

In all this, Being's happening as primal showing-Saying, ruling as language, is determinatively in play. From out of man's language-configured existing every work of art, whether it itself be wrought in

---

[130]"The Origin of the Work of Art," UK 41ff, PLT 41ff & UK 28ff, PLT 32ff. In the case of the painting discussed, the community in question is much narrower, but the implied reference to it is clear.
[131]UK 44f; PLT 44f. On "world" as thus used to speak of what-is as grasped as a meaningful entirety, in any particular instance of its being met, cf. Ch. II, pp. 61-63.
[132]Cf. Ch. I, p. 48.
[133]See "The Origin of the Work of Art," UK 41-45, PLT 41-44.

words or not, has its inception and comes to completion. And it is always via language, through whose transpiring meaning is carried into play, that the response to the work and the taking-place of the illumining of what-is and the disposing of a world that the work permits come foundingly to pass.

All the arts are possessed of the capacity directly to answer to the dictates of the Being of what-is in its happening as the showing-Saying that lets appear. But Being happens changefully, and it happens first of all as self-concealing. Not all ever apprehend the unconcealing vouchsafed in a work of art.[134] And, in time, that unconcealing always dims, as Being's happening as concealing gains upon it. When new peoples arise, the determinative power of former works vanishes, as does the milieu to which they spoke.[135] And, moreover, the same self-withholding in its self-sending-forth that characterizes Being in its self-manifesting by way of thinking from age to age rules in art as well.[136] The powerfully illumining outreach accomplished in particular art works

---

[134]Cf. UK 75f, PLT 66f.

[135]Cf. UK 39f, PLT 40f. In this context a puzzling question and a quite tentative answer to it suggest themselves. The question concerns the fact that in his discussions of the arts and their works Heidegger concentrates no attention on works of music, an abstention surprising in a highly educated German of his obvious artistic sensitivity and interest. Might it be that his silence in this regard arises to some significant degree from the fact that centuries-old musical works can, when performed in the present, meet us, assuredly, with a vitality that is very far from being diminished? Is it credible to think, to take a most salient example, that a fugue of Johann Sebastian Bach, whose historical, communal milieu was certainly markedly different from our own, is experienced now by us as a work bereft over time of intrinsic beauty and power? One cannot but wonder whether the peculiar character of music, as an art whose works possess for many of us, as surely as they did for Heidegger, an immediate capacity to make meaningful, may not have rendered music difficult for Heidegger to assess in accordance with his understanding of a "work." If this were the case, might that difficulty perhaps have served, presumably in company with other factors, to place musical works, together with whatever complexities of interpretation their inhering vitality might have raised, beyond the purview of his manifest concern? With regard to Heidegger's sensitivity to the musical, cf. Ch. IV, pp. 154f, and n. 46.

[136]Cf., e.g., Heidegger's remark concerning the art of the Middle Ages in Appendix 7 of "The Age of the World Picture" (HW 94, QT 143), which relates directly to his discussion of the understanding of the Being of what-is prevalent in that period (HW 83, QT 130).

has changed over time, until now it has virtually withdrawn. Pervasively strong and decisive among the Greeks, it is narrow of scope and impact among ourselves, touching only the few.[137] For among us a detached viewing of a work of art for the sake of individual aesthetic experience[138] has come to replace the ready yielding of a community to the disclosive sway of a work whose disclosing of the Being of what-is can meaningfully open to view and dispose as a forward directed unity its on-going life.

For Heidegger, human life as ultimately understood, i.e., human life as "historical," as ordained from out of Being, is thoroughly undergirded and carried forward via language as a pivotal locus of the happening of Being. In language, all thinking, all art, indeed all human accomplishing whatever moves.[139] Language as elucidative naming that, under Being's governance, opens what-is to view affords the primary arena and the decisive medium of self-undertaking by way of which thinking and art can, by achieving the determinative disclosure peculiar to them, further, via specific centered events, the pervasive ruling of Being. And language is always already given. It is offered to man in his need to speak. In language Being finds utterance. But just as Being repeatedly changes in its self-sending-forth,[140] so language changes. Accordingly, it comes to people after people quite uniquely. Indeed, those peoples are directed and molded by it. For the comprehensions and conceptions that inform the on-going life of a historical people, allowing it to orient itself and to pursue with significant awareness its appointed way, are already stamped out in the language it receives.[141]

Thus does Being, ruling historically via its changeful self-sendings-forth, configure and configure yet again the vast manifold of what-is, in happening as the ever distinctively particularized appearing by way of

---

[137]This is evident in Heidegger's detailed discussion of the Van Gogh painting ("The Origin of the Work of Art," HW 43ff, PLT 32ff), which is powerfully revelatory after the manner of an art work, but which can surely not be taken to play a primary, formative role for any historically discernible community or people.
[138]Cf. "The Age of the World Picture," HW 69, QT 116.
[139]"The Turning," TK 40, QT 40f.
[140]Cf. Ch. I, pp. 48f.
[141]Cf. "The Origin of the Work of Art," UK 84, PLT 74.

which pure Happening, as the single Twofold, in ceaselessly renewed disclosure and restraint from disclosure maintains itself via man as he who lives out time.[142]

▨   ▨   ▨

Via language, the Being of what-is says, i.e., shows, itself as the unfolding Twofold to man and by way of man. Via language, man, in an intricacy of responsiveness that involves every aspect of his individualized corporate life, lets the Being of what-is manifest itself rulingly in its unfolding as the Twofold. In maintaining itself by way of man's disclosive manner of existing, where time, under the leap of projective, orienting apprehending, ceaselessly opens to arriving and departing, lighting and obscuring, the accepted and the new, Being as self-unconcealing self-concealing so happens as primordial showing-Saying as to come to utterance precisely in the elucidative communicative speaking that, in permitting meaningful appearing of whatever kind, is ever fundamentally constitutive for man as man. It flashes forth here and there—now brightly, now dimly, now in the flickering intricate pattern of man's continual speaking, and now in the sudden brilliance of a foundational word or work. Here it shines forth with a radiance that illumines all it touches. There it glows hiddenly, able only to bathe with an obscurity that conceals.[143] Always man offers the arena of its happening. However penetrating or misinterpretive his receptive response to Being may be, man, uniquely poised in knowing awareness within the Twofold—apart, yet ever held within Being's pervasive sway—via language ceaselessly lets Being as unconcealing happen wherever and however it will.

Being and man belong together, not merely as Being and what-is belong together as the single Twofold but in a reciprocity that accords to man by way of the centrally constitutive dialogical happening of language

---

[142]On this role of man, cf. Ch. II, pp. 69-72.
[143]A discussion of Heraclitus' identifying of the Being of what-is, in its happening, as fire affords Heidegger an opportunity to describe vividly this ever-varied flashing of unconcealing. (*"Alētheia,"* VA 275f, EGT 117f.)

the special dignity of a cognizant partner in the accomplishing of the self-manifesting self-particularizing unfolding as the Twofold, as which the Being of what-is, happening most fundamentally as self-concealing, unconceals itself and so maintains itself as itself. Only by way of this unique belonging can the Being of what-is show and preserve itself as such, and only by way of it can man exist as man.

Here the pure Happening of which the word Being speaks brings itself most fundamentally to pass. And here that Happening may be glimpsed most primally as an initiatory happening into relation, sui generis and wholly unique, that rules complexly in the unfolding Twofold as which the Being of what-is accomplishes its self-unconcealing through man by way of time. Through it the decisive belonging together of man and Being comes to pass. Through it the needed manifesting achieved via showing-saying is wrought out. For only through it are the multitudinously nuanced aspects and concrete shapings of the happening that it pervades and quickens ever allowed to come into their own.

CHAPTER IV

EREIGNIS

At this juncture we come to the heart of Heidegger's thinking, i.e., to his direct portrayal of pure Happening, in its self-unfolding, in words that display the peculiar impetus most central to it. For this he uses the verb *ereignen* and the noun *Ereignis*. These words ordinarily mean, respectively, to come to pass or happen, and event. Heidegger repeatedly insists that these usual denotations must be set aside.[1] For him *Ereignis* and *ereignen* do not allude at all to the kinds of events that we ordinarily know, which are to us identifiable occurrences capable of being isolated and pointed to precisely in their specificity and which, if they are to be understood, must each be seen as an outcome (the literal meaning of event) wherein some prior agency or congeries of agencies has brought itself determiningly to bear. Connotations of active coming to pass remain, but they are raised to quite another level of thought. *Ereignis* and *ereignen*, as Heidegger uses them, speak immediately of what from the beginning of our discussion we have been calling pure Happening. But, most importantly for him, their connotations are so probed as to let them manifest directly the most determinative manner of that Happening.[2]

---

[1]See "The Principle of Identity," ID 101, 36. See also "Time and Being," SD 21, TB 20.

[2]Often in English translations of Heidegger's writings *Ereignis* is rendered with "event," "event of Appropriation," or "Appropriation," and *ereignen* with "to appropriate." The suitability of the word "event" is clearly questionable, especially since no effort is made to separate it from its usual connotations. The employment of "Appropriation," which presumably owes much to the use in French translations of "Appropriement" and "approprier," is also inadequate, and will not be pursued here. The verb appropriate does originate from Latin *proprius* (own). The prefix *ap-*, however, parallels Latin *ad-* (to); and therein lies the major difficulty inhering in both verb and noun. "Appropriate" and "Appropriation" speak basically of a taking, a taking to oneself or to a given use or purpose (cf. *The American Heritage Dictionary of the English Language*, Boston, Houghton Mifflin Company, 1971, p. 64). *Ereignis* and *ereignen* do speak, for Heidegger, of a gathering toward the Bringing-to-pass that is in question, i.e, of its "owning" of that toward which it comes to pass. But they

Heidegger's use of these crucial words began quite early in his work. In a later context (1959), he witnesses to the consistency of their use, remarking, in a note to "The Way to Language":[3] "The author has for more than twenty-five years used the word *Ereignis* in his writings for the matter thought here."[4]   He brought them to overt discussion, however, only relatively late. Misunderstanding arose. For attempts were made by those who considered his thought to interpret his *Ereignis* as a designating of Being that could properly be placed on a par with other concepts familiar from the history of metaphysics. Heidegger rejects any such interpretation.[5]   His use of *Ereignis* is to him not at all a continuation of metaphysical thinking, and it has little directly to do with the conceptions of Being that he finds there. Rather, it stands in closest conjunction with his own thinking of Being. It might, in fact, be said to be a crystallization of that thinking, which, to his mind, itself lies so radically outside traditional philosophical usage.

1

When the Being of what-is is understood first of all as the presencing of what presences, other characterizations of Being—e.g., as self-unconcealing concealing and self-concealing unconcealing, as the Twofold, or as self-withholding self-sending-forth—must come into play in order that the meaning of "presencing" be adequately explicated.[6]

---

speak equally, and indeed more decisively, of a bestowing, a providing, a giving-into-its-own, brought to pass with respect to that which is so gathered. "Appropriate" and "Appropriation" are not words that properly manifest this all-important aspect of meaning.
[3]"The Way to Language," US 260, WL 129.
[4]Cf., e.g., "The Origin of the Work of Art,"   UK 85, PLT 74 (1936), and "Overcoming Metaphysics," VA 71 & 72f; EP 84f & 86 (passages which, since they stand early in a work whose contents come from the years 1936-1946, are presumably early). For an illustration of consistency of use, see this chapter, pp. 149f, and n. 29.
[5]See "Time and Being," SD 22, TB 21, and "The Way to Language," US 260, WL 129.
[6]On self-unconcealing concealing and self-concealing unconcealing, see Ch. I, pp. 24-26 & 43f; on the Twofold, see Ch. I, pp. 27-29; on self-withholding self-sending-forth, see Ch. I, pp. 47-49.

Ultimately, in a usage that goes beyond but bears upon all those depictions of Being, Heidegger speaks of "Being as *Ereignis*."[7] In the word *Ereignis* a fresh range of meaning opens to view. Being as presencing, glimpsed most profoundly, partakes of and is wholly determined by the complex initiating named with the word *Ereignis*.[8] Here, despite the misleading specificity inevitably introduced through the use of ordinary language structures, where nouns must serve to name this and that, there must be no thought of entities of any sort. Here we have to do with pure Happening. To speak the word *Ereignis* is, for Heidegger, to display as clearly as may be the manner of that Happening in its carrying of itself forth in self-disclosure.

Being is pure Happening, and *Ereignis* is the peculiar disclosing that informs to its depths the Happening of which the word Being so often speaks.[9] So primal and so pivotal is *Ereignis* that Being itself, when considered with respect to the complexity of interrelating[10] entailed in its

---

[7] "Time and Being," SD 22, TB 21. The singleness of Happening that is intended in the words Being and *Ereignis* is well shown by the fact that the self-lighting self-opening (*Sich-Lichten*) that is for Heidegger central to the portraying of the happening of Being, which brings itself to pass as a light-bringing clearing (*Lichtung*) (cf. Ch. XIII, pp. 505f; Ch. I, pp. 37f; and Ch. II, pp. 72f), is also predicated of *Ereignis*. ("Overcoming Metaphysics," VA 99, EP 110. Cf. Ch. XIII, p. 516.) The pure, all-provisive Happening that the words "Being" and "Ereignis" name from diverse perspectives (see this chapter, pp. 166-168) brings itself to pass as the light-permitting clearing of whose happening both words can speak.

[8] Cf. "Time and Being," SD 22f, TB 21f.

[9] Cf. the discussion of Being in "The Turning," TK 42-44, QT 43-45, which culminates in the statement, *Einblitz ist Ereignis im Sein selbst* ("In-flashing is *Ereignis* in Being itself").

[10] In this study we frequently use the verb relate together with its nominal and adjectival counterparts to speak of the happening-toward-one-another that characterizes the elements constitutive of Heidegger's Twofold. But we use it with hesitation and with regret, for want of any other adequate single verb. Those "elements"—Being, what-is, man—are ultimately the same. They are the same Happening. Yet they happen as differentiated, as the provisionally sundered, whose sundering is transcended in a counterpoised happening-toward-one-another that is always in play among them. Cf. Ch. I, pp. 29-33 & 53, and Ch. II, p. 108, and n. 190. It is this happening-toward one another that our word "relate" is intended to connote. Our reluctance to use "relate" springs from the fact that Heidegger himself specifically alludes to the Latin noun *relatio* as speaking of a drawing together of distinct elements that involves a conforming (*Richten*) of one element to the other. His German word for such relating

self-unfolding as self-sundering, can be presented by Heidegger as secondary to *Ereignis*.[11] But whatever intricate nuances of manner and primacy Heidegger may explore in so speaking, it is pure Happening and pure Happening alone that he has in view.[12]

The pure Happening as which Being happens is a self-concealing that intrinsically carries itself out as self-unconcealing. It is an utter Stillness, an implosive withholding-to-self, that is, precisely as such, an on-thrusting movement that, as the self-differentiating Same, in its very self-bringing-to-pass carries itself forth as the multitudinous unfolding of the Twofold via the manifold of what-is.[13] The self-sundering Same as which the Being of what-is brings itself rulingly to pass is itself what we may perhaps call—with an image that is admittedly far too static—the point of inception for the happening-forth of concealing as unconcealing. For this "Same" which itself rules as a differentiating is precisely "the play (*Spiel*) of Stillness," the provisive forth-going of self-containedness, as which concealing happens as unconcealing.[14]

In the extended context in which Heidegger portrays the happening

---

is *Beziehen*. The participants of the Twofold cannot properly be thought of as distinct somethings, and the relating subsisting among them has nothing to do with a conforming of the commensurate. For Heidegger *relatio* speaks of a conceptualizing assessing (cf. Ch. XI, pp. 419f). We intend "relate" to speak of the happening-toward intrinsic to the Twofold as such. For this relating-toward, Heidegger uses *Verhältnis*, whose root suggests holding, retaining. See this chapter, pp. 160f, and n. 67. In translating or presupposing *Verhältnis*, we have sometimes found it possible to evidence this meaning through the use of "holding-relating," but usually clarity of presentation has required the use of some simple form of "relate." When forms of relate are encountered in our discussion of what we have called the elements of the Twofold, the peculiar meaning that we intend such words to carry in the Heideggerian context must continually be kept in mind.

[11]"Time and Being," SD 19, TB 18f.

[12]In this connection we should note that the structuring found by Heidegger for *Ereignis*, namely a provisive out-going accomplished as a drawing-toward (see this chapter, pp. 148-151), which happens by way of an openness arising from a self-withholding in self-granting (for this latter element see this chapter, pp. 163-165) is fully accordant with the structure shown us for the happening of Being as the Being of what-is. See Ch. I, pp. 29-33, and n. 36.

[13]Cf. Ch. I, pp. 26-33.

[14]"The Nature of Language," US 214, WL 106.

of the Same as pertaining specifically to the bringing-to-pass of pure Happening itself from out of its innermost heart, concealing named as Stillness, he makes the word *Weg* (way) and the verb *be-wegen* and the noun *Be-wegung* akin to it primary in his speaking of that concealing,[15] thereby bringing pointedly forward the connotations of opening and directing-toward intrinsic to his thinking. *Bewegen* ordinarily means to move or to alter something, and *Bewegung* means "movement." Heidegger, citing the Schwabian verb *wëgen*, which means to open a way—as across a land deep in snow—accords to *Be-wegen* the meaning "to prepare a way," where "way" is to be specifically understood as that which permits arrival.[16] Identifying *Weg* (way) as thus understood with the *Tao* of Lao Tse, he finds the word to speak in an ultimate manner of provisive Happening as such.[17] This is "the Way that opens a way for everything, that rends open for everything its course."[18] It is a letting arrive that is, as such, an opening up for the permitting of arriving.

Here, where we very evidently are speaking of pure Happening in a manner wholly similar to that in which we have found Being to be portrayed,[19] the "Same" that gathers together even as it differentiates, as which concealing—surmounting the quiescence and silence that its identification as Stillness bespeaks—so carries itself out as to open the light-bringing clearing (*Lichtung*) that permits arrival and the happening of the Being of what-is via disclosure, itself so happens as to open the provisive ways that that clearing alone can in fact yield.[20] That Same, the accomplishing "play of Stillness," in a pivotal instancing of the distancing "nearing" that we have seen to rule in the on-happening of time as reaching forth the presencing of what-presences—pure Happening in its ruling as the Being of what-is[21]—in its happening endows with ways (*be-wegt*), and therewith permits accomplishment to, the "play-of-world (*das*

---

[15]US 196-214; WL 90-106. On *Be-wegung*, see also Ch. III, pp. 123-125, and n. 63. On the movement (*Bewegung*) in question, cf. also Ch. IX, p. 353.
[16]"The Nature of Language," US 197f, WL 91f.
[17]US 199; WL 93.
[18]US 199f; WL 93.
[19]Cf. Ch. I, pp. 47f & 29-33, and Ch. III, pp. 123-125.
[20]Cf. Ch. I, pp. 29-33, and "The Nature of Language," US 197, WL 91.
[21]Cf. Ch. I, p. 39.

Weltspiel)."[22]   With a question that in its context clearly anticipates an affirmative response, Heidegger suggests that way-providing *(Be-wegung)*—the self-bringing-to-pass of Way as the all-ruling Happening that, itself ruling as the Nearing that draws the distinguished toward one another, permits to everything the course appointed for it and its arrival therethrough—may be called the very "Bringing-to-pass of Stillness *(das Ereignis der Stille).*"[23]   As such it can, then, be seen, in its ultimacy, as the outgoing Bringing-to-pass of the concealing that precisely as keeping-to-self goes unconcealingly forth, that unconcealing concealing which the word "Being," with like connotations of an all-ruling providing in specificity, likewise names.[24]

In this phrase, *das Ereignis der Stille,* the manner of self-disclosure intrinsic to Stillness—concealing as keeping-to-self—is pronounced in a single word.   In its pivotal "Ereignis" the manner of ruling of that concealing as an initiatory disclosing—as the particularizing, forth-gathering providing-of-ways-that-lets-arrive which "way" names and to which "Being" hiddenly alludes—is spoken forth.   All the nuances of pure Happening must be seen in its light, for this is the naming of that pure Happening's bringing of itself to pass that most centrally bespeaks the manner of the unconcealing that is, as the happening-forth of concealing, everywhere ever underway.[25]

Ereignis, as the immediate forth-happening of self-maintaining self-concealing Happening, is disclosive Bringing-to-pass as such, sui generis,

---

[22]"The Nature of Language," US 214, WL 106.   For a discussion of the meaning of "world" intended in this passage, see Ch. V, pp. 173-179.
[23]"The Nature of Language," US 214, WL 106.
[24]Cf. Ch. I, pp. 43f.
[25]In this complex portrayal of pure Happening as a self-withholding concealing—i.e., as Stillness—that, as *Ereignis,* brings itself disclosingly to pass as a self-surmounting happening-forth-from-itself that is a gathering toward disclosure of whatever is, in its particularity, brought to light, there is displayed again at what we might call the point of its deepest discernibility the intrinsic structure that we found for Happening in first considering Heidegger's portraying of the Being of what-is. Cf. Ch. I, pp. 29-33, and see n. 36, in which the ubiquity of that structure in the instancing of pure Happening via the unfolding of the Twofold is extensively detailed.   On *Ereignis* itself as evincing this same structure, cf. this chapter, p. 146, n. 12, and p. 165, n. 89.

deriving from nothing beyond itself.[26] And, ultimately speaking, it alone brings to pass. It is a supremely outgoing *relating*, which, as a primal disclosing, preempts to its sway every nuance of the multifaceted happening that it informs and, precisely in so doing, rules as a spontaneous bestowal powerfully in play throughout the latter.

In order to make manifest the character of this Bringing-to-pass, Heidegger, characteristically, examines the meaning that he finds hidden in the word *Ereignis*, often through references to its companion verb *ereignen*. *Ereignen* is formed from the verb *eignen*, akin to *eigen* (own), and the prefix *er-*. *Er-* is regularly used to lend specificity to the action of a verb. It ordinarily modifies its verb so that the latter speaks either of the beginning of an action or of the attainment in which the action culminates. In our day, *eignen* is an intransitive verb that means to be one's own, to befit, to behoove, or to belong to. Originally the verb was transitive and meant to take possession of. Citing lines of Goethe, Heidegger also finds for it a further transitive signification. Goethe, using *eignen* as a reflexive verb, gives to it, he says, a meaning close to that of verbs meaning to show oneself (*sich zeigen*) and to mark, to designate, or to characterize (*bezeichnen*).[27] *Eignen*, then, speaks also of an identifying disclosure.[28]

In drawing out the full import of *eignen* as a component of *ereignen*, Heidegger emphasizes this latter aspect of its meaning by placing it in close conjunction with *äugen*, to gaze intently or to look about carefully. Bringing out the full connotations resident in *ereignen*, in

---

[26]"The Way to Language," US 258, WL 127.
[27]US 259; WL 128.
[28]Heidegger frequently uses words closely related to *eignen*, especially *eigen*, own; *eigentlich*, authentic, genuine, actual (as adjective or adverb); *eigens*, expressly, particularly. In *Being and Time*, *eigenste* ("ownmost") is often used. These words speak also for Heidegger in one way or another of that which pertains most definitively to whatever is in question and point always to the latter as disclosed and apprehended as itself. In chronologically later contexts, from a time when Heidegger's understanding of *Ereignis* can be seen to have come to full development (see this chapter, p. 169), the presence of such words must always be taken to suggest that ultimate Bringing-to-pass ruling in the happening of the Twofold. On the translating of such words in this study see Ch. I, p. 37, n. 46, and Ch. II, p. 85, n. 102.

a striking sentence in which he hyphenates verbs so as to manifest their full connotations (a sentence in which we, through multiple translation, endeavor to make those connotations directly evident) he can say, *"Ereignen* means originally: to look intently in such a way as to bring that looking to fruition (*er-äugen*), to disclose in catching sight of (*er-blicken*), in glimpsing (*im Blicken*) to call to oneself (*zu sich rufen*), to render something one's own, to see something as appropriate and so take possession, to let belong unto (*an-eignen*)."[29] With similar emphasis, he writes that *Ereignis* is "disclosing looking-at which, bestowing appropriateness on what is beheld, accomplishes being-one's-own and belonging-to (*eignende Eräugnis*)."[30]

In these statements and in the various discussions of *Ereignis* that they epitomize, Heidegger accords to the verb *eignen* the dual directionality—toward the initiator of accomplishing and toward the recipient—that springs from the verb's simultaneous carrying of intransitive and transitive force, even as he utilizes the entire range of connotations found in its varied uses. As a component of *ereignen*, *eignen* speaks of a taking to itself achieved by the unique, disclosing Bringing-to-pass that intrinsically lays claim to and arrogates to its sway that which it discloses, and it speaks equally of a bringing-into-its-own of

---

[29]"The Onto-Theo-Logical Constitution of Metaphysics," ID 100f, 36. The German sentence that this very free and highly complex translation renders runs in full: *Ereignen heisst ursprünglich: er-äugen, d.h., er-blicken, im Blicken zu sich rufen, an-eignen. Aneignen* ordinarily means to appropriate, to adopt to acquire. An attempt has here been made to account for Heidegger's exposing of its verbal stem, *eignen.* In such a sentence as this, any translation is bound to be fraught with inadequacy, for undoubtedly Heidegger, as always, intends that in one way or another every nuance of meaning resident in the words he uses be heard, and the wealth of connotation here repeatedly in play all but defies presentation. Similar attempts to display through extended, highly nuanced translations the complex connotations clustering within *ereignen* and *Ereignis* will be met with throughout the remainder of this chapter and elsewhere, whenever the need imposes itself to present those connotations as fully as possible at once in their complexity and their unity. For a considerably earlier portrayal of *Ereignis* in which a striking number of words found in the sentence just quoted appear, see "Overcoming Metaphysics," VA 99, EP 110.

[30]"The Turning," TK 44, QT 45. See also, "The Way to Language," US 260, WL 129.

what is disclosed and laid claim to. It speaks, that is, of a befitting of the disclosed that entails its happening precisely as *itself*, and of a bringing of the disclosed into belonging. The English verb "own," cognate with German *eigen*, which means both to possess and to acknowledge, suggests something of this dual force of *ereignen* as an owning that, precisely as a disclosive claiming, enowns.

The owning of which *eignen* speaks is a primal accomplishing that, ultimately and pervasively, and in a wholly unique manner, so brings forth as to provide (*erbringt*). Spontaneously and simultaneously it brings to pass both itself and that toward which it directs itself. It is this owning fraught with spontaneous giving (*das erbringende Eignen*) that Heidegger means when he uses the word *ereignen*. The latter speaks of a Bringing-to-pass-that-brings-forth that as an owning, he says, is "more granting than any working, making, and grounding."[31] To it all interrelated aspects and nuances of the unfolding Twofold owe ultimately their happening as they do. And the Bringing-to-pass, *Ereignis*, of which, in the last analysis, Heidegger is willing to say only that it brings to pass[32] has, precisely as a disclosive claiming that enowns, the character of an owning forth.

2

The Bringing-to-pass-that-owns-forth is a taking into relation that is, as such, a disclosive, befitting giving. Accomplishing itself, it gives culminatingly (*er-gibt*) the opening that is the light-bringing clearing (*Lichtung*) via which, as the Being of what-is, pure Happening unconceals and so maintains itself as itself.[33] The German phrase that literally means "it gives" (*es gibt*) regularly carries the idiomatic meaning, "there is." By way of this double meaning, Heidegger approaches Bringing-to-pass-that-owns-forth at this central point. There is, i.e., it gives, Being; there is, i.e., it gives time. And the "It" that gives, which in fact can only be

---

[31]"The Way to Language," US 258, WL 127.
[32]"Time and Being," SD 24, TB 24.
[33]Cf. "The Way to Language," US 258, WL 127. Cf. Ch. I, pp. 32f & 37f.

hinted at but cannot be denoted by a component of an ordinary sentence such as this wherein a subject is set forth as performing an isolable act toward an object, is that Bringing-to-pass-that-owns-forth.[34]   Only by reason of it, although not in a causal manner, are Being and time provided at all.   Thanks to it, they are given into their own, i.e., into that which pertains most closely to them.   This "own" is precisely their belonging-together.[35]

Genuine (eigentliche) time is the ever-opening realm whose on-goingly interrelating extendings reach forth Being as presencing, i.e., as the letting presence that lets presence whatever presences.[36]   Reached forth as presencing by way of ever-opening time, Being sends itself forth as self-adaptive destining, repeatedly transforming its manner of happening, but ever ruling in whatever is at any particular time.[37] Inextricably, time and Being happen.   In their self-reaching- and self-sending-forth can be discerned a giving that is in fact owning forth.   It is a "devoting (Zueignen)," a "consigning (Übereignen)" that so gives each as to surrender it wholly to the other in a mutuality that alone lets it happen as itself.[38]

The enowning Bringing-to-pass-that-owns-forth, disclosing time and Being via its sovereign sway, brings them to pass into the belonging-together that is the sine qua non of the on-going happening of unconcealing.   The arena of that happening is the existing vouchsafed to man.   There the light-permitting clearing—time—ceaselessly opens, as

---

[34]"Time and Being," SD 18ff, TB 17ff. *Demnach bezeugt sich das Es, das gibt, im "Es gibt Sein," "Es gibt Zeit," als das Ereignis.* SD 20, TB 19. The English sentence we have just used to parallel the phrases distinguished by quotation marks in Heidegger's sentence cannot but be misleading. Heidegger here is very deliberately avoiding the use of the verb is (ist) with either time or Being as subject. The usage we have been forced to adopt could seem, quite wrongly, to make of either a particular entity.
[35]SD 20; TB 19.
[36]Cf. Ch. I, pp. 39f. On Being as letting-presence see SD 12, TB 12. Heidegger's phrase "genuine time (eigentlichen Zeit)" (SD 18; TB 17) directly names time as thus given into its own, as the eigen of eigentlich echoes the eigen implicit in Ereignis.
[37]Cf. Ch. I, pp. 29f.
[38]SD 19f; TB 19f.

having-been, yet-to-come, and present reach forth Being as the complex presencing that lets man presence toward what presences to him precisely as the one who, overleaping and so consummating in himself the light-bringing clearing of time, so draws what meets him toward himself as to let it come meaningfully to light and *while* in that clearing, even as he himself, going beyond his thrownness, yet orienting himself from out of the latter, so directs himself in his intent-fraught apprehending of what presents itself to him as to disclose himself, i.e., the existing after whose manner he pursues his way, to himself in his very disclosing of his world through that fundamental awareness via which, whether it be heeded or shunned, he maintains himself as man, as the one who sees, who meaningfully gathers and elucidates that which displays itself to him.[39] Happening as the unfolding Twofold, to which man belongs and which he also encounters, the Being of what-is as the presencing of what presences ceaselessly approaches man as that which matters ultimately and uniquely to him.[40] In himself, in his individualized corporate life, in whatever he meets as it meets him, Being as presencing, in the guise of one destining or another, encroaches demandingly upon him.

In the existing as which he carries himself out man, whether in suppressed or in overt awareness,[41] receives and takes upon himself this incessant approach of Being that continually claims him. Precariously maintaining himself via the ever on-going extendings of genuine time, which so constitutes him as to find consummation through him, he provides for Being as the presencing of what presences the sole avenue via which it can fulfill its ruling and can, through the disclosure in specificity that he offers, accomplish its self-maintaining self-unconcealing.[42] In an inauthenticity of self-carrying-out, man may conceal from himself the impinging of Being. Yet, in authentic acceptance of his role as man, he may also rise to full acknowledgement

---

[39]Cf. Ch. I, pp. 39 & 49f, and Ch. II, pp. 68-75. See also "Time and Being," SD 23f, TB 23.
[40]See SD 12f, TB 12f.
[41]Cf. Ch. II, pp. 91-93.
[42]Cf. "The Turning," TK 38, QT 38f. Cf. also Ch. II, pp. 71-73.

of it.   When he thus undertakes his required relation to Being man
achieves the distinctiveness that most intrinsically belongs to him as
man.[43] But however that relation may be in play, its accomplishing is
never attained in virtue of any power inherent in man himself. Always it
is provided in virtue of the Bringing-to-pass that gives into relation in
owning forth. Thanks to that Bringing-to-pass, and to it alone, Being and
man, via the ever-opening clearing that is time, belong together. Even as
it is in play in Being's self-sending-forth by way of time, that Bringing-to-
pass is in play also in man's disclosive manner of existing, as, via time,
he opens himself on-goingly to Being's importunate claim.   Out of the
sovereign bestowal as which it rules, Being is given over (*zugeeignet*) to
man's peculiar manner of existing, by way of which alone it can
unconceal and therewith maintain itself; and man is given to belong
(*vereignet*) to Being, so that he does indeed allow that unconcealing to be
underway.[44]

3

Central to this belonging-together of man and Being granted by
way of time is language. That which stirs hiddenly in primal Saying as
showing is nothing other than owning (*eignen*) which, as disclosive
claiming that enowns, bestows into relation. That is, it is nothing other
than primal Bringing-to-pass as such.[45]   Primal showing-Saying is,

---

[43]Cf. Ch. II, pp. 99-101.
[44]"The Principle of Identity," ID 100, 36. Heidegger uses the verb *vereignen* with
some frequency. He himself formed it by coupling with *eignen* the prefix *ver-*, which
often serves to give causative force to German verbs. Clearly he intends something
similar for *vereignen*. But the verb speaks of the unique relating connoted by *eignen*
and *ereignen*, which is never to be thought of as causal. The translation "to give to
belong" is an attempt to preserve something of this peculiar force.   For passages
illustrative of the use of *vereignen*, cf., e.g., "The Question Concerning Technology,"
VA 40, QT 32; "The Turning," TK 47, QT 49; "The Principle of Identity," ID 100,
36; ID 103, 38, where *eignen* and *zueignen* are also used; and "Time and Being," SD
19, TB 19, with *zueignen*. On the relating of Being and man here in question, Cf. Ch.
II, pp. 72-77.
[45]See "The Way to Language," US 258, WL 128. On Primal Saying, cf. Ch. III,
pp.119f.

Heidegger says, itself "the mode *(Weise)*"—not so much in the sense of kind or manner as in that of melody or song *(melos)*—"in which the Bringing-to-pass-that-owns-forth *(Ereignis)* speaks."[46] And human language, in which that Saying makes its way to its fulfillment in concrete utterance,[47] is the pivotal support *(Schwingung)*, delicate yet enduring, of that structuring poised only in itself that is spun out as the disclosive Bringing-to-pass-that-owns-forth ceaselessly rules.[48] It is that language that accomplishes all relating in the latter's outreaching dominion *(Bereich)*, as, through that dominion, Being and man, in disclosive self-maintaining, reach and belong to one another.[49]

Heidegger can speak of language as the "house," the "sheltering place," "of the truth of Being,"[50] i.e., of Being in its happening as unconcealment.[51] For the disclosive happening of Being as presencing— by way of the unfolding Twofold—takes place via language. That happening is "entrusted" to the showing as which primal Saying accomplishes itself through the spoken word.[52] Thus it is that its

---

[46]US 266; WL 135. Cf. "The Nature of Language," US 196, WL 90. The character of language as a singing forth, an immediate self-bringing-into-play, of *Ereignis*, is shown us in the fact that Heidegger, in speaking of "the word" in an ultimate sense, can use phrasing that recalls his understanding of *Ereignis*, guardedly, as the "It" that gives in the phrase "It gives Being *(Es gibt Sein)*" or "It gives time *(Es gibt Zeit)*." He can say, in a passage in which he is emphatically distinguishing the word from anything whatever that the word might name forth, "It, the word, gives *(Es, das Wort, gibt)*. He does not there specifically allude to *Ereignis*, but he does suggest that perhaps "the disposing enduring *(Wesen)* of the word"—i.e., primal Saying, which holds sway therein—"conceals in itself that which gives Being." (US 193f; WL 88). It is worth noting here that for Heidegger what we may call the musicality of language as finally spoken forth as uttered words is as intrinsic as is the carrying of meaning. US 204f; WL 98.

[47]Cf. Ch. III, pp. 120f.

[48]"The Principle of Identity," ID 102, 38. See also "The Way to Language," US 267, WL 135.

[49]See "The Principle of Identity," ID 102, 37f.

[50]Two Heideggerian statements are conflated in this sentence: *Die Sprache verweigert uns noch ihr Wesen: dass sie das Haus der Wahrheit des Seins ist.* ("Letter on Humanism," W 150, BW 199) and *Die Sprache wurde das "Haus des Seins" genannt. Sie ist die Hut des Anwesens,* . . . ("The Way to Language," US 267, WL 135).

[51]On truth as unconcealment, cf. Ch. I, p. 38, and n. 49.

[52]See "The Way to Language," US 267, WL 135. Cf. Ch. III, pp. 118-121.

happening as showing-Saying allows Being as presencing to come into its own, i.e., allows it in reaching and claiming man to happen as the presencing of what presences.

Man, in his individualized corporateness, harkens to primal Saying and *says* in speaking. He apprehends what presences *in presencing*, and his apprehending involves intrinsically his speaking. By way of whatever communal context may be in play he elucidates; he interprets; he communicates—he shows whatever, in its drawing near, he gathers concernedly toward himself. And this he does always, finally, through spoken words.[53] Reaching away from out of the provisional autonomy that characterizes his existing, man receives Being as presencing and, whether wittingly or not, brings it to manifestation. He says; he names; and therewith, bringing primal showing-Saying to utterance, he lets the appearing of what presences come to accomplishment.[54]

It is this sounding of the word, which springs out of an attending upon showing-Saying and which fulfills his belonging to Being, that is man's "own," i.e., it is this that pertains most intrinsically to him, defining and distinguishing him as man.[55] The disclosive Bringing-to-pass-that-owns-forth, bringing itself to pass as showing-Saying, grants man entry into the manner of existing that marks him as the one who speaks. Thus it grants him his belonging to Being, precisely in order that he may so surrender himself to the latter as to safeguard it, preserving it in its happening by letting it maintain itself on-goingly as self-unconcealing.[56]

---

[53]Cf. Ch. III, pp. 112-117.
[54]Cf. Ch. III, pp. 125f.
[55]"The Way to Language," US 260, WL 129.
[56]In striking testimony to this interconnection, Heidegger uses asseverations unmistakably suggestive of one another in averring that the disposing enduring as which provisive Happening brings itself to pass as language "needs" the human response of listening—which indeed alone permits speaking—and in stating that Being "needs" the manner of man's enduring, in order rulingly to accomplish itself as disposing enduring. See "Language," US 30, PLT 207: . . . *insofern das* Wesen *der* Sprache, das Geläut der Stille, das Sprechen der Sterblichen braucht, um als Geläut der Stille für das Hören der Sterblichen zu verlauten; and TK 38, QT 38f: . . . *das* Wesen des Seins das Menschenwesen braucht, um als Sein nach dem eigenen Wesen

Via language the belonging-together of man and Being is given and brought to consummation. The revelatory unfolding of the Twofold happens as man sustains his complex, on-going existing through the spoken word. Language and language alone, as the concretely spoken words as which showing-Saying accomplishes itself, frees every entity of the manifold of what-is to its appearing.[57] For each such entity in its very particularity, the Bringing-to-pass-that-owns-forth—as disclosive bestowing that accomplishes belonging—is in play. Primal showing-Saying is the singing-out of that Bringing-to-pass that happens as a gracious summons. Disclosing and claiming what presences precisely as itself, it "praises *(lobt)*" it; it so commends it as to permit *(erlaubt)* it, via the light-bringing clearing that is time, into the specific enduring *(Wesen)* belonging peculiarly to it. Ceaselessly it brings what presences, as what appears, to the shining-forth *(Scheinen)* that pertains to presencing as such.[58]

What presences is the manifold of what-is, given in particularity. The presencing of its particulars is inherently and fundamentally presencing-together. Presencing, the particulars come unitedly into the open of the present and depart out of it, compliantly yielding place to one another.[59] Only in their thus belonging-together does each come genuinely into its own. Here the disclosive Bringing-to-pass-that-owns-forth rules. Bringing itself to pass via primal Saying that shows, it disclosively, approvingly calls forth what-is *as* a manifold. Its owning-forth, as a sovereign granting that gives into belonging, brings whatever presences or absents into its own, into its proper enduring by way of a particular opening present, in compliant interrelation with its fellows.

---

inmitten des Seienden gewahrt *zu bleiben und so* als *das Sein zu wesen,* . . ." Cf. Ch. XII, p. 479.

[57]Cf. US 257, WL 126. Cf. Ch. III, pp. 119f.

[58]*Denn die ereignende Sage bringt das Anwesende aus dessen Eigentum zum Scheinen, lobt, d.h., erlaubt es in sein eigenes Wesen.* "On the Way to Language," US 266f, WL 135. Although disclosure is here in question in this use of *scheinen*, one should surely not forget wholly the verb's dual meaning: to shine and to seem. Cf. Ch. I, p. 44. On the meaning of *Wesen*, cf. Ch. VI, pp. 252-257.

[59]Cf. Ch. I, pp. 49f.

Out of this belonging-together, each particular, one of a manifold, "shows itself in itself and," whether presencing directly or absenting, "tarries (*verweilt*) according to its kind (*Art*)."[60]   That is, out of the belonging together into which it is given, the particularized manifold of what-is disclosively gives itself to man.   Arriving into Being happening as presencing, what-is as what presences, from out of the relation to Being primordially granted to it, fulfills Being's happening as unconcealing. Happening thus, as the manifold of the self-differentiating single unfolding Twofold, what-is as what presences, in disclosively offering itself to man, therewith lets Being as presencing reach him.   Whether discerned as such or not, Being as presencing—in whatever guise—comes to man immediately by way of everything that, coming to appearance, meets him.   That presencing rules also in himself; and man surrenders himself to it in giving himself to the particularized manifold of what-is, i.e., in launching forth to meet that manifold via his disclosive on-going existing and drawing it concernedly toward himself.[61]   The belonging-together of man and Being is, then, accomplished via the belonging together of man and what-is, which latter itself belongs together with Being.

Primal showing-Saying directingly gathers all entities of the manifold of what-is—including man with all the distinctiveness of his existing—into the particular enduring by way of some particular present allotted to them.[62]   Thus it *says*, it *shows*, what-is as a whole and, in its giving of the one peculiarly singled out, it says, it shows, man's apprehending. It discloses that apprehending;[63] it brings it to happen as itself, i.e., as a responsive attentiveness that can itself show through saying. That apprehending is thinking, whether as a cognizant answering or as a response made deviant by oblivion.   Always it is underway. Always it is gathering what meets it to itself and letting it appear in its

---

60"The Way to Language," US 258, WL 127.
61Cf. Ch. II, pp. 78-81.
62Cf. Ch. III, pp. 122-125.
63Cf. Ch. III, p. 120.

appearing via the spoken word.[64]

Given to belong to what is in Being, man speaks, he disclosively summons forth what-is in the particularity of its manifold, and in this belonging-together of man and what-is his speaking fulfills the happening of the primal showing-Saying that, unitingly governing and directing every particular of that manifold, brings itself as a Saying that is precisely a showing to final accomplishment in his words. Speaking, man fulfills his belonging-together with what-is; and in so doing his language, wherein primal showing-Saying brings itself decisively to pass, takes place continually and centrally as that via which his disclosive belonging to Being is itself given.

Language, viewed in its fullest scope, accomplishes all bringing into relation whatever. Taking place by way of man's unique manner of existing in individualized corporateness, it ceaselessly, by many avenues and in many ways, lets those who speak, who communicate with one another, belong on-goingly, self-maintainingly together in community in a world.[65] Happening at the core of man's subsisting as himself, language lets man reach out to gather to himself in his meaning-discovering interpreting that which presences to him, therewith letting him overleap and consummate in himself the ever-opening clearing that is time, via which disclosure may be achieved.[66] And by way of that clearing, what-is, its particulars gathered to belong together as a manifold, is given to belong to man and he to it as his gathering speaking fulfills its showing of itself by letting it, in its coming toward him, arrive into appearing. Here language consummates the relating back toward one another of Being and what-is, the self-sundering single Twofold. In so doing, language consummates the belonging together of Being and man. And yet, at the same time, that latter belonging-together is, as given always antecedently in the given language that man speaks, ever also the sine qua non of the existing vouchsafed to man as the one who—one with the Twofold that he encounters and discloses as a self-disclosing manifold—lets Being happen

---

[64]Cf. Ch. III, pp. 129f, and Ch. XIV, pp. 552-560 & 573f.
[65]Cf. Ch. III, pp. 115-117.
[66]Cf. Ch. II, p. 69, and Ch. III, pp. 113f.

self-maintainingly in happening as unconcealing. Precisely through that belonging, given, yet ever reaching consummation, does Being maintain itself as itself, i.e., as the Being of what-is as what comes to disclosive self-manifestation via the happening of manifesting as such.

Language puts underway and brings to fulfillment this intricate relating of man to the Being of what-is, from within the latter as the Twofold, and by way of time, in virtue of the disclosive Bringing-to-pass-that-owns-forth. Spontaneously, generously singing itself forth via primal showing-Saying, that Bringing-to-pass brings its ruling to culmination in the spoken word where primal Saying bodies itself forth as man answers to it.

Centered here, happening in relation on every level comes to pass. Primal Bringing-to-pass—as the sovereign disclosing that takes to itself, which is simultaneously the freeing disclosing that enowns, i.e., that spontaneously gives into belonging—grants precisely in holding in relation. It itself is—in a way that at once sustains owning in the dual sense and accomplishes a keeping-to-self—as such the holding-relating that gives into play all holding-relatings.[67] And primal Saying, as the very self-singing-forth of that enowning Bringing-to-pass, bears a like character. As the mode of its ruling as which pure Happening brings enowning specifically to pass, primal Saying too is the holding-relating that provides all holding-relatings that are anywhere in play.[68] Happening as the self-accomplishing of the provisive way-providing-that-lets-arrive that is the very forth-happening of self-concealing Stillness itself, primal Saying, as which language originatingly holds sway, happens as an instancing of the very Nearing—the distancing drawing-toward of the distinguished in their relatedness—that is everywhere in play in the complex unfolding of pure Happening as the Twofold.[69]

---

[67]*Das Ereignis ist, eignend-haltend-ansichhaltend, das Verhältnis aller Verhältnisse.* "The Way to Language," US 267, WL 135.
[68]US 215; WL 107.
[69]See Ch. III, pp. 123f, and this chapter, pp. 146-148. Speaking of the neighborliness of thinking and poetry as preeminent "modes of saying (*Sagen*)," which stems "from their nearness (*aus ihrer Nähe*)," Heidegger concludes, giving *Nähe* its full force, " . . . then Nearing (*Nähe*) itself"—which accomplishes the interrelating as

Primal Saying and that Nearing, whose ruling lies implicit in the enowning-Bringing-to-pass-that-lets-belong named as *Ereignis*,[70] come to pass in a differentiable oneness—a oneness that displays with peculiar trenchancy the utter centrality for Heidegger of language—from out of the gathering, differentiating Same as which Stillness, self-withholding concealing brings itself into play.[71] Everywhere, ceaselessly, Nearing-accomplishing primal Saying, as the disposing that gathers everything whatever forth to light, carries the ever-disposing, enowning, belonging-accomplishing Bringing-to-pass of concealing named as Stillness[72] into play by way of the distinguishing gathering-into-relation that lets appearing transpire that is brought to fulfillment solely through the accomplishing of human speaking.[73]

The disclosive Bringing-to-pass-that-owns-forth, which carries out its rule centrally via primal Saying, is nothing other than the provisive self-bringing-to-pass of pure Happening itself. Pure Happening happens rulingly, self-consummatingly, as primal Saying, and in so doing it happens as the directing, enabling showing that is immediately and powerfully expressive of the Bringing-to-pass as which it itself as self-initiating bestowal brings itself disclosively to pass. Happening self-

---

which any nearness is in play—"must hold sway in the manner of primal Saying (*Sage*)." ("The Nature of Language," US 202, WL 95.) It must always be borne in mind that an active meaning is intrinsic to Heidegger's use of "Nearing (*die Nähe*)." Cf. Ch I., p. 40, n. 53.

[70]This evident congruence between the Nearing that distances in an integrity of the particularized that which it draws-toward and the enowning Bringing-to-pass that accomplishes belonging together is not made explicit by Heidegger. It can, however, readily be discerned. See, e.g., "On the Essence of Language," US 214, WL 106: *Die Be-wëgung des Gegen-einander-über im Weltgeviert ereignet Nähe, ist die Nähe als die Nahnis. Sollte die Be-wëgung selber das Ereignis der Stille heissen?* "The way-providing that lets arrive pertaining to the protective juxtaposing of the world-Fourfold brings Nearing enowningly to pass, *is* Nearing as nighness. Should this way-providing itself be called the enowning Bringing-to-pass of Stillness?"

[71]This explication is based on Heidegger's statement that if Nearing holds sway in the manner of primal Saying (see this chapter, p. 160, n. 69) "Then Nearing and Saying would be the Same." "The Nature of Language," US 202, WL 95.

[72]*Ereignis der Stille.* "The Nature of Language," US 214, WL 106. Cf. this chapter, p. 148.

[73]Cf. Ch. III, pp. 119-121.

unconcealingly, self-maintainingly as the Twofold, pure Happening, holding sway as the unconcealing self-differentiating Same that is in play as the forth-happening of concealing, of Stillness as keeping-to-self, happens itself away as the intricately particularized manifold of shaped happening, and in the provisional autonomy of the latter it itself in all its complexity ever rules.[74] In words that point strikingly to the oneness of Happening here in question, Heidegger can allude to the originating of "Difference (*Differenz*)," i.e., of distinctiveness of happening, from "identity (*Identität*)" (in which that Same shows its ruling) as a stemming-forth that can be illumined precisely by a harkening to the "unison (*Einklang*)" holding sway between the enowning Bringing-to-pass (*Ereignis*) that rules as pure Happening's ultimate mode of disclosure and the differentiating carrying-out (*Austrag*) as which, as the Same, that Happening accomplishes its self-unfolding as the single Twofold, the Being of what-is, in all the multiplicity of its taking place.[75] In that ever-transpiring "carrying-out," the Bringing-to-pass-that-owns-forth brings itself to pass ceaselessly throughout every particularized aspect and nuance of the Twofold as such. The belonging-together of those nuances and aspects that finds consummation by way of primal showing-Saying through spoken language pertains to them already also from their inception. Happening as given at all, they happen as on their way toward the disclosive freeing into mutual self-maintaining that language ultimately brings to consummation. In human language and in all the doings to which that language is prior and for which it opens the way as man gathers to himself what meets him, by way of the on-going openness that time provides, unconcealing—the self-unconcealing of pure Happening—takes place. Word and deed alike are, in whatever way they transpire, dictates of that unconcealing.[76] And here everywhere, even to the least instance of disclosing, the Bringing-to-pass-that-owns-forth is in play.[77]

---

[74]Cf. Ch. I, pp. 27-33 & 53.
[75]"The Principle of Identity," ID 84, 21f. Cf. Ch. I, pp. 32f.
[76]Cf. Ch. III, pp. 132-135.
[77]Cf. "The Way to Language," US 257f, WL 126f.

4

The disclosive Bringing-to-pass that pervasively owns forth brings itself to pass as a sovereign, disclosive granting that gives maintainingly into belonging. But the belonging-together that happens variously throughout the unfolding Twofold involves equally and simultaneously a holding-apart. The happening of Being as presencing is intrinsically keeping-to-self, even to the point of disruptive self-assertion;[78] likewise, Being's every self-sending-forth in historical governance is, as such, a self-withholding.[79] In unconcealing itself, Being withdraws as self-concealing.[80] Always Being happens changingly. Always it rules as a destining that adapts itself again and yet again.[81] In like manner, time, via which Being is reached forth in one guise or another as presencing, extends itself as interrelating dimensions that, even in their mutual nearing, are held apart.[82] Only so do they offer the open clearing that allows unconcealing, by way of which the Being of what-is as presencing lets presence whatever presences, as Being and what-is—the single sundered Twofold that in its coming to pass most fundamentally manifests belonging-together, in its unitariness accomplished as differentiation[83]— happen unitedly to and by way of man's existing wherein man, happened away and responding to Being, withstands and lives out ever-opening time as he constantly and concernedly launches himself forth via its open clearing, draws what presences toward himself even as in the thrownness of his existing he is, by time, ever being withheld from himself, and through his participation in language, finally, brings what-is to appearing and so lets Being maintain itself via its happening as self-unconcealing.[84]

It is nothing less than the disclosive Bringing-to-pass, which, as owning-forth, holds sway in the intricate belonging-together accomplished

---

[78]Cf. Ch. I, pp. 24 & 50-52.
[79]Cf. "Time and Being," SD 9, TB 9. Cf. Ch. I, pp. 47f.
[80]Cf. Ch. I, pp. 44-46.
[81]Cf. Ch. I, pp. 48f, and Ch. XIII, pp. 496f.
[82]Cf. Ch. I, pp. 39f.
[83]Cf. Ch. I, pp. 29f.
[84]Cf. Ch. II, pp. 68-72 & 75f.

in the unfolding Twofold, that holds sway also in the withholding that is, in that unfolding, variously and ubiquitously in play.   In words that clearly show us the Stillness, the self-withholding concealing—wellspring of all unconcealing—as which and from out of which enowning Bringing-to-pass happens initiatorially forth in its sovereign provisiveness, "to the Bringing-to-pass-that-owns-forth (*Ereignis*) as such," Heidegger says, "there belongs dispossessing (*Enteignis*)."[85]

To that Bringing-to-pass, which happens forth as a sovereign disclosing granting that so takes to itself and to its sway as to give that which it discloses into its own, i.e., into the revelatory self-maintaining in belonging proper to it, there pertains a countervailing tendency, a withdrawing that restricts the full play of owning-forth. Primordial Bringing-to-pass (*Ereignis*) happens simultaneously as a withholding, as a not taking into its sway, as a not giving to the uttermost of that which it discloses into its own.   The Bringing-to-pass that graciously[86] takes possession of so as to own forth dispossesses itself even in so doing. It so restrains itself as to limit the given gift that it takes to itself.   And it dispossesses that which it gives, denying to it complete, immediate self-manifestation and letting it come into its own in mutual belonging only via the restrictedness of a continual going-forth.   Happening thus as a withholding, in its very possessing through disclosure of that which it rules and informs and gives, which is "most its own (*sein eigenstes*)," it "withdraws" the latter "from limitless revealing."[87]   Accordingly, everywhere it, precisely as a granting enowning disclosing that gives into belonging, initiates the holding-apart that obtains variously throughout the happening of the unfolding Twofold.

In bringing itself to pass thus as a withholding, the Bringing-to-pass-that-owns-forth does not cease to be itself.   Rather, precisely in so

---

[85]"Time and Being," SD 23, TB 23. On *Ereignis* as the Bringing-to-pass of self-concealing named as Stillness, cf. this chapter, pp. 147f.
[86]For this nuance in the meaning of *Ereignis*, see Ch. V, pp. 214f.
[87] . . . *es [Ereignen] sein Eigenstes der schrankenlosen Entbergung entzieht.* "Time and Being," SD 23, TB 22.

doing it maintains itself as itself.[88] For that Bringing-to-pass is a holding-in-relation that so takes possession of as to give into that belonging-together wherein self-disclosive self-maintaining transpires throughout the unfolding Twofold. And were this giving unrestrained, that which it gave would be but a single, total, undifferentiated gift, revealed, but without on-goingness, without change, without coming into relation, without belonging-together. Only as a withholding-holding-apart can primal Bringing-to-pass be an enowning holding-in-relation that lets belong together.[89]

The balance obtaining between the juxtaposed tendencies in the Bringing-to-pass-that-owns-forth as a self-witholding-granting itself shifts and changes. Variously that Bringing-to-pass "reveals itself or withdraws."[90] Hence, throughout the Twofold, the manner of belonging-together of whatever is disclosively given to belong in owning forth varies also. In the diverging of Being and what-is even as they unfold in interrelation as the single Twofold;[91] in the divisive self-isolating of the particulars of what-is even as they transcend their fragmenting assertiveness so as always to presence in some way as a manifold;[92] in the precarious relating of man to what-is and to Being where mis-seeing and misinterpreting may take place even as that relating continues to transpire and ceaselessly allows in one way or another the unconcealing of the Being of what-is[93]—in all this, the withdrawal of the Bringing-to-pass-that-owns-forth predominates over its self-revealing.[94] Yet its self-

---

[88]SD 23; TB 22.

[89]SD 23; TB 22. In the self-bringing-to-pass of *Ereignis* as *Enteignis*, the structure that is intrinsic to pure Happening in its maintaining of itself via its forth-happening as concealing that unconceals, and that is everywhere in play in the happening of Being and of whatever *is*—i.e., the structure evinced as a self-happening-away that comes to pass self-surmountingly precisely as a gathering-toward—is to be caught sight of at an ultimate point, namely, in the very manner of happening of provisive Bringing-to-pass as such. On the structure in question and on its ubiquity, cf. Ch. I, pp. 29-33, and n. 36.

[90]"The Way to Language," US 263, WL 131.

[91]Cf. Ch. I, pp. 44-46.

[92]Cf. Ch. I, pp. 50-55.

[93]Cf. Ch. II, pp. 92f, and Ch. III, pp. 113-115.

[94]Heidegger does not follow out into all these realms of relating the implications of his

revealing remains underway, and wherever genuine disclosure takes place in the unfolding of the Twofold it comes decisively to the fore.

The shifting interplay of these two tendencies in the Bringing-to-pass-that-owns-forth happens at once broadly and subtly. It is directly manifest in language. The primal showing-Saying (*Sage*) from out of which our language is provided, as the self-singing-forth of enowning Bringing-to-pass, like that Bringing-to-pass vouchsafes through fulfilling distinction on-goingly in interrelation, in that it withholds itself, keeps itself to itself.[95] And in the language in whose multitudinous specificity primal Saying brings its ruling to fulfillment, that withholding is continually in play.[96] Given differently to people after people, and offering by way of individuals' speaking, thinking, and doing in community disclosures of diverse revelatory power, language immediately conforms as the saying-forth of Bringing-to-pass-that-owns-forth to the manner in which the latter now reveals itself, now withdraws.[97] Every belonging-together, be it in harmony or in contention, is decisively achieved as man's apprehending, ever accomplishing itself by way of language, lets what-is appear in Being—whether genuinely as itself or as deceptively veiled—via the open clearing that time provides. And in every belonging-together that man, as the one who speaks, thus brings to culmination the Bringing-to-pass-that-owns-forth is immediately holding sway.

Heidegger's portrayal that depicts the disclosive Bringing-to-pass-that-owns-forth as a spontaneous sovereign granting that, precisely as such, withdraws itself and just in so doing preserves the unique manner of its ruling shows to us from a crucially significant perspective that which we have called pure Happening and which we have seen to be a self-concealing that maintains itself as itself precisely in happening as self-

---

identifying of *Ereignis* as *Enteignis*.

[95]*Sage*, like *Ereignis*, vouchsafes in that it withholds. "The Nature of Language," US 215, WL 107.

[96]US 186; WL 81.

[97]On this conforming, see "The Way to Language," US 263, WL 131. On the diverse happening of the language, cf. Ch. III, pp. 139, 116f, & 126f.

unconcealing, i.e., in happening, via the unfolding of the Twofold, as the Being of what-is. When pure Happening is viewed from the vantage-ground provided by consideration of the Being of what-is, a priority attaches to concealing vis a vis unconcealing. The unconcealing that brings itself to pass, as the Being of what-is ever unfolds itself as the Twofold, arises out of and itself partakes of concealing.[98] When, on the other hand, consideration of the owning-forth that can be discerned in the unfolding of the Twofold is central to our viewing of pure Happening, disclosive granting that gives into belonging holds priority vis a vis withdrawal.

These views are not in conflict. Rather, the one is enriched and undergirded by the other. Happening pervasively as self-concealing, Being as pure Happening maintains itself as itself through happening itself away as shaped Happening that happens in unity and in relation with it by way of the open clearing of time lived out by man. The sundering obscuring pull of self-concealing ever encroaches on the unconcealing as which that concealing accomplishes its self-preserving via the responsive awareness of man that alone can let it, in its ruling, disclose itself. But "self-concealing" is never the ultimate word to be spoken of pure Happening. Self-concealing maintains itself through happening as self-unconcealing. Heidegger's use of *Ereignis* and its companion *ereignen* allows him to identify this Happening, precisely in its ultimacy, specifically as a "letting-belong (*Gehören Lassen*)."[99] However strong and pervasive concealing may be in that Happening, it is to be seen finally as the self-withholding that happens as self-disclosive and that, precisely *as* withholding, itself enables the out-going giving-into-belonging that is its very bringing of itself into play. Before all else, pure Happening happens as an utter self-containedness that goes forth as that giving to belong together. That is, it happens self-withholdingly as the Bringing-to-pass that, in one way or another, ever disclosingly, consummatingly, gives and gathers the variously differentiated participants of the single,

---

[98]Cf. Ch. I, pp. 24-26 & 43f.
[99]"The Principle of Identity," ID 103, 38f.

sundered Happening into interrelation as belonging together. Such giving into relation shares priority, we may say, with the concealing, the ultimate Stillness, whose self-bringing-to-pass it is. Indeed, when it itself is at the center of our thinking, all withholding, all sundering obscuring concealing, can even be seen as its way of preserving its manner of ruling as the bestowing that gives the parted to belong to one another, and all unconcealing requires to be understood as an immediate manifestation of that ruling.

This primal role that pertains to the Bringing-to-pass-that-owns-forth which gives disclosively into belonging can be heard in Heidegger's refusal to say more concerning that Bringing-to-pass, seen in itself, than this: "The Bringing-to-pass-that-owns-forth brings to pass so as to own forth *(Ereignis ereignet)*."[100] This is the sui generis initiating, the sui generis self-carrying-out of withholding concealing as vouchsafing disclosing, that lies at the heart of all the complex happening as which pure Happening brings itself to pass.

Here Heidegger glimpses, he believes, something not glimpsed before by thinking. The Greeks knew the belonging-together *(Zusammengehören)* of man and Being[101] under the ruling of Being as presencing. Repeatedly, in elucidating Greek thinking Heidegger finds there a portrayal of that belonging and of the various interrelatings involved in it, in terms of an unconcealing that surmounts concealing.[102] But neither the Greeks nor their successors in the history of thinking caught sight, he believes, of the "letting-belong" ruling in that belonging.[103] Only now, through another sort of thinking that leaps away beyond the metaphysical thinking that has for centuries been underway, has that letting-belong come to light in his disclosive Bringing-to-pass-that-owns-forth, *Ereignis*.[104]

---

[100]"Time and Being," SD 24, TB 24.
[101]"The Principle of Identity," ID 103, 38.
[102]See, e.g., "The Anaximander Fragment," HW 310, EGT 25f; HW 329, EGT 43; "*Alētheia*," VA 129ff, EGT 104ff. Cf. EM 122, IM 134, and ID 93ff, 30ff.
[103]"The Principle of Identity," ID 103, 38f. For a similar judgment on all previous philosophical thinking, cf. Ch. III, pp. 130f.
[104]ID 96ff, 33ff. Cf. "Seminaire du Thor," in *Questions IV* (Paris: Gallimard), 1976,

The theme of belonging-together comes to the fore in Heidegger's later treatments of *Ereignis*. This fact suggests that, although he had used the words *Ereignis* and *ereignen* during many years and could declare that he had done so with a consistency of meaning, Heidegger came finally to an understanding of them that opened a further dimension for his thinking and shed fresh light on the prevailing orientation out of which, in questioning after Being, he spoke throughout his work.[105]

⊞   ⊞   ⊞

In the happening of the Twofold, the disclosing Bringing-to-pass-that-owns-forth in happening as letting belong together (*Ereignis*) pervasively rules. And in glimpsing this letting-belong, our thinking reaches to the inmost wellsprings of the pure Happening that, disclosively opening itself as ever on-going time, brings itself to pass self-maintainingly by way of that openness as lived out by man, as the Being of what-is. In the multifaceted specificity via which Being unfolds itself as the Twofold, that letting-belong shows itself. It stirs in the holding-in-relation as which Being keeps what-is as a whole and man, the special participant in what-is, in hand even in its happening of them away to provisional autonomy that permits them to present themselves as themselves.[106]   It stirs in the enduring-toward that is the definitive happening of Being as presencing.[107] And it stirs in the differentiative gathering-forth, first named as *logos* and bespeaking the primordial way-opening Way[108] as which Being, happening pivotally by way of language, at once disposes what-is to its presencing and governs the on-going

p. 302. (The German translation of the account of this seminar appears in VS. See page 104 there. There is, to date, no English translation.) On the difference between the two sorts of thinking in question, see ID 93ff, 30ff, and "Conversation on a Country Path," G 31ff, DT 58ff. Cf. Ch. XIV, pp. 544f.
[105]That expanded understanding appears significantly, though without explication, in "The Thing" and "The Turning" (1949-50). It comes to overt statement in "The Principle of Identity" (1957) and "Time and Being" (1962).
[106]Cf. Ch. I, p. 53, and Ch. II, pp. 108f, and n. 190.
[107]Cf. Ch. I, pp. 35f.
[108]Cf. Ch. III, pp. 123-125. See "The Nature of Language," US 198, WL 92, and this chapter, pp. 146-148.

existing of man as openness that, in its instancing of Being, in the dual movement constitutive for it, simultaneously, in man's own presencing, gathers itself forward and gathers what-is, as that which meets it, toward itself and to disclosure.[109]

At this point we must be careful not to suppose that we have here touched upon something that is so "inward" to the Happening of which we are speaking as to be "distant" from the utter immediacy of happening that Heidegger intends to name with the phrase "the Being of what-is."[110] For no such inward initiating, detached, as it were, from Being, is in question.   As the Bringing-to-pass that we have just described, the primordial, sui generis, uniquely initiatory letting-belong that is named as *Ereignis* is itself in play, whether directly or as withholding itself, even to the last taking-place as which the unfolding Twofold brings itself to pass.

As Being-in-the-world, as living out in specificity the openness-for-Being incumbent upon him, man, in the belonging-together that is his belonging into Being, has ever to do in one way or another with this gathering enowning-initiating.   If we would understand fully the way in which man lives out this peculiar belonging-together in the concreteness of his Being-in-the world, in the concreteness, that is, of his making of his way among the things with which he has to do, we must have an eye just to this letting-belong, as the latter shows itself precisely in meetings with such things of the world as are most familiar to us and therefore most in need of being freshly seen.

---

[109]On *logos*, cf. Ch. III, pp. 124f.
[110]Cf. Ch. I, p. 23.

CHAPTER V

THE THING

To us it could easily seem at first glance as though it would be hard to find a word more devoid of content than the word "thing." How vast is the word's usage! How inclusive, especially in its compounds—anything, everything, something, nothing—and therefore how indefinite. And yet "thing" does have its specificity. We use it readily of inanimate objects or of intangible matters or facts or events. We use it as a class noun for animate beings, in phrases like "living things," "growing things," "creeping things." But except when alluding to something surprising or unknown, when words like "What's that thing?" easily spring to our lips, we are unlikely to employ it directly as a designation for particular plants, and we are even more unlikely to apply it to animals. Always we feel that it is wholly inappropriate to use it of human beings unless in phrases of endearment or belittlement, such as "pretty little thing" or "poor thing," phrases which, if subtly, assign to persons a less than human status.[1] We would not use "thing" of God.

Our use of the word thing or our abstention from its use, then, is complex and at times inconsistent. But both usage and abstention are significantly conditioned by our relationship to the entity in question, by the way in which the latter concerns us, and, most clearly, by the degree to which it has or does not have for us an integrity of presence and/or action that lets it stand over against us so as to suggest or manifest a selfhood akin to our own.

In turning to Heidegger's consideration of the meaning of thing, we must be prepared for an outlook sharply different from that implicit in our own employing of the word. For us entities that concern us least closely or that are, in one context or another, readily susceptible of abstract designation are most apt to be named with "thing." For Heidegger, on

---

[1] "The Origin of the Work of Art," UK 13, PLT 21.

the contrary, the word "thing," rightly understood, far from being a word so ubiquitously employed as to be largely drained of specificity and force, bespeaks a centrality and immediacy of relationship with us, the human beings who speak of things and deal with them. For it names an elaborately identifiable participant that is crucial to the complexity of happening as which the unfolding Twofold that is named as the Being of what-is brings itself to pass via the openness that only our human existing by way of time can afford.[2]

1

In a lecture entitled "The Thing," the first in a series on modern technology delivered in 1949, Heidegger takes up afresh a theme that had concerned him earlier and that was now to receive the emphasis of repeated consideration.[3] He begins his inquiry by asking after that which marks a thing as a thing.[4] In 1936, having, in a discussion in which the meaning of "thing" is of central importance, considered and rejected three traditional interpretations of thing—a thing as the substance to which attributes pertain, a thing as a unity constituted of a manifold of sensory impressions, and a thing as "formed matter"[5]—he, in contrast, found the distinguishing character of a thing in a self-containedness, a resting-in-self, that belongs intrinsically to it and that gives it an innate steadfastness in its self-presenting that must not be gainsaid.[6] When he returned to the consideration of the thing, this emphasis on a thing's gatheredness and puissant integrity remained. But his thinking had obviously advanced to fresh insights. For we find a complex portrayal of the thing that is in most respects new.

Heidegger's way of defining a thing, when he takes up the matter

---

[2]Cf. Ch. I, pp. 27-29, and Ch. II, pp. 59f & 72f.
[3]See, "Building, Dwelling, Thinking," an essay closely akin to "The Thing," VA 145ff, PLT 145ff.
[4]"The Thing," VA 166, PLT 167f.
[5]"The Origin of the Work of Art," UK 14ff, PLT 22ff.
[6]UK 20, 23; PLT 26, 29. In a passage written four years later, Heidegger identifies a "thing" as "something standing in itself, *ein In-sich-ständiges.*" See "The Being and Conception of *Physis* in Aristotle's *Physics* B, 1," W 322, Ph 232.

in detail, is to say that a thing "things." The German word *Ding* (English: thing) stems from the old German *thing* or *dinc*, which denoted, he says, a gathering convened in order to deal with an affair of concern to a group. These words were well suited to translate Latin *res* (thing), which originally meant "that which concerns (*das Angehende*)." Behind *res* lay Greek verbs *eirō*, with related nouns *rhắtra, rhắos, rhắma,* signifying to deliberate about something through speaking.[7] In this complex of meaning belonging originally to *Ding*, and still, for Heidegger, resident within it, lies for him the only proper point of departure for the understanding of a thing and of what is its provisive carrying out of itself as such, its thinging.

A thing, in thinging, approaches us as something of closest concern. For via the thinging (Dingen)—the genuine happening—of a thing, in an intricate reciprocity of accomplishing, the worlding (*Welten*) of world takes place.[8] In this context "world" is not understood first of all either as the manner of happening of what-is that is brought into play via the self-carrying-forward of human existing per se or in terms of the events and on-going undertakings that, in their lived out interrelating, constitute the meaningful milieu in which human existing in community carries itself out.[9] Rather, world is seen in terms of a happening via the particularity of what-is of certain primary determinants whose interrelating[10] gathers unifyingly together and provides meaning. Those determinants are four: earth and heavens,[11] divine ones and mortals.[12] "World" names the provisive, uniting interplay and

---

[7]"The Thing," VA 173, PLT 174.

[8]VA 179f, 176; PLT 180f, 177.

[9]Cf. Ch. II, pp. 61-63, and Ch. III, pp. 136f.

[10]Here and elsewhere in this study the reader should be careful rigorously to separate "relate" from its usual meaning. Here "relate," in contexts where the Being of what-is is in any way under consideration, does not speak at all of a connecting of inherently discrete elements, one of which is to be subordinately referred to the other. The relating that is now in question is a *holding-toward-one-another* of participants that, while they are distinct from one another, are ultimately one. For we are using "relate" to speak of the happening of the single Twofold. See further, Ch. IV, p. 145, n. 10.

[11]The most obvious translation for *Himmel* would be sky. Since, however, Heidegger intends that the word speak immediately of all the phenomena that show themselves via the sky, we have chosen to translate with "heavens" a word far better able to carry this range of meaning.

[12]"The Thing," VA 178, PLT 178. Heidegger's phrase is *die Göttlichen*, which is

instantiation according to which these four, each after the manner of its peculiar mode of happening, are gathered as one and, thus caught into a singleness of meaningful transpiring, come into disclosure as the particularized happening of what-is. It is precisely that interplay that thus accomplishes itself in the happening of the Four in question which comes to full accomplishment by way of a thing. As a locus of gathering, a thing, in its thinging, centers by way of itself the self-carrying-out in concert of the Four that is the bringing to pass of world.[13]

The Four, earth and heavens, divine ones and mortals, come to disclosure each as itself, yet they are not discretely separated from one another. They are a unity possessed of a complaisant simplicity (*eine Einfalt*), which happens precisely as a gathered Fourfold (*Geviert*).[14] Via that Fourfold, as the fouring that disposes it, indeed even as the Fourfold itself,[15] world brings itself to pass,[16] happening as a specific instancing mode of that pure Happening that self-disclosingly happens as the Twofold. Holding sway as a gathering in which disclosing, enowning bringing-to-pass (*Ereignen*) is preeminently in play, world, in worlding, accomplishes itself as the "mirror-play" of the Four[17]—a mirror play

---

usually translated "divinities." However, in the passages where the Fourfold is specifically characterized (VA 177, PLT 178, and "Building, Dwelling, Thinking," VA 150, PLT 150), *die Göttlichen* are obviously distinguished from *der Gott* (the god). The translation "the divine ones" is intended to maintain this distinction.

[13]"The Thing," VA 179ff, PLT 180ff.

[14]VA 176; PLT 177. As a noun *Geviert* means square, as an adjective, quartered or squared. Heidegger's infusion into the word of connotations of provisiveness and movement make it hard to translate. The accepted translation, "Fourfold," seems most adequate, although it suffers from a serious drawback in that it suggests quite erroneously an immediate kinship of the word with Twofold (*Zwiefalt*). Such a parallel in form does in fact exist with *Einfalt* (simplicity, silliness), here translated by "unity possessed of a complaisant simplicity," which, expressive of a onefold simplicity, is used to characterize the *Geviert*. On the prefix *ge-* as bespeaking a gathering, cf. Ch. VI, pp. 258f. Heidegger's identification of the participants in the Fourfold sometimes varies. See this chapter, p. 178, n. 34.

[15]VA 176; PLT 178. *Wir nennen das im Dingen der Dinge verweilte einige Geviert von Himmel und Erde, Sterblichen und Göttlichen: die Welt.*" "Language," US 22, PLT 199.

[16]"The Thing," VA 179, PLT 180.

[17]*Wir nennen das ereignende Spiegel-Spiel der Einfalt von Erde und Himmel, Göttlichen und Sterblichen die Welt.* "The Thing," VA 178, PLT 179.

initiatory as pure Happening in its ruling and carried to completion in that Happening's self-particularization by way of the manifold of what-is. Ruled structuringly from out of that mirroring, and gathered via the genuine thinging of a thing, which indeed brings itself to pass (*ereignet sich*) therethrough, each of the Four plays its part freely, but its very freedom is a compliant according with and mirroring of the comportment of its comrades.[18]

As such provisive mirroring, world holds sway as an agile "ring-dance of enowning bringing-to-pass (*Reigen des Ereignens*)." Its ruling is a gathering roundelay that—as the holding-apart inherent in distinguishing is opened up and the distinguished yield themselves toward one another—"lightingly clears the Four into the shining of their onefold simplicity."[19] Brought to pass as a giving-into-belonging (*Vereignung*), that ruling simultaneously frees each of the Four into self-disclosure and binds it into the ever-turning figure of a fourfold mutual coming-to-pass-as-one. Therewith each, just in happening as itself, is given into the thus united Four. So gathered beyond itself and given to belong in compliant oneness with its partners, it is given into its "own," i.e., into that which is its genuine mode of transpiring. Brought to pass after the manner in which, primally, enowning Bringing-to-pass (*Ereignis*) ever rules, this bringing of each of the Four into its own happens precisely through its being dispossessed (*enteignet*), i.e., through its being so withdrawn from self-sufficiency in the special character that is its alone as to find its proper fulfilment as itself solely in that simple, onefold coming into disclosure in company with its partners in quadrapartite taking place which a thing's thinging carries to fruition.[20]

Each of the Four—earth and heavens, divine ones and mortals—that are gathered into the unitary Fourfold as which world rules can rightly be named and identified, but if that one is to be properly thought as thus brought forward, the other three must also be held immediately in

---

[18]"The Thing," VA 179f, PLT 180f.
[19]. . . *lichtet er die Vier in den Glanz ihrer Einfalt.* VA 179; PLT 180f.
[20]"The Thing," VA 178, PLT 179. On *Ereignis* as *Enteignis*, see Ch. IV, pp. 163-166.

mind.[21]

Earth is that which supports and fruitfully provides, "spreading itself out into water and stone, rising up also toward animal and plant," and cherishingly protecting all these.[22]

The heavens are the courses of the sun and moon, the stars' brilliance, and the year's changing seasons, the light and the dusk of a day, the darkness and the brightness of a night, the weather's mildness and severity, the scudding cloud and the azure depth.[23] Through their manifesting rife with changefulness in regularity, they delimit and structure, orient and measure out.[24]

The divine ones are "the beckoning messengers of Godhead (*Gottheit*)." Out of the latter in its concealed ruling that, as holy, partakes of a beneficent wholeness (*Heil*), the god—whose mode of enduring (*Wesen*),[25] determined from thence, withholds him intrinsically from any comparison with what presences—appears into the peculiar present-time presence (*Gegenwart*) that accomplishes his manner of holding sway, or else withdraws himself into the veiling that likewise belongs to him.[26]

---

[21]"The Thing," VA 177, PLT 178f.
[22]"Building, Dwelling, Thinking," VA 149, PLT 149; "The Thing," VA 177, PLT 178.
[23]VA 150, PLT 149; VA 177, PLT 178.
[24]VA 150f; PLT 150. The role that is here accorded to the heavens is strikingly foreshadowed in a passage from *Being and Time* already cited (Ch. II, p. 67, n. 45), in which the rising and setting of the sun are spoken of as "regions," whose orienting ruling bestows meaning upon whatever, having to do with life and death, pertains to human existing in its transpiring as Being-in-the-world.
[25]For a detailed discussion of *Wesen*, see Ch. VI, pp. 252-257.
[26]In this sentence, as indeed throughout our identifying of the Fourfold, we have conflated closely similar passages from "The Thing" and "Building, Dwelling, Thinking." *Aus dem verborgenen Walten dieser [der Gottheit] erscheint der Gott in sein Wesen, das ihn jedem Vergleich mit dem Anwesenden entzieht* (VA 177, PLT 178). *Aus dem heiligen Walten dieser [der Gottheit] erscheint der Gott in seine Gegenwart oder er entzieht sich in seine Verhüllung* (VA 150, PLT 150). In these statements Heidegger seems completely to differentiate "the god" from what presences (*das Anwesende*), and so, one would suppose, from what-is (*das Seiende*). But later in the same series of lectures he states that " . . .the god is, when he is, a being and stands as a being in Being and its enduring (. . . *der Gott ist, wenn er ist, ein Seiender,*

"Mortals are men," specifically as those who are able to die. They alone are possessed of the possibility of submitting themselves to death, i.e., to that uttermost possibility wherein Being, concealing its happening in happening as Nothing, encounters and rules men most fully and most inexorably.[27] In such submitting, mortals carry out directly the holding-relating that lets them belong to Being as openness for its self-disclosive happening. "Mortals are," Heidegger says, "the holding-relating to Being as Being, the holding-relating that gatheringly, disposingly endures ([*Die Sterblichen*] *sind das wesende Verhältnis zum Sein als Sein*)."[28] Mortals, that is, let come to pass immediately in themselves, and indeed *as* themselves, the holding relating obtaining between Being and man,[29] a relating, namely, that so endures as to let endure, i.e., that so endures as to bring into play via ever opening time, which man lives out, the initiatory happening of Being *as* enduring, that governs in whatever *is* by way of the openness that man provides.[30] Here the belonging together of Being and man transpires in ultimacy, happening overtly in its unseverability. For it transpires *in extremis*, as a relating that holds firm

---

steht als Seiender im Sein und dessen Wesen . . .)." ("The Turning," TK 45, QT 47.) It seems most probable that the key to this apparent contradiction lies in the word "holy," which appears in the parallel passage from "Building, Dwelling, Thinking." (VA 150; PLT 150.) To the holy (*das Heilige*) pertains *Heil* (wholeness, well-being). The god endures as present directly from out of a ruling marked by completeness (*das heilige Walten*). Whatever presences as not immediately determined by such wholeness has instead another manner of enduring that is, even at its best, kept back from completeness. Cf. Ch. I, pp. 49-52. Heidegger never clarifies the relationships obtaining among what might be called the levels of divinity and the divine beings that manifest them.

[27]Cf. Ch. II, pp. 98-101.

[28]"The Thing," VA 177, PLT 179. On Verhältnis, cf. Ch. IV, p. 145, and n. 10.

[29]Cf. Ch. II, p. 99. On the transitive use of the copula, which is here assumed, cf. "A Heidegger Seminar on Hegel's *Differenzschrift*," translated by William Lovitt; Martin Heidegger, *Questions IV*, Paris, Gallimard, 1976, p. 258; *The Southwest Journal of Philosophy*, Vol. XI, No. 3, Fall, 1980, p. 38. Cf. also Ch. I, pp. 30f. In "On the Essence of Truth," Heidegger identifies human freedom, the disclosive openness vis a vis the happening Twofold that is constitutive of man as man, directly with holding-relating (*Verhältnis*). W 90; BW 133f. On such freedom cf. Ch. II, pp. 105-107.

[30]Cf. Ch. II, pp. 59f.

even toward Being in its happening as Nothing, in utter concealment, from out of the protective fastness (*Gebirg*) that is death.[31]

Heidegger's specific depictions of earth and heavens, divine ones and mortals, the Four, show us a particularization of entities that bespeaks what-is. Yet these Four are, equally clearly, modes of Happening under whose particularized sway groupings of entities are gathered and provided. Prior to their particularization they can be considered as themselves. "From out of themselves," Heidegger can say, "these Four belong together in one." "Antecedent to all that presences," antecedent, that is, to their own particularization, they are "clasped (*eingefaltet*) into the unique Fourfold as which they come to pass."[32] Pointing to this latter character of the Four, Heidegger, using the word "region (*Gegend*)," which in the determining context he defines as the disclosure permitting "clearing (*Lichtung*)"[33]—as which pure Happening brings itself to pass as Being by way of time—speaks of "four world regions (*Welt-Gegenden*): earth and heavens, God and Man."[34] Here, in a portrayal that very evidently bespeaks the complaisant "ring dance" that, bringing to pass the

---

[31]"The Thing," VA 177, PLT 178f.
[32]VA 172; PLT 173.
[33]"The Nature of Language," US 197, WL 91. Cf. Ch. I, pp. 32f & 37f. On *Gegend* as provisive Region, cf. Ch. XIV, pp. 546f.
[34]US 214; WL 106. The words "God (*Gott*)" and "man (*Mensch*)," used here of "world regions," are words fuller in meaning than "divine ones (*die Göttlichen*)" and "mortals (*die Sterblichen*)," usually employed by Heidegger as definitive of the Fourfold. Here their singleness of reference surely renders them appropriate, as "divine ones" and "mortals" could not be, for the naming of provisive openings-up of initiatory Happening for the taking place of what-is. The phrase "God and man" may here best be understood as offering a defining identification that is paralleled in limiting specificity in Heidegger's original naming of that Fourfold, but that should by no means be thought to supersede it. This view is supported by the fact that this use of "God" and "man" has a counterpart in "The Turning," the concluding lecture in the series that opened with the presentation, "The Thing", in which Heidegger's "Fourfold" of earth and heavens, divine ones and mortals, was first set forth. There, in a context where that Fourfold is clearly presupposed, "the god" and "man" are spoken of pivotally. ("The Turning," VA 45, QT 47.) This pairing, again with reference evidently to the particularized entities in question, appears also in "The Nature of Language" (US 211, WL 104). On the relation between "the divine ones (*die Göttlichen*)" and "the God (*der Gott*)," see this chapter, p. 173, n. 12, and pp. 176f, and n. 26.

ruling of world as mirroring, disclosively gathers forth the single Fourfold from out of enowning bringing-to-pass, these "regions of the structured configuring of world (*Gegenden des Weltgefüges*),"[35] attuned as one, are presented as mutually opposite one another in such a way as protectingly to guard each other in their inter-happening.[36] Thus, wholly after the manner of the Twofold, as the unitarily fourfold Four what-is carries immediately into particularized shapedness the one ruling Happening—elsewhere named as Being,—that as disposingly ruling "world" in fourfoldness happens it unitedly forth.

In the unitary happening together of the Four of earth and heavens, divine ones and mortals, as thus complexly understood, as each carries itself out in the manner peculiar to itself, world worlds.[37] Gathering into unifiedness transpires. The intricate, meaningful disposing of on-going communal life brings itself into play. Meaningful ordering opens out. And whatever is comes to genuine appearing; it presents itself in compliant, on-going interrelating with its fellows precisely through the clarifying determining that is the complexly simple governance of the Four brought into play as the worlding of world.

Via the thinging of a thing that worlding of world brings itself to accomplishment. Every genuine thing belongs within the manifold of what-is. It itself appears from out of world.[38] Yet, as "thing," it fulfills simultaneously a unique role vis a vis world. In that it gathers into itself the fourfold determining as which the worlding of world accomplishes itself, each genuine thing presences uniquely as a focal entity, as a fundamental centering in the happening of what-is that, as such, through the peculiar manner of transpiring as itself—its thinging—accorded to it, permits the intricate self-carrying out of the Twofold as the quadrapartite happening of world to bring itself to pass.[39] In a uniqueness of role that

---

[35]US 208; WL 101.
[36]US 211; WL 104. Heidegger's phrase is *gegen-einander-über*, a variant of the usual *gegen-einander*, "opposite one another," "face to face." He attributes the amplified phrase with its connotation of safeguarding to Goethe.
[37]"The Thing," VA 178, PLT 179.
[38]"Language," US 22, WL 200.
[39]See this chapter, pp. 185-187. In "The Origin of the Work of Art" (1936),

has an analogue, if but distant, only in the role of human beings, whose peculiar manner of existing as openness-for-Being (*Da-sein*), allows all unconcealing whatever of pure Happening in its self-maintaining as the Being of what-is to come to accomplishment,[40] the thing shows us a manner of happening that, wrought out in distinctiveness via a particular center of disclosure, permits self-maintaining Happening as such—now in its happening as the Twofold in a fourfold manner as world—to bring itself to disclosure.

As in the gathered Fourfold as which world comes to pass and in the intricate relating of world and thing the transpiring in oneness yet in distinction that ever marks the wholly self-initiatory Happening that brings itself to pass as the Twofold is shown us in certain of its instancings, many of the deep-lying lineaments that we have found Heidegger to present for that Happening come also directly into view. The protective juxtaposing of disclosive regions that accomplishes itself in the "world

---

Heidegger presented a formulation that bears some relation to the one now under discussion. In a work of art, which, in the subsisting-in-itself that belongs to it, has in its own peculiar way the character of a thing, there takes place a contending of earth and world. Earth, embodied in stone or metal or pigment, is "the self-closing." It resists the disclosure forced upon it as, through its inclusion in a work, it is drawn into the exhibiting of meaning and order that takes place as the work in and through itself manifests world. And world—the disclosive complexity of order and meaning—resists being drawn into and displayed by way of the radically specific, stubbornly self-contained embodiments of earth. The work gathers these two elements into the fruitful strife that, bringing them into disclosure, allows it to transpire as a work. (UK 44ff; PLT 44ff.) Now, in the context with which we are chiefly concerning ourselves, world has become primarily the Twofold itself, as that Twofold gathers itself forth after a specific manner of its happening and therewith brings itself to pass as an open realm of meaningful relationships. The thought of demarcation and ordering is connected now with "the heavens," earth's direct partner that itself is seen as inherently bestowing itself and entering into relation. And the divine ones and mortals have also been newly included as participants in the bringing to pass of world. (VA 178; PLT 179f.) The role of the work of art that gathers earth and world into fruitful self-disclosure through itself is significantly similar to that later portrayed by Heidegger for the thinging thing. We would therefore surely be justified in seeing in that role, in retrospect, an anticipation of "the later, fully developed, thingly character (*das Dinghafte*)" of any "thing" when, as is the case, in "The Origin of the Work of Art" that character is ascribed to such works (UK 10; PLT 19).
[40]Cf. Ch. II, pp. 72f.

Fourfold (*Weltgeviert*)" that finds unconcealment through the thinging thing arises in the disclosively enowning self-bringing-to-pass of pure Happening as the self-differentiating Same that is the very "play of Stillness"—the self-opening of concealing—that gathers forth the complex provisive proffering of time and space via which that Happening unfolds itself as the Twofold.[41]   And it is precisely that puissant self-containedness, that Stillness, that "stills Being"—as which pure Happening is so centrally named—"into the disposing enduring as which "world," as a gathering of the single, quadrapartite Twofold, holds sway.[42]

As that Same as which pure Happening at once withholds itself in self-containedness and brings itself to pass unconcealingly as the Twofold, that Happening rules as the differentiating sundering that unitingly structures forth (*der Unter-Schied*) that is itself the primal carrying-out (*Austrag*) as which the Twofold accomplishes its self-disclosure.[43]   And that very sundering is to be seen as in play when, ruled by the fourfoldness according to which world brings itself to pass, the happening Twofold gathers itself into the open, into unconcealing, through centering itself via the singular specificity of some particular thing.   For it is that uniting-sundering, happening as the primal carrying-out that carries toward in carrying apart,[44] that at once rends asunder world and thing to their distinctiveness and gathers them toward one another intimately as thus sundered,[45] therewith dispossessing them into a mutuality that alone

---

[41]"The Nature of Language," US 214, WL 106.
[42]*Die Stille stillt. Was stillt sie? Sie stillt Sein in das Wesen von welt.* "The Turning," TK 47, QT 49.
[43]Cf. Ch. I, pp. 29f & 32f.
[44]"Language," US 25, PLT 202.
[45]US 27; WL 204f. In Heidegger's complex discussion of the *Unter-Schied* that is now in question ("Language," US 24-29, PLT 201-206), this nuance in the latter's happening is presented through an elucidation of the meaning of the word *Schmerz* (pain), which is found in the poem that in this context serves to elicit and inform his thinking.   "Pain," as a rending asunder that is itself a structuring forth that gathers together, "structures the delineating rent that is the uniting-sundering.   Pain is the uniting-sundering itself. (*Der Schmerz fügt den Riss des Unter-Schiedes. Der Schmerz ist der Unter-Schied selber.*") (US 27, PLT 204.)

brings them into their own.[46]

As the self-accomplishing of pure Happening that can, when otherwise viewed, be named as the provisive Same,[47] that uniting-sundering is also itself nothing other than that primal, stilling Stillness which, intrinsically provisive precisely as self-withholding self-concealing, happens itself forth in opening for all happening its course and, so doing, enowningly brings itself to pass as unconcealing.[48] Accomplishing "stilling," even as "Stillness" itself, the uniting-sundering holds world and thing apart, in urgent self-restrainedness poising them in themselves.   And precisely in that enabling distinguishing, it bestows them toward one another, carrying them beyond themselves into the togetherness that alone fulfills their happening disposingly as themselves. Stilling, "it lets things rest," in the intensity of their self-presenting, "in the gift of world," and it lets world, gathering itself intensely forth via one or another specificity, "suffice itself in the thing."[49]

As unitingly distinguished via the ruling of this provisive uniting-sundering, world, as an instancing of initiatory Happening itself, "grantingly allows (gönnt) things," happening them, in its holding sway as the bringing into play of the Fourfold of earth and heavens, divine ones and mortals, as the things that they are.   And things, as shaped, centered happenings that specifically receive the ruling of world, "gesture (gebärden) world."[50]   Slight though they be, each in itself,[51] in their intense specificity they bring world, as a luminous bringing to pass of the quadrapartite Twofold, immediately and, we may say, concretely into play.   Via their thinging they "carry world out," bearing it forth,[52] as they focus forth the happening of the Twofold through their self-

---

[46]US 29, 28; PLT 205, 204.   "Dispossessing" translates uses by Heidegger of *Enteignen*. On dispossessing as intrinsic to enowning bringing-to-pass (*Ereignen*), cf. Ch. IV, pp. 163-165.
[47]Cf. Ch. I, pp. 29f.
[48]Cf. Ch. IV, pp. 146-148.
[49]"Language," US 30, PLT 207. On Heidegger's understanding of "rest," cf. Ch. IX, pp. 352f.
[50]US 24; PLT 202.
[51]"The Thing," VA 181, PLT 182.
[52]"Language," US 22, PLT 200.

presenting as things, from out of "world in its worlding," within the manifold of what-is.

Thus does the thinging of a thing show us afresh the self-bringing to pass of the unitary Twofold. From out of its depths pure Happening as that Twofold, as an intricate, single yet fourfold Happening shapes itself forth unconcealingly in a multitude of happening entities gathered, via that fourfoldness, into a close mutuality of reciprocal interrelation; and— precisely through centering itself in some one entity, some one thing, wherein, in ways that ramify away widely but are drawn into meaningfulness through that centering, it embodies its self-disclosing—it therethrough accomplishes its self-maintaining as self-unconcealing.

Via pure Happening's self-opening as time, that self-maintaining is fulfilled.[53] Always a "thing" is unique. It is "this or that thing." This means that it is, as thing, a particular something that whiles (*weilt*) in its appointed manner for its appointed duration (*ein je Weiliges*) by way of the on-going openness of a particular present.[54] And it is precisely in thus taking place that a thing, whiling in its thinging, gathers forth the happening of the single Fourfold. As such, it "stays (*verweilt*) the Fourfold," bringing the four, in the illumining potency of their intricate holding sway, to while—to present themselves on-goingly—centeredly in itself.[55]

Here the "distancing Nearing" that is, as such, an accomplishing from out of disclosing Bringing-to-pass that enowns of the drawing close—in opened out oneness—of the diversified, as which pure

---

[53]Cf. Ch. I, pp. 37f.

[54]"The Thing," VA 179, PLT 181. For a like use of *je* ("at all times," "at any given time," "each") as a component of *Jeweiliges*, to bespeak this particularization specified via time, cf. Ch. I, p. 49, n. 89. Cf. also, apropos of the use of *je* by Heidegger, Ch. II, pp. 58f, and n. 4. On "whiling" as an on-goingness that is ever a coming that is a going-away, cf. Ch. I, pp. 49f.

[55]"The Thing," VA 179, PLT 181. The verb *verweilen* is ordinarily intransitive, meaning to tarry, delay, or stay. (See its related use in Ch. I, p. 50.) Heidegger's transitive use of *verweilen* in the present context relies on the prefix *ver-* —which frequently has causative or enabling force—and speaks of a staying that is a bringing-to-while.

Happening brings itself complexly to pass,[56] carries out its pivotal governance at once in the provisive opening to differentiation that yields the protective over-against-one-another in play in the Fourfold of world, and in the gathering-forth via which that differentiating fulfils itself in the thinging of a whiling thing. For in the mirror-play that is the worlding of world and in the thinging of the thing that gestures world as the quadrapartite Twofold forth into accomplished unconcealment, it is precisely that distancing nearing that immediately holds sway.[57]   This ruling of Nearing is complexly fulfilled via time when, as the four— intrinsically remote from one another in the distinctiveness belonging to each—are brought near to while via the thinging of the thing, in that thinging, "nearing" likewise holds sway in withholdingly gathering toward one another the ever on-going, ever interrelating dimensions of time and by way of them reaches forth the presencing of whatever presences from out of the fourfoldness therewith brought directly to bear.[58]

The staying, the bringing-to-whiling, that nearing-thinging accords the Fourfold brings to bear by way of the while granted the thing in question a gathering that accomplishes belonging-together with respect to the constitutive Four, and this means also with respect to the given particulars as which, via that while, the Four present themselves in their interrelating with one another. In the staying that the genuinely thinging thing brings into play, the sui generis disclosing that enowns in letting belong together—as which pure Happening initiates itself as self-maintaining—centers itself rulingly through its centering into the presencing, in-gathering thing of the manifold taking-place as which pure Happening bodies itself forth. "Verweilen ereignet," Heidegger says succinctly; genuine staying that lets "while" so brings to light and takes to itself as to bestow the disclosed so claimed into its own, i.e., into the belonging proper to it, the belonging-together in which alone it is

---

[56]Cf. Ch. IV, pp. 146-148. Cf. "The Nature of Language," US 214, WL 106.
[57]"The Thing," VA 176, 179f; PLT 178, 181.
[58]"The Thing," VA 172, PLT 173 & VA 179ff, PLT 180ff. Cf. Ch. I, pp. 39f.

genuinely itself.[59]   Stayed, brought to while in relational unity via this or that thing, the Four—earth and heavens, divine ones and mortals—come into their own.   And happening thus, and taking place in ever shifting particularity, these four constitute what-is.   Hence every particular of the latter's manifold as gathered forth by way of the while traversed by a thinging thing is governed by such staying.   It too is brought into its own through this governance, by the same disclosive letting-belong-together— in this case its belonging together with all its presencing fellows.   For it is the staying of the Fourfold via the thinging of a thing that allows the presencing-together of the particulars of what-is, which is itself an accomplishing of whiling, to come genuinely to pass.[60]

Things are of two sorts.   There are things brought forth of themselves, such as tree or brook, bull or horse; there are things that depend upon man for their bringing-forth, such as jug or clasp, picture or crown.[61]   In choosing things to portray in their thinging, Heidegger draws from those of the latter sort. In one portrayal that shows at length in specificity the thinging of a thing, he depicts a bridge.[62]   Vaulting over the stream, the bridge unites and discloses the latter's banks as banks and, through this joining, gathers together toward itself the expanses of land stretching away behind them.   It defines the landscape, bringing its varied features to stand forth in unique identity precisely through the interrelating that it, drawing them together into a unity, confers upon them.   Borne up by its piers, in every season the bridge gathers the changeful stream to itself.   Sustained in its suitability, it focuses upon itself the onslaught of the storms that can alone maintain the stream.   Ceaselessly it receives the flowing water, taking it in beneath its arches and giving it forth again unhampered to mirror the open heavens.   Unifying the landscape by its central standing, the bridge draws men to itself.   It offers them a way from bank to bank or town to town.   And

---

[59]"The Thing," VA 172, PLT 173.   Cf. Ch. IV, pp. 149-151.
[60]Cf. Ch. I, pp. 49f.
[61]"The Thing," VA 181, PLT 182.
[62]"Building, Dwelling, Thinking," VA 153ff, PLT 152ff.   In "The Thing" (VA 166ff; PLT 168ff), Heidegger considers in like detail the role of a jug as a thing.

when they follow it as *mortals*, for them the bridge bespeaks that "last bridge" that conducts toward the most extreme possibility, the *Nothing* of death, toward which they are "always already underway." In its soaring over the gulf it spans, it betokens the determined vaulting beyond the commonplace that must mark that perilous course. Likewise, in its sure overleaping, the bridge discloses to those who pay heed the striving fundamental to them as mortals to mount beyond "that which, belonging to them, is ordinary and unsound and harmful (*Unheiles*) in order to conduct themselves via that which, belonging to the divine, is sound, whole, saving (*vor das Heile des Göttlichen*)." Carrying them via its overarching of the endangering gulf, the bridge gathers mortals, in the midst of their pursuance of their wonted ways, into proximity with the divine ones. The latter's beneficent presencing manifests itself to them in its otherness precisely there in their hazardous crossing. The patron saint of the bridge, messenger of the fullness and serenity of the divine,[63] beckons those who pass to consider and give thanks in acknowledgement of the fulfilling sustaining vouchsafed to them.[64]

The single Fourfold of earth and heavens, divine ones and mortals, can come into its own, i.e., can happen interrelatedly in genuinely self-disclosive unity, only by way of the singular concreteness of this or that thing. Just as time, as the self-opening of Happening via the threefold outreaching that proffers presencing, is focused uniquely and provisively in the whiling of such a thing, so space as such—for which time's self-opening makes room and which, together with time, permits presencing to accomplish itself—as the opening-up of Happening that permits every sort of emplacing,[65] finds a decisive locus of self-accomplishing in the

---

[63]For this nuance of meaning cf. "Remembrance of the Poet," EHD 16ff, EB 248ff.
[64]"Building, Dwelling, Thinking," VA 153f, PLT 152f. At this point in his presentation Heidegger pursues his discussion with mention of many bridges. We might easily infer that he is displaying to us the essence common to them as bridge-things. This inference would, however, be mistaken if it implied to us that there is an essence (*Wesen*) that is an abstractable, universal bridgeness that is exemplified in each. The shared *Wesen* discernible among the bridges must be understood rather as the manner of accomplishing itself as bridge that, wholly intrinsically, governs each in its distinctive milieu. VA 154; PLT 153. On *Wesen* see Ch. VI, pp. 252-257.
[65]See "The Nature of Language," US 213, WL 106.

thinging thing's self-carrying-out. Via such a thing, originative "space" brings itself rulingly into play. The bridge as a thing is a "place (*Ort*)." It itself permits (*verstattet*) to the gathered Fourfold a specific abode (*Stätte*). Thinging thus as place, it is a center of disclosing defining. Only in relation to it as "place" in this sense do the locations and the ways that it gathers about itself appear as themselves in disclosive interrelation. And only with their appearing does space—genuine, bounded, defined space—open among them. In permitting the Fourfold an abode through its thinging, the gathering bridge first constitutes about itself the realm of spaces that in any particular on-going present is the fundamental arena for the unfolding in concreteness and in discernible interrelation of the manifold of what-is.[66]

2

The thinging of the thing is the pivotal locus for the self-presenting of what-is that is its encountering of man. There world, the inseparable milieu of man's living-out of his existing,[67] transpires immediately; it "worlds." The thing is indeed that which, in its gathering forth of world, concerns man directly and decisively. For their part, men are summoned to encounter things in the very yielding of themselves to the vouchsafing of openness that is, in their existing, incumbent upon them as men.[68] In this context, where human comportment in relation to things is in question, Heidegger's central word for such self-carrying-forward via openness is "dwelling (*Wohnen*)." For him, to dwell means in the living out of ever-opening time "to be brought to peace." It means, that is, via the ruling of provisive Happening to be brought to remain freeingly, protectingly, circumscribed into a freeness (*Freie*) that, with no admixture of self-assertive demanding, from out of a "peace" that is contentedness

---

[66]"Building, Dwelling, Thinking," VA 153ff, PLT 154f. For a discussion in which Heidegger, quite similarly, sees any notion of space as derivative and understands "place" in terms of constitutive interrelatings fraught with intending see SZ 102-110, BT 135-144. His portraying of ordinary measuring as secondary to the measuring-out that is a differentiating, relating seeing (see this chapter, p. 200) also belongs within this circle of understanding.
[67]Cf. Ch. II, pp. 61-63.
[68]Cf. Ch. II, p. 60.

with permitted enduring husbands anything met with into the mode of enduring (*Wesen*) belonging peculiarly to it. Dwelling, as such, precisely as delimited to be itself, inherently leaves whatever is met with to happen after its own manner or, again, countering any deviance from full self-presenting that may beset it, so meets it as protectingly, preservingly to harbor it back (*zurückbergen*) to enter anew into genuine transpiring as itself. This sparing husbanding (*Schonen*), which bespeaks the genuine meaning of freeing, is Heidegger emphasizes, the "fundamental tendency (*Grundzug*)" pervading dwelling as such.[69]

---

[69]"Building, Dwelling, Thinking," VA 149, PLT 149. In the paragraph underlying the above presentation, Heidegger builds his thought forward via uses of *zufrieden sein* (to be satisfied or content), *Friede* (peace), *Freie* or *Frye* (nominal adjectives meaning freeness), *freien* (to free), and *einfrieden* (to enclose), and he specifically states: "peace (*Friede*) means freeness (*das Freie, das Frye*)." English equivalences necessarily obscure the pervasiveness of the unity that is thus given to his portrayal of dwelling as a circumscribedly freed husbanding that frees, i.e., that, out of the protectedness thus vouchsafed it, is empowered protectingly to free whatever, met by it, is via it preserved to happen as itself. The happening of freeing precsiely by way of a protective enclosing, which is shown to us in *Zurückbergen*, sheltering, harboring back, and in *Einfrieden*, a peace-providing enclosing, is characteristic of Heidegger's view that delimiting circumscription is ever the sine qua non of any happening-as-itself. See UK 95f, PLT 82f, and "The Question Concerning Technology," VA 17, QT 8. Cf. also, Ch. IX pp. 347f & 350, and n. 91, and Ch. X, p. 383. Human existing, as which dwelling is carried out, is intrinsically limited to transpiring by way of a particularity of openness vis a vis whatever via such openness is particularly provided. As genuinely carried forward, it accepts such limitation. (See Heidegger's description of Greek man in "The Age of the World Picture," HW 96f, QT 145. Cf. also Ch. IX, p. 350.) For a similar thought regarding human freedom, see Ch. II, pp. 107.) Presumably just such compliant contentment with restriction is implied in Heidegger's allusion to the circumscribing of dwelling into the freeness, *das Freie*—a word that itself suggests openness (the meaning of *Freie* as a full-fledged noun being "open place" [see "The Question Concerning Technology," VA 32, QT 25, and n. 23])—that permits it to free anything met with to be itself. What is in this passage said by Heidegger of the sparing husbanding (*Schonen*) accomplished as a freeing in dwelling is paralleled very closely in his defining of *Retten* (saving). ("The Turning," TK 41, QT 42; "The Question Concerning Technology," VA 36, QT 28. Cf. also, Ch. XII, p. 488f.) The parallel between freedom, as a letting-be, and dwelling should also be noted. (Cf. Ch. II, p. 105-107.) The verb *zurückbergen* again plays a central role when a bringing to fully disclosive transpiring by way of an accomplishing of openness is in question from quite another perspective, in Heidegger's portraying of provisive Happening named as the Region (*die Gegend*). Cf. Ch. XIV, p. 547. On the meaning of *bergen* and on Heidegger's using of it, cf. Ch. I, p. 24, n. 12. On the meaning of *Wesen*, cf. Ch. VI, pp. 252-257.

As a sparing husbanding, dwelling is a self-disregarding dedication to that with which the dweller has to do, wherein a commitment to the precariousness—immediate and ultimate—intrinsic to man's fully authentic mode of existing is ever decisive.[70] Men dwell as *mortals*,[71] i.e., as those who surrender every claim in acknowledging and accepting the ineluctable provisionality of their course: its culmination in death.[72] Accordingly, dwelling takes place as a sojourning *(Aufenthalt)*.[73] It is a pausing for a while in-the-world, a transient residing here and now, which carries out the ever-opening present that human existing, as such, both accomplishes and withstands as it takes its way, individual by individual, toward death.

The sojourning of mortals, as which genuine dwelling takes place, is the correlate of the "staying," the bringing to whiling, as which the thinging of things gathers forth the Fourfold in the belonging-together that is the intrinsic unity of the four and therewith lets world world. Mortals sojourn "in close company with things, *(bei den Dingen)*."[74] Their

---

[70]Cf. Ch. II, pp. 99-101.

[71]"Building, Dwelling, Thinking," VA 150, PLT 150.

[72]Cf. Ch. II, p. 99.

[73]"Building, Dwelling, Thinking," VA 151, PLT 151.

[74]VA 151; PLT 151. *Bei* (beside, near, at home with) carries a strong connotation of closeness. We should here be reminded of Heidegger's very deliberate use of this preposition to speak of man's existing, as Being-in-the-world, as a being in company, authentically *(sein bei)* or inauthentically *(aufgehen bei)*, with that which is there met. Cf. Ch. II, pp. 99 & 89f, and n. 118. Here and elsewhere we have hesitated to use "at home with" as a translation for *bei*, because of the connotation of comfortable closeness that the phrase can scarcely fail to convey. It is never Heidegger's intention to speak of a comfortable, a wholly easy state for man. His stress on mortality alone, which we have just been pursuing, attests this. More than this, we must remember that any being-at-home-with always remains that of those whose authentic manner of existing involves the acknowledging of a homelessness entailed by the nullity intrinsic to thrown Being-in-the-world. Cf. Ch. II, pp. 94f. In a passage in "The Turning," the last in the series of lectures to which "The Thing" belongs, Heidegger makes a very strong statement of at-homeness (TK 46; QT 49). Cf. Ch. XIII, pp. 519f. Writing of that which the turning in the happening of Being as the manner of holding sway of modern technology would bring to pass, in a context where men's rapport with things is in question he says that men will dwell "as those native to Nearing." Here the word used is *einheimisch*, which means indigenous, native. Men, as the very ones who they are, will dwell as those who pursue their way as home-born in Nearing.

transient abiding is pursued in an immediacy of reciprocity with the things with which they have to do. Accepting and affirming the on-going opening-out of time that transpires via their unique mode of existing, mortals meet things in attentive receptiveness. Drawing near to things, in the integrity of their self-presenting via a lived-out present unclouded by their own self-assertion, mortals allow those things to come to light and to have free play as the things that they are.[75] And where such meeting takes place, under the governance of the distancing Nearing in play in the worlding of world the things via which and in which world holds sway and brings itself to disclosure are, like the ruling world regions themselves, protectively juxtaposed to one another wholly as themselves. Drawn toward one another precisely in their distinctiveness, they are, Heidegger says, "open to one another in their self-concealment." In the very integrity of their self-withholding in self-presenting, they shield one another in proffering themselves together.[76]

Caringly, enablingly met with, things thing. Mortals and things, abiding together for the while allotted them, bring to accomplishment the worlding of world. Like the bridge-thing of which we have spoken, everything that genuinely takes place gathers to disclosure in ever

---

But even here no easy comfort obtains. In the context in question the thought of men as mortal again lies close at hand (TK 47, QT 49). And more importantly, Nearing is, as such, a distancing. This nativeness, this being-at-home, partakes once again of the precariousness ever inherent in man's existing as provided and governed from out of Being—a precariousness that forbids any facile being-at-ease, any simple being-at-home.

[75]The parallel between this relating of mortals and things and the transpiring of *Being*-there, as Being-in-the-world, in its concerned meeting of that which proffers itself as ready-to-hand is striking. Cf. Ch. II, pp. 64f. See also the description of the authentic self-opening of *Being*-there as bringing the latter "precisely into its concernful Being-in-company-with (*Sein-bei*) that which is ready-to-hand," Ch. II, p. 100.

[76]"The Nature of Language," US 208, WL 101, and US 211, WL 103f. Heidegger here uses of things the same phrase, *gegeneinander über*, that he uses elsewhere for the four world regions. (See this chapter pp. 178f and n. 36.) We should here surely be reminded of what-is understood as what whiles in presencing, i.e., as a gathered grouping of particulars compliant in on-going happening together. (Cf. Ch. I, pp. 49f & 55.) For the yielding to one another that is fundamental to the transpiring of what whiles as what presences is shown us now in greatly enriched specificity in this mutual interplay of things.

unfolding particularity the Fourfold of earth and heavens, divine ones and mortals. And mortals, in their sojourning in closely concerned relation with things, carry out a dwelling that "preserves (*verwahrt*) the Fourfold" in thing upon thing heedfully met.[77] Via the openness of on-going time, which they dedicatedly live out, they allow the "staying" of the Fourfold, which the thinging of the thing brings to pass, to come to full accomplishment. Transpiring as attentiveness to things, their dwelling is as such the preserving husbanding-forth of the constitutive Fourfold. It is an enabling permitting, wrought out caringly in acknowledgement of intrinsic character, that lets the Fourfold in each of its components met in unity show itself and bring itself determiningly to bear *as* itself.

"Mortals dwell insofar as they save (*retten*) the earth," letting it manifest itself as itself in its myriad embodiments. They dwell "insofar as they receive the heavens," leaving to the seasons and the changeful weather their determinative roles and letting day and night, light and darkness rule unchallenged. "Mortals dwell insofar as they expectantly await (*erwarten*) the divine ones as divine ones," not seeking to reach or compel them and not attempting in their absence to make divinities for themselves. Dwelling, they simply wait trustingly in undemanding openness even when unsoundness and harm (*Unheil*) beset them. Steadfastly they maintain themselves in an expectant relation to the divine ones that they can themselves in no way fulfill. Therewith they hold toward the divine ones "the unhoped-for (*das Unverhoffte*)," that which, although hidden and unhoped for in itself and requiring to be given, yet informs the waiting-hoping that, grounded in prior relation, looks toward it and to which it may come. "Mortals dwell," finally, insofar as they themselves "accompany and guide" the enduring that is their own peculiar mode of existing, i.e., insofar as they, authentically living out openness for the happening of Being, rightly assess and direct their own unique capability for death. They neither impose a meaning on death nor orient themselves by death as though it were "empty Nothing." Rather, they willingly direct themselves into the ever encroaching sway of death as

---

[77]"Building, Dwelling, Thinking," VA 151, PLT 151.

death, i.e., of death as that sheltering fastness wherein Being, wholly concealed as Nothing, meets man and manifests itself by way of him.[78]

It is as dwelling that men as mortals carry out in specificity the unique role that Heidegger sees for man in his intricate belonging within, and belonging together with, the unfolding Twofold that is named as the Being of what-is. The centrality of "dwelling" as a name for definitive human comportment is tellingly given in the statement: "Dwelling is *the fundamental tendency* (*Grundzug*) of Being in accordance with which mortals are."[79] The "Being" of those who dwell answers immediately and without intervening disruptedness to the happening of Being as such. Through it the belonging-together of Being and man in immediate relating finds consummation. The Four of the single Fourfold—earth and heavens, divine ones and mortals—bespeak, each in its peculiar manner of carrying itself out, primal modes of Being's protean, self-initiating happening that are, while unitary, likewise constitutive of these four primary determinants of what-is.[80] And it is human dwelling as genuinely undertaken that so corresponds to that primal fourfold happening of Being as to permit the latter to accomplish its governance as it presents what-is as gathered to disclosure via the *thing*, which, in the genuine self-presenting that dwelling permits to it, lets the four determinants come into their own, in unity, i.e., lets them, in their changeful particularity and in their indissoluble belonging-together, come to light determiningly, in a single happening, in the opening-out of a world.[81]

---

[78]VA 150f, PLT 150f; "The Thing," VA 177, PLT 179. On the meaning of *retten* (to save), used here of mortals' dwelling vis a vis earth, see "The Turning," TK 41, QT 42.

[79]"Building, Dwelling, Thinking," VA 161, PLT 160. This statement should bring us to see dwelling as a direct, specifically named and delineated carrying out of human existing (*Dasein*) as such in its unvitiated mode of transpiring. On *Dasein* as an instantiation of Being, cf. Ch. II, pp. 59f.

[80]In *The Question of Being* (pp. 80f) Heidegger symbolizes this fourfoldness in the Happening named as Being by superimposing crossed diagonal lines on the word *Sein*. See ZS 80ff, QB 81ff.

[81]On the role of man as such vis a vis the Twofold, which is here made specific in a particular way, cf. Ch. II, pp. 61-63, and n. 22, and Ch. IV, pp. 152-154.

The happening of Being that rules thus complexly is precisely the letting presence that lets presence whatever presences[82] in traversing the open clearing of time, together with its fellows, in the gathered, ordered unity that "presencing" connotes. Every *thing*, together with whatever it gathers to itself in disclosure, presents itself by way of a particular on-going present,[83] even as do the mortal human beings who dwell caringly in company with it. It arrives into a particular while and departs. Every *thing* has its inception and runs its course. Hence, the dwelling of mortals in company with things has in no sense to do with that which offers itself in the static integrity of something already constituted and self-sufficient. The on-going meeting of mortals and things, of which the word dwelling speaks, is a meeting of the changing and vulnerable. Thus the receptiveness that inheres in dwelling as a husbanding is a receiving of that which requires so to be met as continually to be allowed to become itself.

The protective self-offering in openness as which, as dwelling, that husbanding takes place as a presencing (*Anwesen*) of man toward that which presences toward him[84] is an enduring unto ever new transpiring, ever new enduring-unto, on the part of that which is being met. And the gathering forth of the Fourfold to disclosure, which is accomplished via the husbanding dwelling of mortals in their sojourning in company with things, is ever wholly dynamic. It is an accomplishing marked always by the on-goingness toward what-is-yet-to-come in its arriving into its while that is entailed in Being's holding sway in reaching itself forth as presencing by way of ever-opening, ever-on-going time.[85]

Heidegger's word for the dwelling of mortals as viewed specifically with respect to this accomplishing of coming-to-disclosure, which is coming remainingly to be, is *bauen*. The root meaning that lies within this verb is, he says, to dwell, to remain. Earlier kindred verbs

---

[82]Cf. "Time and Being," SD 12, TB 12.
[83]Cf. Ch. I, pp. 49f.
[84]"The Age of the World Picture," HW 83, QT 131. On *An-wesen* as enduring-unto, see Ch. I, pp. 35f.
[85]Cf. Ch. I, pp. 39f.

194

Chapter V

evinced this meaning. It has been lost in the modern use of *bauen* itself, but it should be discerned as still present in related verbal forms—namely, in the first and second person singular indicative *bin* and *bist*, found in the complex conjugation of the German verb *sein*, to be.[86] *Bauen* as now ordinarily employed means both to construct or build and to till or cultivate. For Heidegger these meanings of the verb, which evince only implicitly the underlying notion of a bringing-to-be in the sense of a bringing to remain,[87] can be properly understood only from out of the meaning of dwelling.

Genuine dwelling, as a heedful remaining in rapport with ever-arriving, ever-departing, whiling things, carries itself out in the dual manner of which *bauen* now speaks. It transpires both as the nurturing of whatever grows of itself and as the erecting or fashioning of that which is "built."[88] Here it is that the dedication of mortals, in their dwelling, toward the permitting of things to be the things that they are brings itself concretely to bear, in the fostering of things that come forth of themselves and in the careful fashioning of those that require to be brought forth by another.

The reciprocity obtaining in the sojourning of mortals in company with things is most sharply focused at this point. Things gather to light the Fourfold of earth and heavens, divine ones and mortals, therewith bringing to pass by way of themselves the worlding of world. Mortals, toward whom, as those who are-in-the-world, that disclosure is offered via the nurturing building as which their dwelling takes place, permit or themselves bring to accomplishment the disclosive self-presenting of the things in question, through specific caring and specific deeds. Thus they, in company with things, carry out the staying (*Verweilen*) of the Fourfold, the bringing of the four to tarry and while in the worlding of world.[89]

---

[86]"Building, Dwelling, Thinking," VA 147, PLT 147; EM 54f, IM 59. English "be" and "been" are, clearly, from the same roots. Cf. *The American Heritage Dictionary of the English Language*, Boston, Houghton Mifflin Co., 1971, p. 1509.
[87]On Being as remaining, cf. Ch. I, pp. 35f.
[88]"Building, Dwelling, Thinking," VA 148, PLT 148.
[89]See "Building, Dwelling, Thinking," VA 151, PLT 151.

It is precisely in nurturing building (*Bauen*) that the dwelling of mortals husbands the Fourfold, protecting it forth into disclosure. In that in this dual manner it brings things to light and permits them to present themselves as the things that they are, dwelling brings to consummation the gathering-forth of the Fourfold that comes to pass via things; that is, it brings to consummation the whiling of the unitary four. It husbands the Fourfold—in husbanding things—into its own. On-goingly, it "preserves the Fourfold into things."[90]

3

The dwelling of mortals carries itself out by way of the nurturing building that lets things present themselves *as things*. Nurturing building fulfills dwelling; yet by the same token it permits dwelling to come to accomplishment, for through it the things are brought forth and cared for in company with which alone dwelling can take its on-going way. Only insofar as man builds and nurtures (*baut*) does he dwell.[91] The specific undertakings of building or nurturing that are accomplished vis a vis particular things are, however, not themselves of such a character as to be constitutive for dwelling as such.[92] The origination of dwelling, the opening up of the possibility of dwelling as the fundamental tendency of Being in accordance with which mortals are, lies, rather, in an exceptional undertaking that is also a bringing-remainingly-to-be (*Bauen*) but that as such carries itself out uniquely; via it the belonging-together of Being and man brings itself to pass with peculiar directness and peculiar initiatory power. This constitutive accomplishing is the province of the poet, whose poetic speaking is here intrinsically precedent to all the other achievings of men.[93] "Poetic composing (*Dichten*)," Heidegger can say, "is . . . original bringing remainingly to be (*Bauen*)."[94]

In accordance with this understanding and in testimony to it,

---

[90]VA 152; PLT 151.
[91]" . . . Poetically man dwells . . . ," VA 202, PLT 227.
[92]VA 191; PLT 217.
[93]VA 192; PLT 217f.
[94]*Das Dichten . . . ist das anfängliche Bauen.* VA 202; PLT 227.

Heidegger, in explicating the origination of dwelling, takes as his specific starting point the words of a poet, Friedrich Hölderlin: "Poetically man dwells upon this earth."[95] Dwelling, as which man's unique mode of existing is caught sight of in its transpiring as a bringing to light of the unitary Fourfold of earth and heavens, divine ones and mortals, takes place, whenever it does take place, in a poetical manner. Here the word "poetic" does not allude first of all to the production of poetry, which we

---

[95] . . . *dichterisch, wohnet/ Der Mensch auf dieser Erde.* VA 191, PLT 216.   In his essay, ". . . Poetically Man Dwells . . . , " Heidegger considers in detail lines 24 through 38 of Hölderlin's poem "In lieblicher Bläue blühet mit dem metallenen Dache der Kirkenturm (In lovely blueness blooms the steeple with metal roof) . . . " :

> *Darf, wenn lauter Mühe das Leben, ein Mensch*
> *Aufschauen und sagen: so*
> *Will ich auch seyn? Ja. So lange die Freundlichkeit noch*
> *Am Herzen, die Reine, dauert, misset*
> *Nicht unglüklich der Mensch sich*
> *Mit der Gottheit. Ist unbekannt Gott?*
> *Ist er offenbar wie der Himmel? Dieses*
> *Glaub' ich eher. Des Menschen Maass ist's.*
> *Voll Verdienst, doch dichterisch, wohnet*
> *Der Mensch auf dieser Erde. Doch reiner*
> *Ist nicht der Schatten der Nacht mit den Sternen,*
> *Wenn ich so sagen könnte, als*
> *Der Mensch, der heisset ein Bild der Gottheit.*
> *Giebt es auf Erden ein Maass? Es giebt*
> *Keines.*   (VA 194)

> "May, if life is sheer toil, a man
> Lift his eyes and say: so
> I too wish to be? Yes. As long as Kindness,
> The Pure, still stays with his heart, man
> Not unhappily measures himself
> Against the deity. Is God unknown?
> Is he manifest as the heavens? I'd sooner
> Believe the latter. It's the measure of man.
> Full of merit, yet poetically, man
> Dwells on this earth. But no purer
> Is the shade of the starry night,
> If I might put it so, than
> Man, who's called an image of the godhead.
> Is there a measure on earth? There is
> None."   (PLT 219-220)

ourselves might think of as something that could enrich human dwelling simply by making manifest the ranges and nuances of its occurring. "Poetic" bespeaks, rather, the primal accomplishing in receptiveness via which a human composing so brings language to the spoken word as directly to bring to consummation in the peculiar immediacy and historical specificity of particular utterance the dictation, the directing happening, of the Being of what-is and hence to disclose the latter's unfolding as the Twofold.[96]

Genuinely poetic utterance is, we may say, alone genuine speaking. In a sense it is incommensurate with all "everyday" speech, for the latter is not the standard by which to measure the poetic but is, Heidegger avers, itself only "a forgotten and used-up poem."[97] In poetic speaking, whether that speaking be found in poetry proper or in prose,[98] "language (*Sprache*)" as the mode as which—as primal showing-Saying (*Sage*)—the self-initiatory Happening centrally named as Being accomplishes itself as self-disclosing brings its ruling immediately to fulfillment via humanly spoken words.[99] Via such speaking the pivotally significant gathering forth of world and thing in the intricacy of their sunderedness as intimately united is centered, for via it there takes place a peculiarly pure achieving of that "naming" that ever summons into appearance.[100]

Language in its primal happening, as that which confers Being, carries pure, initiatory Happening directly into play in the latter's bringing of itself to pass as the unfolding Twofold.[101] The self-contained self-concealing, the Stillness, that, in its self-maintaining, happens itself forth as self-unconcealing—after the manner of the disclosive, enowning Bringing-to-pass that brings into belonging (*Ereignis*)[102]—through ruling as that uniting-sundering (*Unter-Schied*) that both allows the openness of differentiation and gathers into unitary disclosure, in thus disposingly

---

[96]Cf. Ch. III, pp. 133f.
[97]"Language," US 31, PLT 208.
[98]US 31; PLT 208.
[99]Cf. Ch. III, pp. 119-121.
[100]"Language," US 30f, PLT 208. Cf. Ch. III, pp. 99-101.
[101]Cf. Ch. III, p. 119.
[102]Cf. Ch. IV, pp. 146-148.

bringing itself to pass governs pivotally as "language" in the ultimate sense of the word. That language "is"—i.e., it happens provisively as disclosive of Being—"in that the uniting-sundering" that accomplishes stilling, the intensely powerful, self-restrained forth-happening as which Stillness accomplishes itself, "brings itself disclosively, enowningly to pass."[103]   The uniting-sundering   as   which   initiatory   Stillness accomplishes itself is itself a gathered, gathering calling-forth (*Geheiss*), a summoning into disclosure. It rules as the "gathered sounding of Stillness (*Geläut der Stille*)." And it is this gathered sounding, this forth-happening, word-intending providing, as "the soundless gathering call through which primal Saying opens the way for and brings to arrival the relating of world," that in the first instance the word *Sprache* (language) names.[104] In primal Saying, as the disposing enduring as which language holds sway, rests that way-opening-letting-arrive that holds the four world regions in "the single Nearing pertaining to their protective juxtaposing." And it is primal Saying that, in happening thus as the be-waying that disposes world's constitutive Fourfold, likewise gathers all things together into the Nearing that governs in just such juxtaposing among them.[105]

    Precisely this complex Happening focuses itself to fulfillment in the human speaking of language that, as a saying-after, speaks forth Being as named in the "is" and, as the final accomplishing of language, gathers to light whatever proffers itself in initiatory, pure Happening's self-unfolding.[106] Ruling as primal showing-Saying so as to dispose into its appearing whatever is, and as such coming to pass as the very mode, the cadence, as which enowning Bringing-to-pass (*Ereignis*), the self-

---

[103]*Die Sprache . . . ist, indem sich der Unter-Schied ereignet.* "Language," US 30, PLT 207. Since the word "Sprache" here names language solely as primally provisive and not as particularized as spoken words, we have taken this "ist" as transitive. Language, that is, "ises"; it happens initiatorially as the mode via which Being allows whatever is to arrive into unconcealing. Cf. Ch. I, pp. 30f. On uniting-sundering (*der Unter-Schied*) as a bringing to pass of stilling Stillness, see this chapter, pp. 180-182.
[104]"Language," US 30, PLT 207. Cf. also "The Nature of Language," US 215, WL 208. On the prefix *ge-*, as denoting gathering, cf. Ch. VI, pp. 258f.
[105]"The Nature of Language," US 216, WL 108. Cf. this chapter, pp. 183f & 189f.
[106]US 216; WL 108.

bringing-to-pass of Stillness itself, speaks, this "language" happens as the focal mode via which pure Happening fulfills its self-maintaining self-disclosure in its self-accomplishing as the Twofold, as it summons forth the answering speaking by way of which human beings proffer that which is the sine qua non of all disclosing whatever, the spoken word.[107]

Only via human speaking can any disclosure whatever take place, and the originative disclosing for which human speaking is thus "needed" centers peculiarly in the gathering forth of world by way of the world-embued self-presenting thinging of a thing.[108] It is just this self-presenting that is preeminently accomplished in poetic speaking. As found there, the thing's self-proffering may not be like that of entities with which we have ordinarily to do, which are concretely and presently present; for it may, instead, take place in a presencing that happens as absenting.[109] Nonetheless, as that wherein disposingly ruling "language" brings itself into play with peculiar immediacy, the gathering forth of world by way of things named that is achieved through poetic speaking accomplishes with special clarity the illumining both of whatever is called forth and of whatever crucial inter-relatings pertain to the disclosed.[110] For it is given to poetic speaking to utter words via which alone the self-bringing-to-pass of pure Happening, in its determinative quadrapartite bodying of itself forth as the Twofold, can directly achieve enowning disclosure such that the dwelling of human beings as mortals, i.e., as those who dedicate themselves to the husbanding of things and hence to the gathering forth into puissant enduring of the Four of earth and heavens, divine ones and mortals, that constitute world, can find fulfillment.

When Heidegger considers the poetic with specific reference to dwelling, both the immediacy within it of the ruling of pure Happening as

---

[107]Cf. Ch. III, pp. 119-121, and Ch. IV, pp. 154f & 147f.
[108]Cf. this chapter, pp. 174f & 179f. On the "need" for human speaking, see "Language," US 30, PLT 208.
[109]Cf. "Language," US 20f, PLT 198f.
[110]For an instancing of this role of poetic utterance, see "Language," US 28, PLT 205, where Heidegger considers the luminous disclosure of the Fourfold of world by way of "things," namely bread and wine, summoned in the poem that is under consideration.

such and the central instancing of that Happening via the transpiring of human existing come directly into view. At the heart of the poetic undertaking we find a human self-accomplishing that carries into play a severing distinguishing that is a genuinely disclosing uniting. In it initiatory Happening, as the primordial sundering-uniting (*Unter-Schied*) that in the happening forth of the Twofold first provides openness and, so providing, gathers meaningfully forth, brings itself to pass.[111]   That which is poetical that pertains to dwelling is a comprehending human seeing which, in bringing to bear discerning differentiation, first gathers the Fourfold to disclosure. On the basis of that seeing and from out of it the particular gatherings to light accomplished in dwelling are possible and specifically arise.[112]

The poetic seeing in question is, Heidegger says, following Hölderlin's words, a thorough measuring (*Durchmessung*). It is a perspicacious surveying of the "open between of heavens and earth" that is apportioned (*zugemessen*) to human dwelling.[113]   With this "between" we are not speaking of a spatial distance, but of a distance that is to be thought in terms of distinction and relationship. It is the surveying, measuring glance of man that opens the between in question, that undergirds and makes possible the turning-toward one another of the heavens and the earth, and that thus makes possible, in the context of their discerned relating, the defining of space and the sort of measuring with which we ordinarily concern ourselves.[114]

This thorough measuring, this comprehending assessing of earth and heavens in distinction and in relation, is apportioned to man as his intrinsic achieving.[115]   As such, it is the dimension (*Dimension*) that disclosingly unites heavens and earth.[116]   As that dimension, it instances

---

[111]Cf. Ch. I, pp. 29-33, and n. 36.

[112]On a discerning seeing as fundamental to the carrying forward of human existing, cf. Ch. II, pp. 173f.

[113]" . . . Poetically Man Dwells . . ." VA 195, PLT 220.

[114]VA 195; PLT 220.

[115]On the pivotal role of the capacity to distinguish as such, as intrinsic to and definitive of the transpiring of human existing, cf. Ch. II, pp. 77f.

[116]" . . . Poetically Man Dwells . . . ," VA 195, PLT 220.   Heidegger's

immediately the originating happening of the Being of what-is that, as the pure Happening that opens and illumines itself by way of time, in making manifest at once differentiates and gathers together. The manner of enduring, the constitutive mode of Being's on-going happening, that puts into play this dimension—this surveying that measures in comprehending—"is the cleared and lighted and thus encompassable accomplishable allotting (*durchmessbare Zumessung*) of the between." It is the opened-out bringing-into-play of the distinguishing-relating discerning which first lets transpire the meting out (the disclosing, that is, which precisely as differentiative shows the differentiate in unity) of "up-thither toward the heavens as down-hither toward the earth."[117]

Man *is* as man "insofar as he withstands the dimension,"[118] insofar, that is, as he withstands the opening of the "between"—which is intrinsic to his spanning seeing—as an opening-up via which at any given time disclosure of the Twofold is vouchsafed, yet for which no fixed, extrinsic support exists. That opening-up transpires as fundamental. It partakes of the primal precariousness in which man in his unique mode of existing surmounts, withstands, and consummates as an arena of disclosure the ever-opening clearing of time by way of which the ever-unfolding Twofold, which complexly unconceals itself as self-concealing, brings itself to light.[119]

Since man's on-going comprehending of the differentiating-gathering-into-relation of the heavens and the earth, while it pertains to

---

understanding of *Dimension* is in accordance with the meaning of *Durchmessung*, a noun of his own formation, from *durchmessen*, to traverse or, with separable prefix, to measure throughout. It is in keeping with the noun's origin from the Latin *metiri*, to measure, plus the intensive prefix *dis-*. For another context in which Heidegger takes "dimension" to speak of the accomplishing of disclosure, see "Time and Being," SD 15, TB 15. Cf. Ch. I, p. 39, n. 51.

[117]*Das Wesen der Dimension ist die gelichtete und so durchmessbare Zumessung des Zwischen: des Hinauf zum Himmel als des Herab zur Erde.* VA 195; PLT 220. On the meaning of *Wesen*, cf. Ch. VI, pp. 252-257. On the structure shown us here—a happening-away that disclosingly gathers toward—as characterizing the happening of the Being of what-is as such and as instanced in the manner of existing of man, cf. Ch. I, pp. 29-33, and n. 36.

[118]VA 198; PLT 223.

[119]Cf. Ch. II, pp. 74-76.

him as man, is nevertheless thus precarious, man must comprehend also and again intrinsically who he himself is and how his disclosive seeing is rightly undertaken. His way of properly being man must itself at any particular time be disclosed, be "measured," be gathered into the scope of his comprehension. The comprehending seeing as which poetic composing (*Dichten*), poetic bringing to language, transpires is, as the primal assessing "measuring" that man undertakes as man on the earth, under the heavens, itself the accomplishing wherein the needed measuring of man's subsisting as man is carried out.[120]

Once again the measuring in question is a gift vouchsafed. It takes place on the basis not of a self-reliant grasping but of a "letting-come (*Kommen-lassen*)."[121] Poetic speaking arises out of a comprehending that takes the measure of man in his uniqueness in that, taking place in responsive openness, it receives the "measure," the illumining disclosure, in light of which and indeed *in* which man is himself genuinely disclosed. It is from out of the sui generis initiating as which the Happening that unfolds itself as the single Twofold disclosively enowns in letting belong together that this measure-taking, which lets man disclose himself to himself, is given. "In poetic composing, the taking of the measure brings itself to pass in such a way as disclosingly to enown and let belong together (*ereignet sich*)."[122]

The measure (*Mass*)—the assessing disclosing—that brings itself to pass in the poetic, in being vouchsafed thus to man from beyond himself, has to do with the whole of that gathering surveying, the opening-uniting "dimension," that is brought into play in this disclosive human seeing.[123] Accordingly, it pertains to the whole range of takings-place as which the Twofold, which we name the Being of what-is, unfolds itself. And it pertains first of all to the unitary Fourfold—earth and heavens, divine

[120]" . . . Poetically Man Dwells . . . ," VA 196, PLT 221.
[121]VA 199; PLT 224.
[122]*Im Dichten ereignet sich das Nehmen des Masses.* VA 196; PLT 221. The extended rendering of *ereignet sich* here offered as a translation springs out of the discussion of *Ereignis* and *ereignen* presented in Chapter IV, on pp. 149-151.
[123]" . . . Poetically Man Dwells . . . ," VA 198, PLT 223f.

ones and mortals—that is the primal instantiation of pure Happening in its bringing of itself to pass as shaped happening in the constituting of what-is.[124]   Hence, when in poetic seeing and poetic speaking the assessing disclosure vouchsafed is carried into play in an assessing disclosing that opens the way genuinely for human dwelling, man comprehendingly sees himself as the mortal upon the earth, under the heavens, vis a vis the divine.   And in this he sees himself, ultimately, in relation to the Being of what-is, in accordance with which the single Fourfold, in all its manifold particular centerings, can alone come genuinely to light.

The "measure," the defining disclosure, that is received by man and carried into play in the accomplishing of poetic speaking is provided him via his co-partners in the Fourfold.   In his explication of Hölderlin's lines that let us see decisively man's dwelling as poetic, Heidegger sets forth detailedly the interrelating of the Four (before they had as yet been named by him as one) in the accomplishing of the poetic disclosing that is for him there crucially in question.   So central are the Four to Hölderlin's utterance and so intricate and fundamental are the relatings that Heidegger there discovers among them that one must wonder, indeed, whether Heidegger's discerning of the Fourfold—earth and heavens, divine ones and mortals—may not perhaps have arisen largely from his considering of just these poetic words.[125]

"Man measures himself, matches himself," Hölderlin says, "not unsuccessfully against the divine."[126]   The disclosure via which man rightly sees and can undertake his role as man is the disclosure of the divine.   In that disclosure and through it alone primordial relationships and modes of self-presenting come to light.

Hölderlin speaks specifically of God, asking concerning his manifestness.[127]   And yet, as Heidegger emphasizes, "God is, as he who

---

[124]On such instantiation, cf. Ch. I, pp. 28f.

[125]Other sources may certainly have contributed as well. In the *Gorgias* (507-508), Plato, for example, speaks inclusively by referring to "heaven and earth, gods and men."

[126]" . . . *misset/ Nicht unglüklich der Mensch sich/ Mit der Gottheit.*" ". . . Poetically Man Dwells . . . ," VA 194, PLT 219.

[127]VA 194; PLT 219. Although Hölderlin can speak of "God," he thinks of a plurality

he is, unknown for Hölderlin."[128]   The decisive disclosing with which we have here to do is the mysterious disclosing of the unknown *as* unknown.[129]

That manifesting happens by way of the manifestness of the heavens,[130] and it bears in itself the illumining measure that is now in question. The latter is neither God nor the heavens nor the manifestness intrinsic to the heavens. It consists, rather, "in the manner in which the god who remains unknown is *as* such a god manifest through the heavens."[131]   Here the reference to the heavens is not solely to the heavens. In this context where primordial measuring, primordial gathering to light in differentiation and relation, is in view, the oneness of heavens and earth as seen comprehendingly by man is assumed by Heidegger as fundamental. Thus he avers that "the looks, the aspects (*Anblicke*) of the heavens" include all that in any way takes place and manifests itself "in the heavens and accordingly (*somit*) under the heavens, and accordingly on the earth."[132] The whole of reality is participant in this constitutive disclosing.

Disclosure of the god is vouchsafed to the poet, disclosure, that is, of the exalted otherness that belongs to him as God. That otherness is shown by the heavens, whose phenomena bespeak and embody a ruling power whose ways they mirror, which, as it were, clothes itself in them. "The countenance of the heavens," Hölderlin says, "is full of characteristics of him [God]." "Lightnings," fiercely illumining and rife with destruction, "are the passion of a god.[133]" Everything that presents itself in the unified arena as which the heavens and the earth are gathered

---

of gods, (cf. e.g., "Remembrance of the Poet," EHD 19f, EB 253f), and he speaks of "the god" (e.g., EHD 21, EB 255). Therefore his allusions to "God" should never be taken as references to the God of Christian faith.

[128]VA 197; PLT 222.
[129]VA 197; PLT 222.
[130]VA 194, 200; PLT 219, 225.
[131]VA 197; PLT 223.
[132]VA 200; PLT 225.
[133]VA 200; PLT 225.   These lines are, Heidegger says, from a poem closely associated with "*In lieblicher Bläue blühet . . . .*" He does not name it.

to light by comprehending, receptive human seeing declares the god yet shows him as other than itself.

Here there takes place a guarding and a preserving of the god, himself unseen. He confirms his otherness, his alienness, precisely in so yielding himself to those intimately familiar appearances that continually meet man as to present himself on-goingly and to be sustained in that presenting, albeit as hidden and "unknown."[134]

This appearing of the god, his appearing as "the unknown," via the manifestness of all that is inclusively named in the phrase "the heavens," is the poetically apprehended disclosure, the measure, in accordance with which man can rightly assess himself. Comprehendingly, man meets and receptively beholds what-is, in a seeing intrinsic to which is insight into his own appointed role. The key to this insight is the self-authenticating awareness that whatever so familiarly and manifestly presents itself attests the prior, providing self-presenting of a god for whom manifesting is but a cloak to make hiddenness and the otherness of supremacy known.

The appearing that characterizes the god differs decisively from the "mere appearing" that characterizes that which, belonging to the heavens and the earth, presents itself to ordinary, unpoetic seeing.[135] For "the appearing of the god by way of the heavens"—and that means by way of all the manifestations of earth that are gathered into one with those of the heavens—"consists in an unveiling *(Enthüllen)*." This unveiling does not take place as a simple self-manifesting such as that on which man so readily relies in all that is familiar to him. It "lets be seen that which conceals itself," but in so doing it in no way seeks to tear the concealed forth out of its concealedness; rather, it lets the concealed be seen only in that "it guards and preserves *(hütet)* the concealed in its self-concealing."[136] This is an unveiling wherein veiling—protective keeping hidden—is primary in the manifesting that is brought to pass.[137]

---

[134]VA 200; PLT 225.
[135]VA 200; PLT 225.
[136]VA 197; PLT 223.
[137]Although Heidegger does not hyphenate the verb *enthüllen* as he so often does with verbs whose components he wishes to expose so as to draw out concomitantly the

Here we come directly upon the happening of the Being of what-is in accordance with which the god as unknown *is*. Heidegger suggests his allusion to Being only through the capitalizing of a demonstrative pronoun. In the manifold "looks (*Anblicken*) of the heavens," as seen poetically and brought to light in poetic speaking, "the poet summons That which in self-unveiling," i.e., in a mode of happening that so hides as to disclose and preserve *as* hidden, "lets that which conceals itself appear"—appear, that is, "*as* that which conceals itself."[138] In every transpiring of pure Happening as unconcealing, as that Happening brings itself to pass as the Being of whatever is, there is in play a concealing, always prior and ever to be surmounted yet ever instanced in that very surmounting.[139] Just this Happening is here unobtrusively shown us as that in accordance with which the god so presents himself, so appears, as, through his appearing, to provide for man, who sees, the disclosure in whose light alone all self-disclosure can be rightly apprehended.

The god "remains" as god, i.e., on-goingly is as god by way of time,[140] from out of a self-protecting self-concealing that maintains itself only through manifesting itself as such concealing. Thus the god endures most immediately, we might say, from out of Being. That is, the god directly instances the Being of what-is, as Being is caught sight of most fundamentally when an attempt is made to consider it in itself.

Here we must recall Heidegger's asseveration that when the god appears, he appears uniquely "from out of the holy sway, the wholeness-bestowing sway [of divineness (*Gottheit*)]."[141] When pure Happening, as the Being of what-is ruling as divineness, centers and shapes itself particularizingly as a god, it is, as the holy, directly in play as the self-

---

meaning resident in each, he is clearly relying in the context here under discussion (". . . Poetically Man Dwells . . . ," VA 197, 200; PLT 223, 225) on the specific force of the verbal stem and the prefix as these would be discerned separately. The prefix *ent-* means "forth" or "away from." (Cf. Ch. I, p. 24, n. 12.) We have here to do with a veiling (*Hüllen*) that, *as* a veiling, happens as a disclosing, as a veiling-forth.
[138]VA 200; PLT 225. For a similarly subtle use by Heidegger of a capitalized That, see "The Question Concerning Technology," VA 13, QT 4.
[139]Cf. Ch. I, pp. 24-26 & 43.
[140]On the happening of Being as remaining, cf. Ch. I, pp. 35f.
[141]"Building, Dwelling, Thinking," VA 150, PLT 150. See this chapter, p. 176 and n. 26.

providing plenitude that first of all preserves itself in happening as a self- protecting self-concealing but that just in so doing unconceals itself, gathering itself to light via on-goingness, so as to maintain itself in its primordial withholding of itself to itself.[142]

When, via poetic speaking, the god is brought to light *as* unknown, i.e., as the one who conceals himself, the self-unconcealing self-concealing as which Being rules in whatever is, is proclaimed and hence is manifested as the disclosive self-preserving self-concealing of the god.[143]   When the poet "so says the looks of the heavens [inclusively understood]" as to present the appearances in question not in familiar guise but in the strange, the radically unfamiliar mode of that in which hiddenness is manifest as such, so that they themselves stand forth as the alien, the strange (*das Fremde*), it is the self-concealing, unknown god who is presenting himself.[144]   In the peculiarly disclosive looks of such poetically presented appearances, it is precisely the god who is occasioning surprise and consternation; it is he who is manifest as radically unfamiliar, alien, strange.[145]   And just in this surprising manifesting as alien, the god hiddenly consummates his "remaining" in accordance with Being in the full ranges of its happening, i.e., in accordance with the self-initiatory Happening that, in coming to pass self-differentiatingly as self-unconcealing self-concealing, first and above all lets belong together and gives into relation.   For precisely in this

---

[142]Cf. Ch. I, pp. 24-26 & 42f. Heidegger's depiction of "the Holy" can parallel his portrayal of Being closely enough so as to show clearly the identity of the Holy as a mode of pure Happening's bringing of itself to pass rulingly as the Being of what-is. See, e.g., "'Wie Wenn Am Feiertage . . . ,'" EHD 71: *Das Heilige aber, über die Götter und die Menschen, ist "älter denn die Zeiten." Das Einstige, allem zuvor das Erste und allem nachher das Letzte, ist das allem Voraufgehende und alles in sich Einbehaltende: das Anfängliche und als dieses das Bleibende.* "But the Holy, 'beyond gods and men,' is 'older than the ages.' The most unique, first before everything, and last after everything, is what precedes all and encompasses all within itself: what is the original and, as the latter, what remains." (There is no published English translation of Heidegger's essay, "'Wie Wenn Am Feiertag. . .'") For a portrayal of the happening of Being, cf., e.g., Ch. I, pp. 29-36.
[143]". . . Poetically Man Dwells . . . ," VA 200, PLT 225.
[144]VA 200; PLT 225.
[145]*Durch solche Anblicke befremdet der Gott.* VA 201; PLT 226.

manifesting of himself as alien and other and unknown, the god "evidences his unceasing nearness."[146]

Poetic speaking, then, lets the appearances of the heavens—and hence also of the earth—be seen and in those appearances lets the self-concealing, unknown god be seen in his otherness and his nearness. Heidegger uses the word *Bild* (image, figure, picture) to speak of this layered convergence of disclosure. *Bild* is, he says, the familiar word for the aspect or look (*Anblick*) of something. Such a picturing image takes place as a depicting look that lets something presently invisible, which is other than the image, nevertheless be seen. "It thus pictures or images (*einbildet*) it," the invisible, "into something alien to it (*in ein ihm Fremdes*)."[147] Just so do the looks of the heavens, together with those of the earth, disclosively image the invisible, unknown god.

Here we must carefully note that the imaging in question is not resident in those looks themselves, as though it could be immediately discerned by anyone looking upon them. Rather, it is an imaging that pertains to the looks of the heavens *as poetically said*. The measure, the assessive disclosure, that the poet as poet receives consists in such imaging. Therefore his poetic utterance "speaks in 'images.'"[148]

These images are no mere imaginings, no mere phantasies. They are, rather, "imaging gatherings-forth." That is, they are "imagings-into (*Ein-Bildungen*)" that transpire "as visible inclusions of what is unfamiliar, other, strange, in the looks of what is intimately familiar."[149] Thus, Hölderlin's words—the striking image that Heidegger cites— "lightnings . . . are the passion of a god," are no mere empty figure. In them the familiar phenomenon of lightning appears in mysterious guise; it

---

146VA 201; PLT 226.
147VA 200; PLT 226. In regular German usage, *einbilden* is a verb with separable prefix that is used with a reflexive pronoun in the dative to mean to imagine, to think or believe, to be conceited. Heidegger uses the verb with an inseparable prefix and with its object in the accusative. Clearly he wishes to avoid its usual, very subjective, connotations.
148VA 201; PLT 226.
149 . . . *Ein-Bildungen als erblickbare Einschlüsse des Fremden in den Anblick des Vertrauten.* VA 201; PLT 226.

manifests the unseen, unknown god via whom—as a participant in divineness—in urgency and power bestowal comes; and in so doing it brings to light ultimately in speech—the pivotal mode of disclosure—the unfolding of the Twofold, the Being of what-is, as, in the juncture of time wherein the poetic words speak, the latter is most fundamentally bringing itself to pass.

The image of the self-protecting self-concealing god is found decisively, then, not in the appearances of the heavens and in those of the earth that are gathered into one with them, but in the obediently uttered human saying, the spoken depicting, that, in accordance with the immediately governing Happening of Being, gathers all that is encountered forth to light and determiningly into play in the boldly proclaiming poetic image.

Heidegger's consideration of image (*Bild*) moves finally to the citation of Hölderlin's lines that immediately follow Hölderlin's crucial assertion concerning the poetic character of man's dwelling upon earth. "No more pristine (*reiner*)," says the poet, "is the shadow of the night with its stars . . . than is man, who is called an image of deity."[150] The shadow—the night—attests the light. It is, as shadow, cast by the light and ascribed to it. In like manner the "measure" that poetic speaking (*Dichten*) takes, i.e., the assessive disclosing that it brings into play, attests the god. It witnesses to and manifests the god in his otherness. It announces and shows forth reliance upon the divine. It provides itself as what is strange (*das Fremde*). For it is a mysterious and unwonted disclosing into which the god, as invisible, husbands and preserves (*schont*) his remaining as god, precisely through his preserving of himself "into the familiarness of the looks of the heavens" as those looks, and he with them in hiddenness, take place as image—the image, namely, that transpires as the poetic word.[151] Even so can man—man as poet, who speaks and dwells poetically on the earth—be said to be an image of deity, a manifesting witness, in the speaking of language—which belongs to him

---

150VA 201; PLT 226.
151VA 201; PLT 226. For another portrayal of poetic speaking as fulfilling a mediatorial role, cf. Ch. XVI, pp. 651f.

and uniquely characterizes him as man—to the hidden yet self-evidencing sway of the divine.

In the imaging-forth that man as poet accomplishes, the role of the heavens is crucial. It is the heavens as poetically disclosed, and that means as disclosed in immediate relation with the earth, that offer, as imaged forth, the manifesting in accordance with which, alone, man can comprehensively see and assess, in its complex interrelatedness and in the mystery of its happening, whatever is. For the heavens manifest the ruling god, manifest him as hidden, yet as he who bestows and disposes. Brightness and illumining are the hallmark of the heavens.   Yet, Heidegger says, "the heavens are not simply light." In the very brightness of the heavens, which seemingly delivers everything up to thorough visibility, darkness that conceals and preserves as inviolate remains in play. The brilliance of the heavens' shining height is at once "in itself the obscure mysteriousness (*das Dunkle*) of its all-sheltering breadth."[152] The blueness that most immediately identifies the heavens for us is none other than the color of unfathomable depth.   And the heavens' brightness is never fixed and uniform; it is, rather, an on-going, ever-shifting brilliance. As brightness it is the rising and setting that ever belong to the obscuring twilight that, one with the manifesting light, "protectingly harbors everything proclaimable," everything that can be announced and shown.[153]

It is precisely these heavens in all their mysteriousness that the poet images forth in his speaking. His saying of spoken images, which so transpires as to accomplish poetic presenting, "gathers the splendor and the sound of the appearances of the heavens into one with the darkness and silence belonging to what is other and alien and strange," belonging, that is, to the hidden god who in nearness rules. This disclosure, i.e., the heavens as said and shown in poetic image, is most immediately the provided measure that poetic presenting at once receives from out of the happening of the Twofold and proffers as a fundamental manifesting via

---

[152]VA 201; PLT 226.
[153] . . . *die alles Verkündbare birgt.* VA 201; PLT 226.

which the Twofold, in its manifoldness, brings itself to light.[154]

Once again, in the passage just considered, the Being of what-is is not specifically mentioned by Heidegger. Attention centers only on the mystery-fraught disclosing of the god who, *as* disclosed, remains unknown. Nevertheless, the thought of the ruling happening of Being is clearly present in Heidegger's words. An unconcealment (*Unverborgenheit*) that, in the poetic imaging of the heavens, discloses their depths precisely as hiddenness; an unconcealment whose happening as structured by an ever-on-going withholding in the very providing of the disclosure that itself protects by concealing (*verbergen*) even as it manifests—all this speaks plainly of the disposing Happening named as the Being of what-is, which as a harboring-forth, a revealing (*Entbergen*) that is at once a concealing, on-goingly comes to pass as a letting-appear that ever rules as a mystery, the mystery, namely, of keeping to self and arriving and departing.[155]

In imaging what *is*, the poet's words say and show it precisely in its Being. In the manifestness of the heavens as poetically said, a ubiquitous throwing into relief of the light by way of the obscurity and mystery, the darkness and hiddenness, that everywhere interpenetrate and bound it rules as decisive. Such disclosing breaks into and transmutes the inadequate human discerning that sees everything as readily familiar. The insuperable strangeness of the changeful, the undiscoverable, the unplumbable, marks every appearance of the heavens and every appearance of the earth, as earth and heavens are gathered into relation and the former, unlighted of itself, is made manifest in accordance with the latter's variously shadowed light. And in this manifestness of the

---

[154]VA 201; PLT 226.

[155]Cf. Ch. I, pp. 24-26, and n. 12, and p. 41. Heidegger's portrayal of poetic disclosing should remind us of his early juxtaposing of a preoccupation with the familiar to an awareness of mystery and concealedness, in a context where the transpiring of erring is in question (cf. Ch. II, pp. 92f), and also of his juxtaposition of the easy familiarity of the "they" regarding everything known, in the everyday, inauthentic transpiring of human existing (Ch. II, pp. 88f), with an authentic awareness of uncanniness, of nullity, of a being-without-abode in the world (Ch. II, pp. 94f). For clearly, like elements are to be found here in his discussion of poetic disclosing as taking place in contrast with ordinary, undisclosive seeing.

heavens that transpires ever in conjunction with a protective, sustaining hiddenness, the god, the supreme bearer of bestowal, whose remaining near is a self-concealing, in his otherness in which the happening of Being is immediately in play finds a fitting arena of disclosure. Precisely in manifesting himself—his own characteristics—in the spoken images that picture forth the heavens, he makes known his hiddenness, which is imaged forth, spoken, and shown in accordance with the strange mystery of the heavens. The god's announcing of his hidden nearness, too, like all else that is proclaimed, is protected, harbored, and vouchsafed in inviolateness through being uttered in a disclosing that acknowledges and displays the mystery of concealment, even as it brings to light.[156]

Thus does man, in the accomplishing of poetic presenting, receive and bring to bear the assessing disclosing, the measure, that carries inherent in itself a genuine adjudging of what-is considered according to its constitutive particularizations as these latter meet man and are met and received by him directly in obedient openness after the manner of his peculiar capacity for disclosure by way of language. This disclosing permits to the poet, and to any who, in attending to his presenting, see and hear with him, an apprehending that rightly makes manifest, through rightly differentiating and gathering into relation, that which at the juncture of time in question presents itself and is in view. But more than this, this disclosing permits a rightly assessing human apprehending of seeing, listening, speaking man himself. Whoever genuinely receives and brings to fulfillment the intricate disclosing that is vouchsafed to him in poetic accomplishing apprehends and acknowledges his role even as he carries it out. In an awareness that is self-awareness, he comports himself as the one who, via imaging utterance, gathers to genuine manifestation— i.e., lets come to *self*-manifestation—the heavens and the earth, whose on-going taking-place in interrelation opens the arena for human doing, *and* the divinity, peculiarly other and hidden yet self-proffering, whose

---

[156]" . . . Poetically Man Dwells . . . ," VA 201; PLT 226. In this connection one should note Heidegger's use, in a single complex paragraph, of closely allied words: *bekunden*, to speak of the god's declaring of his ceaseless nearness and *das Verkündbare*, to speak of "everything proclaimable that the twilight harbors."

governing bestowals the takings-place of the heavens and the earth show forth. It is in this complex assessive disclosing in which man, in accomplishing the poetic, apprehends and brings into imaging utterance— which is itself a disclosing of himself to himself—that man receives and brings into play the measure, the meting-out, that opens to its proper scope the mode of enduring, the mode of determiningly comporting himself, that is properly his as man.[157]

Here man, accomplishing his poetic role under the heavens, upon the earth, catches sight of and discloses himself via appearings that bespeak hiddenness and mystery. He accepts and surrenders himself to the precariousness of receiving only such disclosure as is bounded and shadowed by a concealing that ultimately withholds from his grasp whatever is manifest. He acknowledges the ruling of a god who, as other, declares his nearness by manifesting himself as hidden. And turning receptively toward that manifestation, he affirms the fullness that it bespeaks and does so in a concomitant awareness that he too is wholly dependent on the bestowings vouchsafed from out of the holy sway of the divine.

Here man, seeing and speaking poetically, catches sight, in this standing-in-relation, of the extremity that belongs inherently to his proper comporting of himself as man. He knows himself as the one who is called upon in assenting awareness to pursue on the earth and under the heavens a way that is bounded by concealment, by a withholding of disclosure. And he knows that that concealing, that eclipsing of self-manifesting, encroaches radically upon himself. He is the one—and he only—who, in radical distinction from the god whose self-manifesting lets him truly see himself, is summoned *in awareness* to abjure every claim to provide for and to secure himself, and who is hence summoned so to pursue his way as to surrender himself to that concealing as it constantly claims him and as it awaits him in ultimacy in the befalling of death.

Man takes his way as man "as the mortal," that is, "he is capable of death as death." Continually, indeed, he dies. But this capability is

---

[157]VA 196; PLT 221.

not simply an intrinsic possession of man. It rests in that poetic seeing
and presenting that, in a given juncture of time, gathers heavens and
earth, the ruling god, and man as the receptive, witnessing mortal forth to
genuine disclosure via the disclosively imaging, mystery-affirming poetic
word.[158] And where that word is spoken and heard, man, precisely in
discerningly assuming his role rightly vis a vis all else that is in the
particular juncture in which he speaks and sees, participates in the
illumining self-disclosure of Being and makes his way in right relation to
Being's ruling sway.

In keeping with the utter contingency of man's genuine pursuing of
his way as man, the possibility of poetic human comportment, i.e., of the
poetic seeing and speaking that thus gathers disclosingly forth, is wholly a
gift. Genuine poetic presenting and self-presenting takes place only, in
Hölderlin's words, "so long as kindness, the pure, stays with his [man's]
heart.[159]" This pure kindness is, Heidegger says, a graciousness (*Huld*,
Greek *charis*), which is such as to evoke a response like to itself. Man
comports himself poetically only inasmuch as this graciousness has come
intimately close to the peculiar on-going mode of taking his way via time,
in relation with the unfolding Twofold, that belongs to man as man. It
arrives as the enabling claim (*Anspruch*) of the assessing disclosure, the
measure, that reaches man so fundamentally that he, in an ultimacy of
response, obediently heeds that disclosure, allowing it to attain
consummation through his own discerning and his own utterance and,
precisely in so doing, taking his way in accordance with it.[160]

In this pure graciousness, *Ereignis*—the sui generis initiating
Bringing-to-pass most interior to the Happening that is the unfolding
Twofold—is, although not specifically named, immediately in play. The
poetic, in accordance with which man comports himself in responding to
that graciousness, "brings itself enowningly to pass (*ereignet sich*)."[161] It

---

158VA 196; PLT 221.
159 . . . *So lange die Freundlichkeit, noch/ Am Herzen, die Reine, dauert.* . . . VA
203; PLT 228.
160VA 203; PLT 229.
161VA 204, 203; PLT 229, 228. Cf. also VA 202, PLT 227. For another allusion to
graciousness in a context in which *Ereignis* is in question see "A Dialogue on

lets belonging-together transpire with primal immediacy and hence lets the constitutive participants in the manifold of what-is—the heavens and the earth, the divine ones and mortals—come, via mutual interrelating, into their own. And in this it lets man genuinely belong to the Happening named as the Being of what-is.

The poetic itself comes to pass, after this manner, from out of what we may venture to call a prior happening of primordial letting-belong-together, i.e., from out of the letting-belong-together of Being and man. Man is only capable of genuinely poetic comportment insofar as the determinative enduring that pertains to him as man "is given to belong (*vereignet*)" to Being, i.e., to that which itself "is inclined toward man," and which therefore, thus happening in relation, "needfully uses" as the vehicle of its self-disclosure that very mode of enduring in openness by way of time that marks man as man.[162] In the pure graciousness now in question, the utter primordial giving as which the initiatory disclosing whose ruling ever gives the ruled into its own in giving it into belonging reaches and intimately touches man in his pursuing of his way.[163] It does so as the claim (*Anspruch*) via which the measure—the manifesting of insuperable hiddenness which genuinely enables human seeing and saying—that is received and accomplished by poetic presenting turns man, himself ruled by gracious giving, obediently toward itself.[164]

The bringing to light accomplished by the poet, then, carries

---

Language," US 144, WL 47.

[162]*Aber der Mensch vermag das Dichten jeweils nur nach dem Masse, wie sein Wesen dem vereignet ist, was selber den Menschen mag und darum sein Wesen braucht.* (VA 203; PLT 227.) A decisively parallel statement with direct allusion to Being (*Sein*) occurs in "The Turning," (TK 38; QT 38f): . . . *zum Wesen des Seins das Menschenwesen gehört, insofern das Wesen des Seins das Menschenwesen braucht, um als Sein nach dem eigenen Wesen inmitten des Seienden gewahrt zu bleiben und so als das Sein zu wesen,* . . . On the "need" spoken of in the verb *brauchen* as used of Being, cf. Ch. I, p. 53, and n. 107.

[163]Cf. Ch. IV, p. 151.

[164]" . . . Poetically Man Dwells . . . ," VA 204, PLT 229. Heidegger often uses *Anspruch* to speak of the way in which ruling Happening, in various modes of governance, meets and lays claim to man. Cf. e.g., "The Question Concerning Technology," VA 27, QT 19. Cf. Ch. VI, p. 259.

crucially to completion via the openness that man's manner of existing by way of time alone provides the ultimate disclosing as which pure Happening so brings itself to pass as to grant all manifesting, all gathering-forth in orderly interrelating whatever. In his poetic presenting, when that presenting genuinely takes place, man is so given to belong to the Being of what-is that the unitary Fourfold as which the latter particularizes itself constitutively in unfolding as the Twofold is gathered to disclosure uniquely and with peculiar determinative clarity. Thus it is that human dwelling, the human comportment to which it belongs to gather the Fourfold to disclosure on-goingly by way of things, rests upon and transpires as an instancing of the poetic. And thus it is that poetic presenting, transpiring as—in this dual sense—"the fundamental capability (*Grundvermögen*) of human dwelling,"[165] takes place in a unique, initiatory manner as the bringing-remainingly-to-be (*Bauen*) that at once permits dwelling and carries it out.[166]

In poetic presenting, things, the erstwhile familiar, various things of the heavens and the earth, are so apprehended and said as to present themselves with singular immediacy as the things that they are and to stand forth as the loci wherein world brings itself to light. In the things imaged in the words of a genuine poet the world beheld appears as for the very first time.[167] Standing forth astonishingly, those things manifest the prior hiddenness attested in their appearing newly brought to pass, and, as possessed of such genuine appearing, they so take place as imaged forth in the poet's speaking as to permit and accomplish in themselves the gathering on-goingly to light of the Fourfold of heavens and earth, divine ones and mortals as which pure Happening, as the unfolding Twofold, brings itself to pass as the worlding of world. As such poetically shown appearances, these "things" take place most immediately and fundamentally as the things that they are. Via these things that it

---

[165]VA 203; PLT 228. For an instance of such poetic presenting, cf. Ch. XVI, pp. 646-650.
[166]Cf. this chapter, pp. 193f.
[167]HHF 25; HFH 96. Cf. "The Origin of the Work of Art," UK 43, PLT 43, where the word "thing" has a somewhat narrower application.

apprehends and says, poetic presenting at once receives and puts into play the assessive disclosure, the measure, whose carrying out is the peculiar province of every genuinely poetic undertaking.

Poetic presenting itself transpires as "measuring (*Vermessen*)" in the most fundamental sense of that word. Unlike ordinary measurings, that presenting carries no ready-made standard against which to measure something whose coming to be it has already planned out. Rather, it itself is a receiving of meaningful disclosure, such that the meaningfulness apprehended so informs the apprehending to which it comes as to render the disclosing in question, as humanly appropriated and brought to utterance, an assessing that manifests as imbued with that meaning that which is being shown forth in words and, in the crucial arena where speaking and hearing take place, decisively brought to light. Heidegger early averred that man receives the standard (*Richtmass*), i.e., the underlying determinant, of his speaking and acting from what-is as disclosed to and through him.[168] The seeing and hearing of the poet, his reflecting upon meaning, and his disclosing of the latter via his imaging words adjudge and present as possessed of exceptional import that which comes to appear in poetic utterance, even as the poet Hölderlin says and shows lightning as manifesting the ruling of a god. In the poetic comportment that transpires as a letting come remainingly to be (*Bauen*)— as, in its saying, it lets things appear genuinely, in the fullness of their meaning as gatherers-forth of the Fourfold—the foundation is laid for the dwelling of man, i.e., for the on-going human comportment via which, in company with and by way of things variously built and nurtured, the Fourfold is, in accordance with the illumining received and vouchsafed through the poetic, in juncture after juncture variously brought to light in the meaningful gathering-forth of what-is.

Transpiring thus, poetic presenting receives and contains in its very transpiring a disclosure of such dwelling itself, as that dwelling takes place in interrelation with all that is, i.e., with whatever at any given time requires to be gathered forth to self-manifestation. Poetic presenting

---

[168]"On the Essence of Truth," W 80, BW 124; W 91, BW 135.

encompasses in the comprehending seeing that characterizes it the scope of man's dwelling. As this disclosive assessing, poetic presenting first lets man's dwelling come to be as dwelling. It first permits it, that is, to take place as a genuine responding—informed by awareness—to the Happening that, as the unfolding Twofold, the Being of what-is, is claiming it, bestowing it, and calling it forth. As such, that presenting most originally puts into play the enabling bringing-to-self-manifestation that the verb *bauen* as most amply understood bespeaks.[169] Every reciprocal relating between dwelling and the nurturing building that at once carries dwelling into accomplishment and allows it to transpire on-goingly, through providing the things via which the disposing, on-going bringing to light of the Fourfold of heavens and earth, divine ones and mortals can bring itself to pass, must spring ultimately, if it be genuine, out of the disclosing of relationships and meaning that is vouchsafed via the speaking of the poet.

In genuinely poetic seeing and speaking, which must be received and carried into play by those who hear and therewith see as the poet sees, there rests ultimately the possibility of man's pursuing his way, on the earth, under the heavens, properly as man, in an obedient attentive dwelling in close company with things. In poetic speaking, the singing-forth of the initiatory disclosing that enowns in bringing into belonging (*Ereignis*) as which language most primally comes to pass[170] accomplishes itself immediately in the spoken word, as man is given to belong, in a fullness and immediacy of responsive commitment, to the Being of what-is, and the latter, via complex shapings and centerings of itself in an intricacy of interrelating, brings itself to pass with a peculiar directness that, in illumining and humblingly claiming man, in face of the concealing that alone grants and unconceals, lays in imaging language the foundation for every genuine disclosing of the Twofold that man, living out openness for disclosure by way of time, can hope to achieve.[171]

---

[169]Cf. " . . . Poetically Man Dwells . . . ," VA 202, PLT 227.
[170]Cf. Ch. IV, pp. 154f.
[171]Cf. Ch. IV, pp. 152-154.

Heidegger's consideration of human dwelling in company with things turns our attention to the sphere that might perhaps be most accessible and congenial to our own thinking, the sphere, namely, of our life among the things, natural and man-made, that surround us. Yet it is hard to find his portrayal commensurate with the life that we ourselves know. Do the words of poets ever determine the manner and course of our existence? Are there any such words to which we accord the power to disclose reality most truly? Is it not the practitioners of the sciences, and certainly not the poets, who can and do give us most relevant and accurate information regarding the character and significance of the things and events of our world? Is it not the sciences and their attendant technologies that most fundamentally determine our life, showing us the possibilities that lie open to us and discovering to us what we may do and what we may become? Moreover, for us as belonging thoroughly to our time, is there any proclamation of the disposing ruling of a god that can win our acknowledgement and bring us to foreswear our claim to be the prime disposers of everything whatever? Knowing what we know of the workings of reality as scientifically disclosed, i.e., of inherently explicable events and causal sequences, how can we suppose that a god plays a crucial role in the universe in which—to borrow Heidegger's word—we "dwell"?

Heidegger himself would in large measure not dispute our response. At the close of his explication of poetically determined dwelling, he indeed raises the question as to whether we today "dwell poetically." Immediately he replies, "presumably we dwell altogether unpoetically."[172] For him, however, this fact does not suggest, as it may for us, that the words of Hölderlin whose meaning he has carefully unfolded should be judged to be irrelevant or untrue. Rather, for him, in the unpoeticness of our present dwelling on the earth, the truth of the poet's pronouncement is hiddenly confirmed. Just as blindness can afflict

[172]" . . . Poetically Man Dwells . . . ," VA 202, PLT 227.

only one who is capable of seeing, so an unpoetic mode of transpiring can pertain only to a human dwelling that carries within itself originally and indeed inextinguishably the possibility that the poetic may provide to it its determining impetus and stamp. We who listen to Heidegger are dedicated to the pursuits of science and technology and are prone to make all judgments in their light. Constantly, even indefatigably, in the scientific spirit, we measure and calculate. To Heidegger, however, our measuring and our calculating are a kind of frenzied travesty. They are but blind pursuits that witness distantly to the genuine measuring of which the poet Hölderlin speaks and which poets undertake, i.e., the intricate disclosing of the Twofold via the illuminingly poetic word.[173]

In all our measuring we know nothing of the measure, the opening out of reality to assessing seeing, that, according to Heidegger's understanding, is vouchsafed via poetic presenting and that is so received by the poet as to be brought on-goingly to bear as a determinative bringing to light. We know nothing, that is, of the self-manifesting, via the appearances of the heavens and the earth, of a god who in otherness rules in those takings-place and who, through that manifesting, which claims us into relation with himself, enables us—and can alone enable us—to see ourselves as and to be the self-forgetful mortals, dedicated to obedient receptiveness and accordant self-surrender to whatever disclosure may be given, that, in the last analysis, we as human beings are.

We are oblivious of such disclosing, for in fact, in the happening as which the Being of what-is now rules, it has been withdrawn from us. Heidegger, again following Hölderlin, sees ours as a time when no god presents himself in determinative manifestation. Hölderlin himself— whose poetic pronouncements are for Heidegger decisively determinative although not so seen by others—speaks of "the time of the gods that are fled *and* of the god that is coming." This is our time, which, Heidegger says, Hölderlin "founds," i.e., summons into view as underway, in bringing it to poetic utterance.[174] In it we live under a double absence of

---

[173]VA 203; PLT 228.
[174]"Hölderlin and the Essence of Poetry," EHD 44, EB 312f.

the gods. The self-manifesting of no god proffers to us through poetic presenting a disclosure, a measure, that provides to us and summons us to a genuine assessing. Assessing has, rather, in the strange excess of our self-reliance when no god meets us, fallen into our hands. Pertinaciously discovering and assigning significance from within our present perspective, we now ever more intently pursue our unpoetic way all but bereft of such insight into the primal constituting of reality. For the absence of a god bespeaks a withdrawing of immediate disclosure, the province of the poetic, and a dearth of that gathering puissantly to light of the Fourfold of the heavens and the earth, divine ones and mortals that a poetically based dwelling in company with things could bring to pass.

Precisely that which militates against our acceptance of Heidegger's portrayal of human dwelling upon earth as a mode of life that, when genuine, is, in his terms, poetically enabled and attuned, i.e., our unquestioned prior acceptance of the superiority and normative validity of our ways and our knowledge, looms for Heidegger as a testimony to the rightness of his own outlook. We shortsightedly make our way as though we had entered upon the culmination of human achieving. He, on the other hand, ponders the unfolding of the Twofold, the on-going happening of the Being of what-is, discerns a complexity of transpiring that quite escapes us, and summons us to step out of our usual perspective so as to see as he sees. It is pure Happening that concerns him, as, self-initiatorially giving and gathering itself restrainedly to light, it intricately shapes and bodies itself forth via the on-going manifold of what-is, through bringing itself self-maintainingly to pass as self-unconcealing self-concealing by way of the disclosing that man, in his living-out of time, provides.

What is it that Heidegger would tell us about our age and about all that we here and now so readily take for granted? The recounting is long. It requires that we commit ourselves in thought to the orientation that has just disturbed us. Yet ours is an age that demands understanding. The scientific and technological methods and presuppositions on which we so readily rely have carried us into crises. The world they have fashioned promises and gives us much, but it also threatens us variously with

destruction.   We can ill afford self-confidently to reject Heidegger's
proffered insights.   What, according to Heidegger, is this unpoetic
dwelling that we now pursue?   What are the technology and the science
that determine it, when seen from the perspective that speaks first of the
Being of what-is?   And what may we expect, what may we best do, in this
problematical age?   Heidegger's responses to such questions cannot but
speak meaningfully to us.   They cannot but extend our perspectives and
help us to think afresh regarding both the deepest questions—those
concerning the real—and the most immediate questions—those concerning
the understanding and action that may be incumbent on us in this critical
time.

PART TWO

The Modern Age

CHAPTER VI

TECHNOLOGY[1]

"Technology is," according to Heidegger, "a mode of revealing." Enigmatic at the outset,[2] these words can now sound for us with significant force. Understood literally, the word "revealing (*Entbergen*)" speaks of a harboring-forth, i.e. of a disclosive providing, that takes place immediately from out of the protective concealing as which Being brings itself to pass.[3] In thus speaking directly of the self-manifesting self-unfolding of the Being of what-is happening as the Twofold, the word *Entbergen* necessarily suggests at the same time the role of man, the special being who, among all that is, allows that self-manifesting to reach fulfillment.[4] Hence, seen from out of Heidegger's own orientation of thought, this brief definitive statement regarding technology can now suffice to open to us implicitly the whole concern of our primary study.

If left to ourselves, we might perhaps be tempted to undertake the consideration of modern technology by examining technological

---

[1]Behind our use of "technology" stands Heidegger's use of the German noun *Technik*. *Technik* has a richer meaning than does technology. It can mean technology, technical science, engineering, skill, technique, or execution. In Heidegger's discussions concerning technology, this full range of meaning is decidedly secondary, but its presence should be noted. For him the very name that he uses suggests an inclusiveness of reference to actions and processes that is not immediately resident in our "technology."

[2]Cf. Authors' Prologue, p. 1.

[3]Cf. Ch. I, pp. 24-26. The German prefix *ent-* in compounds connotes specific activity indicating either the establishing of a new state or else the abandoning of a prior one. *Entbergen* is a word of Heidegger's own coinage. The coupling of this active prefix with the verb *bergen*, to harbor, protect, conceal, provides a word that, with its suggesting of initiatory forth-going movement, contrasts subtly but significantly with the noun *Unverborgenheit* (unconcealedness) and the adjective *unverborgen*, both formed after the past participle *verborgen*, with their similar suggestion of the negating of concealedness and their lack of full verbal force. Cf. Ch. I, p. 24, n. 12.

[4]Cf. Ch. II, p. 60. On the self-unconcealing happening of Being as the Twofold, cf. Ch. I, pp. 27-33.

phenomena, many of which we know well, in the light of the detailed study of the tenets and perspectives of Heidegger's thinking into which we have inquired thus far. For it might seem that we could thereby best discover the place that technology must occupy in that thinking. Such an approach would, however, be diametrically opposed to that which Heidegger demands of us. We have carried out thus far a surveying of salient aspects of his thought which, gathered into structural sequence, offer us some over-all knowledge—opened out in depth at numerous crucial points—of the understanding of reality out of which he himself speaks. But when particular phenomena are to be understood Heidegger can never be content, nor can we, with an understanding of them that fits them into a preconceived scheme. For Heidegger, it is in our direct encountering of that which is, that which immediately meets us, that with which we have here and now to do, that the disclosure that truly permits understanding must come. If we are to gain insight into the character and meaning of modern technology after Heidegger's manner, we must confront technology itself. More than this, we must confront it as an *historical* phenomenon. We must remember that for Heidegger the on-going unfolding of the Twofold that meets us concretely as what-is takes place epochally and changefully.[5]   Our concern is specifically with modern technology, the phenomenon of our present age. Only if, placing ourselves to the best of our ability within the Heideggerian perspective and looking intently from thence, we look first of all at that technology itself can we catch sight with significant comprehension of the "revealing" that, Heidegger tells us, is there underway.

1

When we ourselves speak of technology there comes immediately to our mind the whole host of those machines and devices that have proliferated in our time for the performing of all manner of tasks. And beyond this we are aware of an outlook—an approach to reality, a manner of dealing with problems and persons and things—that we would call "technological." If asked, we might well include this too in our definition

---

[5]Cf. Ch. I, pp. 48f.

of technology. For it is an attitude of mind that takes its rise out of the myriad successes that have been achieved through the use of technological apparatus and that, in turn, works constantly toward the bettering and further dissemination of such apparatus and of the effective techniques made possible by it. Thus the word technology speaks to us of a particular aspect of modern life, in which human thought and action produce and utilize all manner of instruments and machines whose functioning is specifically designed to facilitate every sort of activity undertaken in our world.

Heidegger would by no means wholly disapprove of our defining of technology in this way. For he characterizes as "correct (*richtig*)" a definition of technology that sees it simultaneously as a means to an end and as a human activity, i.e., as a single phenomenon for which the apparatus and tools provided, the action that provides them, and the purposes they serve are all, together, equally constitutive.[6] And yet such an "instrumental" and "anthropological" definition scarcely begins to tell us of technology. Modern technology is indeed for Heidegger one vast "instrument (*instrumentum)"*—an identification that points directly to its "machine" character, in the broadest sense of the word.[7] It is an immense arrangement of material and non-material components, an enormous contrivance (*Einrichtung*), controlled and directed by man.[8] As such, it is ever being honed and extended by him to serve human purposes and perceived needs, even as it itself constantly takes up into its complex network the human activity that disposes and governs it. But such defining can be only preliminary. Modern technology and with it modern man, who both controls and is subservient to it, belongs within the great manifold of what-is. Its true governance comes, therefore, from out of Being. In modern technology there is manifest the destining, the self-

---

[6]"The Question Concerning Technology," VA 14, QT 4f.
[7]For this identification of the instrumental as possessed of what we may call a machine-character, see HHF 35, HFH 99f, and cf. this chapter, p. 238f. The meaning of the Greek word *mēkhos* from which our word machine is descended is means or expedient, that which enables. Cf. *The American Heritage Dictionary of the English Language*, Boston, Houghton-Mifflin Co., 1971, p. 1527.
[8]"The Question Concerning Technology," VA 13f, QT 4f.

sending-forth, as which Being now rules in its self-unfolding as the Twofold.[9] It is, Heidegger believes, precisely this, Being's present manner of holding sway, that the familiar character of technology can, if we are properly attentive, disclose to us.

We have already glimpsed the most salient characteristics of technology that point beyond technology as such, in discerning as fundamental to the latter a single-minded intention that everything technological shall be laid hold of in such a way as to serve some specific end. Intrinsic to technology is an urgent proceeding always toward some confidently foreseen goal *(Zweck)*.[10] Preoccupied with itself and supremely confident in itself as worthy of perpetuation, technology advances from undertaking to undertaking. Always there is in play within it a planning *(Planung)*, an intent, well-informed projecting of ends to be attained and devising of procedures for their attainment,[11] which draws purposefully into its domain and fixes there as available anything whose availability is seen as requisite for the accomplishing of the purpose in hand.[12] Transpiring thus, as ever in train toward some envisioned goal, technology focuses its attention on usefulness. And this concentration and the manner in which it is put into play exhibit at a crucial point the mode

---

[9]See Ch. I, pp. 47f. For the identification of the destining in question, cf. this chapter, pp. 257-259.

[10]Cf. "The Question Concerning Technology," VA 20, QT 12. The urgency with which technology is pursued must remind us of the urgency that characterizes human *Being*-there as such in its self-carrying-forward from out of thrownness. Cf. Ch. II pp. 68f, and n. 50, p. 89, and p. 100, and n. 157.

[11]See "The Principle of Identity," ID 99, 35.

[12]There is evident here, in this structure according to which technology takes place, a close parallel to the structure belonging to the transpiring of man's *Being*-there as Being-in-the-world. Analogously to *Being*-there's casting forward of itself *(Entwerfen)*, to gather toward itself that which meets it, from out of the basis provided in its thrownness as Being-in-the-world, on the basis of whatever is usably available to it technology continually casts itself forward via its planning, gathering toward itself whatever it needs for its undertakings. And in its pursuing of goal upon goal it evinces a ubiquitous tendency to confirm and augment its accomplishments and hence to sustain itself as itself that directly parallels *Being*-there's carrying out of itself, as care, after the manner of a for-the-sake-of-itself intent on self-maintaining. Cf. Ch. II, pp. 68f & 73-76. On the structure here in question as instancing the structure characteristic of the happening of Being as the Being of what-is, cf. Ch. I, p. 33, n. 36.

of happening definitive of modern technology as such.

We have seen that Heidegger early made central to his thinking a notion of usefulness. In its fullest sense, usefulness is a manner of self-presenting that, in a serviceability (*Dienlichkeit*) that belongs intrinsically to it, characterizes the so-called "equipment (*Zeug*)" that, as ready-to-hand (*zuhanden*), proffers itself from out of the intricately circumstanced relational context of a world that is meaningfully met in the intending "concern (*Besorgen*)" constitutive of human beings' existing as that existing, as *Being*-there-as-openness-for-Being that is Being-in-the-world, carries itself forward in sustaining itself as itself.[13]   Proper using (*Brauchen*) transpires as a counterpart to just such self-presenting. It is a responsive intending that must be thought of as freeing the used, in an initiatory way, to the self-presenting that is properly its own.[14]

The technological "using" of which we have just been speaking is clearly far other than this. In the domain of technology, everything— whether natural feature or sophisticated artifact, whether person or art work or idea—is viewed solely in terms of the place it occupies in a continually proliferating complex. Everything possesses significance only insofar as it is seen and is taken charge of as something useful for the serving of an end beyond itself. Technological using is a derogating employment that, arising out of intense purposefulness, leaves no place for true self-presenting. Moreover, since that using instances a technology which, in accordance with the very manner of Being's

---

[13]Cf. Ch. II, pp. 64-68.
[14]See "The Origin of the Work of Art," UK 72, PLT 64: *Dieses Brauchen aber verbraucht und missbraucht die Erde nicht als einen Stoff, sondern es befreit sie erst zu ihr selbst.* For an interesting parallel to this phrasing, with respect to "saving (*retten*)," see "The Turning," TK 41, QT 42. On using as freeing see also SZ 83-85, BT 114-117, and Ch. II, pp. 67f. The predominant meaning of *brauchen* is to need, to require. It can also mean to use. When, as in the primary passage just cited, Heidegger employs it to speak of using, he obviously brings forward the latter meaning. But the "using" in question remains a needful using; and the needfulness suggested, which bespeaks an allowing of the used so to present itself as to play the role toward which it inherently tends, must finally be understood from out of Heidegger's employment of *Brauch* and *brauchen* to speak of Being's governing provisively via the handing-over into relation with itself of that which it allows to be. Cf. Ch. I, pp. 53, and n. 107.

happening that governs it, can never be content with what it has achieved but, quite to the contrary, always strives toward some further goal, it takes place as insatiable and is always, ultimately, a using up. Unquestioningly it claims and in one way or another consumes in the name of technological advance whatever comes in its way.[15]

To illustrate the way of happening that is characteristic of technology, Heidegger adduces several examples, among them that of the Rhine River seen—as it now must be—as one component in a technological nexus. When a power station is built into the Rhine, the river does not, as once it did, present itself amid environs determined and thrown into relief as gathered into conjunction with man's communal life. Formerly that life was accommodative and responsive toward the surroundings of which it availed itself. In erecting bridge or town, it drew the river into its own realm of meaning and so let it show itself, through varied relations, as precisely the river that it was.[16] Now, in contrast, the power station dominates the river. It is a product of technology, whose way of presenting itself and whose function are far different from those of any ancient bridge that might have spanned the river at the same point.[17] The bridge would have determined and disclosed the river amid all the relationships that gave it, there and then, its uniqueness.[18] The power plant, in contrast, with its dam and its turbines, imposes itself on the river in such a way as to strip it of those relationships as factors that are significant for it. And in so doing, it

---

[15]Cf. "Overcoming Metaphysics," VA 96f, EP 106f; VA 91f, EP 103. In these passages, as in "Overcoming Metaphysics" generally, Heidegger's characterizations of technology and the technological are given in the context of his consideration of the metaphysical undergirding of the present rule of technology provided in Nietzsche's metaphysics of the will to will as will to power. (See, e.g., VA 80f, EP 93.) Accordingly, the character of technology as a using up is there considered, not in itself, but with reference to the manner of Being's ruling happening as discernible in the manner of transpiring of that will. For an explication of this "metaphysical" context, see Ch. VIII, pp. 315-319.

[16]Cf. "Building, Dwelling, Thinking," VA 152ff, PLT 152ff.

[17]"The Question Concerning Technology," VA 23f, QT 24.

[18]Cf. Ch. V, pp. 185f.

divests it of its power to present itself as itself.[19]   The power plant discloses the Rhine only as one element in a detailedly designed process. The river is now simply a water-power supplier, and Heidegger tellingly portrays its role as a mere link in a chain of supply.   The hydroelectric plant that has been set into the river's current

> sets the Rhine to supplying its hydraulic pressure, which then sets the turbines turning. This turning sets those machines in motion whose thrust sets going the electric current for which the long-distance power station and its network of cables are set up to dispatch electricity.[20]

And the Rhine itself is lost, swallowed up in a purposeful process, the supplying of electrical power to waiting consumers, which alone, from the technological point of view, is of any importance whatever.

In giving this portrayal of the way in which the Rhine is now presenting itself under the dominant sway of the power plant, Heidegger is not simply speaking of an isolated aspect of the river's way of encountering us here and now, as though the latter could still be seen as a component of its environing world that is discerned and valued in its uniqueness and for its own sake. He himself raises the question whether even in the presence of the hydroelectric plant and its dam the Rhine is not still "a river in the landscape." He admits that it may perhaps be so. But he denies that even when we today see the river thus we are able to focus our attention immediately upon it and hence to let it stand forth and meet us as itself and for its own sake.   The Rhine now lies in the landscape, he says, "in no other way than as an object (*Objekt*) on call for inspection by a tour group ordered there by the vacation industry."[21] With this observation Heidegger brings into play a fundamental tenet of his view of the modern age. The dominance of technology is not to be felt only in the presence of machines and industrial units, even when these are viewed in conjunction with the infrastructure of human work, of

---

[19]In making this point Heidegger contrasts the Rhine as viewed in conjunction with the power plant with the Rhine as shown in a poem by Hölderlin. VA 23; QT 16.
[20]"The Question Concerning Technology," VA 23, QT 16, and n. 4.
[21]VA 23f; QT 16.

planning and skilled execution, that sets up and utilizes plants and equipment of every kind. Rather, the determining power of technology now rules everywhere in the modern world. A river is not only a supplier of water power. It is a supplier of money for travel entrepreneurs and a supplier of pleasurable experiences for tourists characteristically bent on enjoying their vacations to the fullest and seeing all they can in the shortest possible time. Here again the River Rhine is lost to us in what was once the completeness of its relational self-presenting as an entity of unique integrity and intrinsic power. Again it is but one element in a process. Its importance lies only in its being on hand to serve an end beyond itself.

What has been said of the Rhine could be paralleled if we were to speak of all manner of other constituents of what we call the natural world, if we were to speak of works of art or endeavors of scientific or philosophical thought, yes, even if we were to discuss persons and groups in all the varied actions and communications that comprise their life.[22] Everywhere every sort of entity tends to present itself and to be viewed explicitly or implicitly as something useful for the achieving of a goal or as something that, in an overarching schema, plays its part toward the attaining of some assumedly beneficial end. It cannot and does not confront us as and for itself. It does not call us and let us call it into immediate and meaningful mutual relation[23] in the midst of its fellows. This state of affairs is endemic in our Western world; and it is spreading ever more widely and rapidly, like the very apparatus and "know-how" of our technology, with seeming inexorability over the whole earth.[24]

Not surprisingly, the state of affairs here in question displays itself

---

[22]Cf. "The End of Philosophy and the Task of Thinking," SD 64, TB 58.

[23]Here and elsewhere in this study the reader should be careful rigorously to separate "relation" from its usual meaning. Here "relate," in contexts where the Being of what-is is in any way under consideration, does not speak at all of a connecting of inherently discrete elements, one of which is to be subordinately referred to the other. The relating that is now in question is a *holding-toward-one-another* of participants that, while they are distinct from one another, are ultimately one. For we are using "relate" to speak of the happening of the single Twofold. See further, Ch. IV, p. 145, n. 10.

[24]Cf., e.g., "*Logos*," VA 227, EGT 76; FD 11, WT 13f.

pervasively among us in our language. This fact, although often little heeded by us, is, in the context of Heidegger's thinking, one of decisive import. Technology everywhere affects us, and its ways impinge upon us. Our most ordinary speaking shows its impact, for in that speaking there is strikingly evident a failure to meet with things as they intrinsically are, which springs ultimately from a preoccupation with some goal beyond the thing of which we are ostensibly speaking.

Constantly language serves us in this way. It functions as a mere instrument, a means to purposeful understanding and to the garnering of what we consider to be relevant information regarding that with which we have to do.[25] Indeed, so pervasive is this attitude that the very language that we employ in this instrumental way itself suffers from the same disregard that it brings to bear toward that of which we speak. For we fail to engage ourselves seriously with language, even as, via language, we fail to engage ourselves with all else. Habitually language is not so spoken or written among us as to play its proper role through being allowed to exhibit directly and fully that which is being spoken about.

The presence in our employment of language of this dual-faceted failure of engagement evinces the degree to which the determinative power of technology has laid hold upon us. Everywhere we show its influence, often clearly, often in subtle ways. Characteristically—in a tendency that is by no means unique to us and our time but that is, thanks often to ubiquitously disseminated advertising and journalistic parlance, peculiarly urgent among us—we are in haste to "simplify" our speaking. To this end, adverbs constantly give way to adjectives, so that we "go slow," "eat right," and "think big." Prepositions disappear. We now shop Bloomingdale's and debate our opponents. In both our speaking and our writing, seemingly needless intricacy disturbs us and, even at the cost of actual clarity, we regularly omit relative pronouns, verbs, prepositions, or conjunctions that seem to us superfluous and uselessly repetitive in their contexts. Such behavior clearly shows our unwillingness to involve ourselves in full seriousness, through the use of more demanding

---

[25]HHF 35, 37; HFH 99f, 100.

syntactical means, either with our language itself or with that which we are attempting to say.

Often, under the pressure of this tendency in our speech, we borrow directly from terminology that, in its uncompromisingly purposeful terseness, belongs squarely within the technological sphere. Instead of troubling to use a "besides" or "also" or even an "and," we make do all too often with a simple "plus" to join one thought with another. When we wish to make some conclusive or definitive statement or to sum up the significance of something, how often we escape the need to clarify connections by asserting "the bottom line is . . . ." And many a simple "equals" has replaced a more appropriate verb that would speak of the achieving of a consequence or would allude to some other specific interrelation.

Countless other words have a technological ring. Do we not constantly speak of "processing" this or that? At home we routinely operate our food-processors. At work we may process job applications or the applicants themselves, requests for psychiatric help or accident victims brought to an emergency clinic. And everywhere we resort with greater and greater readiness to our *word*-processors, a term that surely must have made Heidegger cringe. In all such instances the verb "process" is itself expressive of the distance, the lack of immediate relation, that subsists between us and that with which we have to do. Always we are looking toward an anticipated result of the process in question, and we are all too apt to be quite unconcerned, for their own sakes, with the things or persons let alone the words that have come into our hands.

Similarly, in speaking of reactions to any sort of intellectual or artistic presentation, we often do not have in mind the direct responses of thinking, feeling persons. Rather, we are apt to be concerned only with comments that show how the presentation in question was received and to refer to the "feedback" that was obtained. Again, the word for a type of conscious awareness of our own organic functions that can open out on greater vitality and greater well-being is "bio-feedback," a term that denotes a structured process viewed only abstractly and that gives no hint of the immediacy of self-awareness that it is employed to name. We do not *listen* to persons and their words. We "tune in" on conversations. If we do not understand a person, we are "not on his wave-length." If we are in sympathy with him, there are "good vibrations" between us. In what passes for religious thinking among us, we may speak of spiritual or psychic "energy" that we can "plug into" through one "technique" or another. And all such words readily show how far we often are, even in our most important relationships, from immediate personal involvement with whomever or whatever we meet, encountered as itself.

The same tendency toward a distancing is shown in the abbreviations and acronyms that litter our speech. We watch TV, visit LA, or attend NYU. We outwit the IRS and watch for news of the UN or the EU. Our doctors are MD's, and our dignitaries VIP's. We take our Labs to the vet's and employ our war vets in labs. Prematurely born babies have become "preemies," and fatally ill persons are "terminal." We catch the "flu," gasoline is "gas," and businessmen and lobbyists expect their "perqs." Assiduously we work out names for groups or organizations whose initial letters will, like NOW and MADD, be terms that can be recognized and easily remembered. Pronounceable acronyms such as NASA and NATO virtually become proper names. Others, like "radar" and "laser," so pass into the general language that their acronymic origin is all but forgotten.

Where these familiar kinds of abbreviating occur, we usually fancy that we are indulging in them simply for the sake of brevity or conciseness. But we also feel, not always consciously, that they display our intimate knowledge of, sometimes even our affection for, that which they name. This naming has, however, a diminished, even a skeletal character. We do not so speak of what we name as to address it in its fullness. And we do not allow it to address us. Familiar words with their connotations and associations are suppressed, and familiar entities are therewith set forth only in a depleted guise. In this way, we may indeed grasp them more readily than when we name them in full. But we grasp, as it were, only a kind of shadow whose substance, the entities themselves in the fullness of their relation to us, does not come into our ken.

For Heidegger, the technological cast of our everyday language,

which shows itself in such varied ways, arises out of a decisive change
intrinsic to language itself, under the dominion of technology, that is far
more radical than we customarily suspect. Since language lies at the root
of the taking place of human existing and hence of every specific human
undertaking that springs therefrom,[26] it must therefore be language that,
in the domain of technology, centrally manifests and determines the
latter's way of carrying itself forward. For technology, however,
language is significant only as a provider of technologically useful
information. To it, the historically determined phenomenon of diverse
languages can appear only as a severe impediment to communication, and
every element in any language that contributes to gathered richness of
meaning must present itself solely as an obstacle to be circumvented.
Technology gladly dispenses with the excess baggage of syntax and
connotation wherever possible, for the sake of clarity and precision in the
storage, communication, and utilization of data. Indeed, it would
abandon ordinary language altogether if contentless signs and numbers
could be brought wholly to suffice for its work.

Here the computer, that machine whose role among us continually
expands with such astonishing speed, cannot but spring to mind as a
prime instance of the technologizing of language. Computer
programming "languages" are formulated in "commands" in which the
basic grammatical structure of an imperative sentence—verb/object (or
objects)—is utilized, sometimes with the inclusion also of a subject. But
in those formulations the range of import and the relational and qualitative
nuances intrinsic to our usual highly varied grammatical units—
sentences—have no place. Recognizable words occasionally appear, but
for the most part the commands in question are written with fragments of
words, numbers, punctuation marks, and other symbols. The messages
that the operating system or word processor provides are frequently so
terse and devoid of explanatory phrases as to be all but incomprehensible
to the uninitiated. All manner of identifications are conveyed simply
through the use of single letters or numbers. Once again, the words that

---

[26]Cf. Ch. III, pp. 112f & 129-141, and Ch. XIV, pp. 573f.

in our customary speaking and writing would make evident interrelationships among elements of the thought that is being expressed are often absent, now in a radical way. A mere space, for example, can have prepositional force, indicating that an operation is to be initiated to move information from one location to another, or a dependency that would ordinarily require the use of an adverbial phrase can be indicated with a simple backward slash. And the computer user must learn that a "word" can be any group of symbols whatever that stands between spaces. Behind these levels of "language" use lie numeric codes whose members are comprised solely of digits or of letters and digits in combination, and behind these lie the open and closed electrical circuits on whose proper patterning the functioning of the computer depends. The rigorously informational character of the "languages" employed in computer use is strictly conditioned by their subservience to the electro-mechanical character of the machine that they are intended to control. Nothing beyond the providing of precise information is either needed or desired. Indeed, anything beyond the narrow confines of the "syntax" of the "languages" in question would completely disrupt the functioning of the machine.

At a time—1958—when the rise of the computer to prominence was but beginning, Heidegger saw with striking clarity the scope of the event that was taking place, and he saw in it the focal point of technology's bringing of itself to bear on the transpiring of modern life. That dominion was centering itself precisely in the sphere of language. There, he tells us, in words that directly recall his original description in *Being and Time* of the everyday,[27] there has followed from the hastiness and mere habitualness that mark our ordinary uses of language in that inauthentic mode of existing a technologically oriented viewing of language that renders it, like everything else with which we now busy ourselves, predominantly an instrument in our employ. Ever more extensively it serves in our intercourse with one another as a purveyor of information and as a means to promote and disseminate facile

---

[27]See SZ 177f, BT 222. See also SZ 371, BT 422. Cf. Ch. II, p. 89.

understanding.[28] And precisely in the bringing to the fore of such a role
for language, the computer has its central place.[29]

Pursuant to the construction of what Heidegger calls the "electronic
brain," in a widely recognized development, the building of specialized
variants—calculating machines, thinking machines, translating machines—
is, he says, underway. All the activities for which such computing
machines are intended—calculating, thinking, translating—have their
inception and are carried out in the element of language. Accordingly,
Heidegger calls the multiplicity of these machines, understood as a single
technological layout (*Anlage*), the "language-machine (*Sprachmaschine*)."
This name itself bespeaks the prior happening of language that carries
itself out in the functioning of the machines in question. For by way of
the latter, he says, "*the language-machine*," i.e., language in its present
happening as an instrument, "has realized itself (*sich verwirklicht*)"; it
has, in concrete centerings, brought itself disposingly into play.[30]

Here the now pervasive role of modern technology comes crucially
to the fore in a way both unobtrusive and unavoidable, which is therefore
all but unsuspected among us. An ever-increasing host of machines
constitutes the "language-machine" of which Heidegger speaks. Those
machines are man-made, and men "program" and "operate" them. Quite
naturally it appears to us that man has mastery over the "language-
machine." And yet the truth of the matter is quite other than this
appearance. Given the fundamental character of language as a disposing

---

[28]Cf. Ch. II, p. 89.
[29]HHF 35; HFH 100.
[30]HHF 35; HFH 100. On the meaning of *wirken* (to work), which is the root of
verwirklichen—literally, to make real—and on *Wirklichkeit* (reality), see Ch. VII, p.
282. In Heidegger's use of *Sprachmaschine* to speak simultaneously of the originative,
provisive happening of language, which is as such a mode of the happening of Being
(Cf. Ch. III, pp. 118-120), and of the single "layout" of machines in which that
happening comes to accomplishment, we have a peculiarly clear and telling instance
that shows us the unitary happening of the Being of what-is as the ruling Happening
that "is" what-is, i.e., that both enables what-is to present itself and itself manifests
itself immediately in that which, in that very presenting, concretely *is*. Cf. Ch. I, pp.
30-32, and n. 32. On Heidegger's use, in this regard, of "is," as transitive, see "The
Onto-Theo-Logical Constitution of Metaphysics," ID 132, 64.

happening that underlies and predisposes every mode of human undertaking, it must be the case that, since the "language-machine," in receiving the happening of language and bringing it into play—in that "machine's" diverse calculative efforts—"puts language into operation," it is language itself that in fact exercises mastery over the way in which human existing now carries itself out. For it is that "language-machine" that now decisively delimits the way in which language grants itself to us. Precisely "from out of its own mechanical energies and functions," it both regulates and providingly metes out antecedently "the manner of our possible usage of language." The *machine*-character of the medium of activity in use via which comprehensible, manageable information is processed and provided by way of rudimentary energy-transfers and basic machine functions, i.e., its character as an instrument inflexible and automatic in its accomplishing of its tasks, both determines the way in which language is, by it, brought into use and brings into currency and ascendancy from out of that usage the only sort of language that now has dominion among us.[31]

Here the manner of happening of modern technology focuses its sway. Thus crucial is this "language-machine," this machine-complex via which the happening of language now brings itself to bear, as a directing mode of technology's taking place. For that "language-machine" now is—indeed, most importantly, it increasingly will become—a "mode" (than which none could be more central), according to which "modern technology takes disposal over the manner and the world," the disclosed structured reality, "pertaining to language as such."[32]

Certainly no such radical depleting of language as that which the computer evinces is required or practiced in most arenas wherein language presently serves us, but the very restriction of language to the conveying of information and the use of it to achieve precise control that is there rigorously in play is that which does in fact everywhere mark our

---

[31]HHF 36; HFH 100.
[32]*Die Sprachmaschine ist—und wird vor allem erst noch—eine Weise, wie die moderne Technik über die Art und die Welt der Sprache als Solcher verfügt.* HHF 36; HFH 100.

attitude toward our employment of it. Since language is now
determinatively disposed via technology and proffers itself from the
technological point of view, and since, accordingly, the purpose that
language is now intended to serve is that of rendering immediately
accessible whatever technology may wish to put to use, the denotative
provision of information must be the primary function that language is
expected to perform. Just this view, indeed, extends now even to the
theoretical understanding of language. So-called "natural language" with
its imprecision is widely considered to be a lesser or incomplete form
whose perfected counterpart would be a language that was wholly
formalized.[33]   In the domain of technology, then, language has a
definitively new character. Connotative, meaning-full language, and with
it everything that only such language can adequately disclose, is under
attack in the interest of the peculiar task that technology so resolutely
undertakes.

   Unwittingly we abet this attrition when we speak as we now so
often do. And we bear ample witness to the technologizing of language to
which Heidegger points. In the very casualness of the habitual kinds of
speaking that have been earlier adduced we display with peculiar force the
ubiquity of the technological attitude. When, in our abbreviated speech,
we pass over individuality and immediacy of relationship, we are, just as
in our use of our computers, if more subtly, in accordance with this
technological outlook, stripping down our language in order to facilitate
our dealings with that with which we are concerned. We are detaching
ourselves from the latter, communicating a minimum of information about
it, and therewith despoiling it and relegating it to a nameable category,
wherein it can be at our disposal. The terms of our verbal shorthand can
be applied indiscriminately and with perfect ease to all the members of a
group or class, or they can summon up with an incisiveness that is often
singularly opaque a particular entity that is being named. In all this, we
are, in a quite technological manner, asserting our dominion over that of
which we speak. We do not let it encounter us as itself. Rather, we

---

[33]"The Way to Language," US 262f, WL 131f.

grasp and present it to ourselves and to others as something standard and commonplace. In so doing we make it subservient to the purpose that, although almost always unacknowledged, so pervasively informs all that we do, namely the easy pursuance of a course of life where demanding involvements, whether of thought or action, can be avoided as much as possible and where everything will be to the greatest feasible extent manageable and under our control. And here no less than in the transforming of a river into a water-power supplier or an adjunct of the tourist industry—and indeed even more deeply and insidiously—the primary tendency informing modern technology is inexorably in play. Bringing itself to bear upon us pivotally in the granted language[34] that we speak, it shows to us crucially, yet in a scarcely noticeable manner, the intimately encroaching pervasiveness of its rule.

2

Technology relentlessly impoverishes whatever it touches. In pursuing its characteristic way, it must be able to plan ahead so as to set specific processes in motion for the achieving of intended goals, and in order to carry out such planning it must necessarily concern itself with resources whose number and potentialities for use it can readily calculate and with which it can work at will. Indeed, such purposeful calculating (*Rechnen*)—which should surely remind us distantly of that reckoning that belongs intrinsically to the intent-fraught concern that characterizes human existing as Being-in-the-world in its having to do with that which is met with[35]—is the only mode of thinking via which technology gives its attention to that which it gathers into the purview of its concern.[36] As that concern carries technology ever forward toward new

---

[34]Cf. Ch. III, pp. 118 & 120-122, and Ch. IV, pp. 154f.
[35]Cf. Ch. II, pp. 65-67, and n. 45.
[36]"The Principle of Identity," ID 98, 34f. See also "Memorial Address," G 14f, DT 46, where Heidegger's overt mentioning of the technology that is clearly in question does not begin until later in the discussion (G 18; DT 50). On the basic character of calculative, reckoning thinking, see "Science and Reflection," VA 58f, QT 170. See also Ch. XI, pp. 417f. On the metaphysical basis of such thinking as presently carried out via technology, see Ch. VIII, pp. 314-321.

accomplishments, that calculative thinking is everywhere brought into play. It directs itself without reservation toward everything on earth.[37] And indeed, building on its successes, it now as it were even overlooks the earth and reaches increasingly toward the conquest of cosmic space as well.[38]

It is precisely because calculation is thus definitive for modern technology that rudimentary information language emerged as the vehicle of technology's advance. Such language is silent concerning intrinsic quality and genuine distinctiveness and concerning the kind of interrelation wherein these are disclosed. It presents only the specific characteristic or connection that is needed in a particular context to serve a projected end and that can, accordingly, be taken up into a particular line of reckoning and calculated upon as the latter is followed out. For technology must, by its very nature, be concerned only with the most expeditious ordering possible for the materials that come under its sway. Technology cannot concern itself with what is unique and therefore unpredictable and unmanageable. Indeed, it must deal always with what is comprehensible within its own preexisting frame of reference, and that means with what can be readily categorized so that its behavior can be foreseen and manipulated to serve the desired end. In pursuing its way thus, technology can have to do, therefore, only with general characteristics concerning which accurate communication can be confidently made. It can give no heed whatever to those special features and relationships that give to anything a singular character.

In keeping with this, its fundamental way of meeting and dealing with whatever comes under its sway, technology suppresses uniqueness and promotes indistinguishability. In a manner that is highly reminiscent of Heidegger's portrayal of the addiction to averageness of the "they" in its dominion in everydayness,[39] technology fosters likeness, for only the latter can provide a basis adequate to its single-minded activity, by

[37]"Overcoming Metaphysics," VA 98f, 80f, & 83; EP 109, 93, & 95.
[38]"The Nature of Language," US 189f, WL 84. Cf. "Memorial Address," G 20, DT 51.
[39]Cf. Ch. II, pp. 87f.

permitting the calculating, amassing, organizing, and reorganizing that modern technology endlessly undertakes in the service of greater and greater proficiency in the utilization for its own peculiar ends of whatever comes into its sphere. Uniformity alone accords with thorough-going calculating and planning out.[40] For this technology, *like* entities of every sort stand to one another in relationships that, in their *likeness*, can be equated and duplicated as the need arises.

Likeness is all that technology can see in what it surveys, and it is what it itself must strive to attain for the products that it so assiduously provides. One river is as good as another, as long as it can provide electrical power or fit adequately into a package-tour schedule. A person matters only insofar as he or she can get a given job done, can be counted on to buy a particular product, or can be polled as one in a representative sampling of voters or consumers. Above all, machines or devices of a given type are like one another. Millions of exemplars of a given model are produced, parts are and must be interchangeable, dimensions must be standardized, and manufacturing techniques and processes aim at greater and greater precision.[41] Where differences between products designed for the same purpose exist, they are frequently small and superficial. And as sophistication in design and means of production grows such differences often tend to diminish at a corresponding rate. Significant advances do occur in these areas, but when this happens competitors immediately strive to duplicate as quickly as possible any newly successful product that has shown itself better able than its predecessor to serve some particular end. Everywhere the drive within technology itself to (as it itself might say) maximize plannability and availability forces it, inexorably, to prefer and produce the *like*. The more readily one entity can be equated with or substituted for another, the more efficiently

---

[40]Cf. "Overcoming Metaphysics," VA 96f, EP 108, where, in a passage in which technology is not specifically mentioned but its mode of transpiring is immediately in question from the preceding context, Heidegger avers: ". . . reality consists in the uniformity pertaining to reckoning that can be planned out."

[41]Cf. "The Question Concerning Technology," VA 28, QT 20f, for an illustrative reference to engine stockparts (*Bestandstücke*).

technology can function. Hence wherever its dominion extends—and it extends to our every sphere of concern[42]—technology, preferring the indistinguishable, actively discovers and promotes substitutability. Mass production flourishes. And everywhere we encounter the alikeness—whether in flashlights or shopping centers, in aspirations and amusements, in tastes or opinions—that is the hallmark of our technological age. On it depends the constant forward march of technology as, alluring us with the abundance and ready availability of its standardized products, it establishes the needed base—material, economic, and ideational—for its own projected advance.

Heidegger calls this permanent fund of the undistinguished and readily available *das Bestand*.[43]   The German noun *Bestand* always carries connotations of durability.   It ordinarily means permanence, stability, or continuance, and so constituents, residue, or stock. Heidegger extends the meaning of the word and significantly changes its primary emphasis.   For him *Bestand* does indeed speak of a reserve whose continuance and stability are of vital importance. But now, in the context of his portrayal of modern technology, *Bestand* speaks—in a manner that surely presupposes the circumstanced, self-intending self-presenting of the ready-to-hand that is intendingly met in concern (*Besorgen*) and the uncircumstanced, public availability of the on-hand that devolves from it[44]—first of all of a reserve that is purposely provided and that is immediately at hand, standing ready for use.

Under the dominion of technology, this is the way in which entities actually present themselves. And they do so far more fundamentally and pervasively than we are apt to recognize. Heidegger cites the example of

---

[42]"Overcoming Metaphysics," VA 80, EP 92.

[43]VA 24; QT 17. The use of *Bestand* presupposes Heidegger's prior use of related words, *Beständigkeit, Beständigung, Bestehen*, to speak of the continuance in the open arena of disclosure provided by time, vouchsafed what-is in the ruling happening of Being as presencing. Cf. Ch. I, pp. 50-52 & 54f, and also this chapter, p. 264, n. 107.

[44]Cf. Ch. II, pp. 64-68 & 88. In *Being and Time* (SZ 153, BT 195), Heidegger, intending the word to connote mere subsisting, uses *Bestand* as elucidative of the meaning of *Vorhandenheit*, onhandness: ". . . circularity belongs to a manner of Being of onhandness (*einer Seinsart von Vorhandenheit* [*Bestand* {*subsistence*}])."

an airliner seen on a runway. Using a familiar conceptual framework, we might well, in observing the airplane, say that it is first and foremost an object (*Gegenstand*)[45] at which we as subjects are looking. Heidegger admits that we can certainly represent it to ourselves in this way. But if we do, our conceiving will be out of touch with the way in which the machine is itself being made manifest to us. The latter is presenting itself directly from out of a technological context. This means that it is being disclosed, it is being put forth and put into play, as an entity ruled and determined by the happening of technology. As thus put forward, the airliner "stands on the taxi-strip only as standing-reserve (*Bestand*)," i.e., as what alone it truly is, namely, something "ordered to insure the possibility of transportation," which must therefore "be, in its whole structure and in every one of its constituent parts, on call for duty, i.e., ready for take-off."[46] This prearranged readiness that is the way of on-goingly presenting itself intrinsic to the airliner in question likewise characterizes every entity that belongs within the realm of technology. There all is *Bestand*, a reserve that stands prepared and stands by precisely in order that each of its elements may fulfill its appointed function.[47]

Whatever comes under the dominion of modern technology is, as so ruled, brought into the standing-reserve. Such resources are, however, not in themselves the first concern of that technology. Modern technology strives after an ultimate disposal over whatever is. It seeks a mastery such that it may be able to produce whatever it purposes to provide and such that it may, to that end, be able at will to draw upon whatever presents itself to its grasp. Wide though the arena of its sway now is, in keeping with the present-day assumption that the "material" is that which is most objective and elemental, technology's foremost concern is with spheres having to do with machine-based readying and

---

[45]It is evident that Heidegger intends a distinction between the *terminus technicus Gegenstand* and *Objekt* as denoting something looked at. Cf. this chapter, p. 231.
[46]"The Question Concerning Technology," VA 24, QT 17.
[47]VA 24; QT 17.

production.[48] And since its machines are power-driven, technology's central need is for energy—abundant, available energy—that can be used to drive those machines and hence to accomplish at the most basic level of achieving the countless processes of production, both existing and newly projected, that it has forever underway.   Modern technology must, therefore, concern itself above all with the procuring of a secure energy supply.   Accordingly, there is, underlying its preoccupation with the available resources that it ceaselessly lays up for itself, a primary concern with nature as "the chief storehouse of the standing energy reserve, (*Energiebestandes*)"[49] that it fundamentally needs. Indeed, under the rule of the technology that now so hungers for energy, it is always thus that nature is first of all confronted.[50]

In its encountering of nature, technology discovers and extracts energy wherever it can, whether it be in the form of coal and uranium from the earth, of oil from the sea bed, or of nitrogen for fertilizer from the air.[51] Routinely, it stores and stockpiles the energy resources that it obtains,[52] in order that it may use them to keep the required supply of power flowing into the unimaginably intricate network of processes and events that constitute the vast out-working of technology among us. This emphasis on the storing up of energy is in fact that which, Heidegger says, sharply distinguishes modern technology from the technologies, with their devices for the producing and immediate utilization of power for specific applications, that belonged to former ages.[53]

Like all amassing of the standing-reserve, this storing up is no static laying by. Rather, it surges with forward-thrusting vitality.[54] In its

---

[48]"Overcoming Metaphysics," VA 80, EP 93.
[49]"The Question Concerning Technology," VA 29, QT 21. Cf. "Memorial Address," G 19f, DT 50.
[50]VA 22; QT 14.
[51]VA 23; QT 15.
[52]VA 23; QT 15.
[53]VA 22; QT 14.
[54]The on-thrusting movement characteristic of the carrying forward of technology should surely remind us of the thrust imparted to human *Being*-there as Being-in-the-world in the very thrownness that lies always at the root of the latter's on-going taking place, which is always discernible in any comportment via which *Being*-there carries

accomplishing, "the energy concealed in nature is unlocked, what is unlocked is transformed, what is transformed is stored up, what is stored up is, in turn, distributed, and what is distributed is shunted about ever anew."[55] Here the Rhine power station that Heidegger so vividly juxtaposes to the once unsubjugated river suggests itself as a paradigm of technology as such. The power plant exploits the river, taking control of the energy that it yields. Under the impetus of its working that energy is retained and channeled, so that it can continuously be made available to empower a multitude of technological undertakings, great and small, that are always in some irreducible way dependent on an assured supply of electrical power. Just so, through the carefully planned garnering and distribution of energy from today's many sources, does technology accomplish its appointed task, the ever more inclusive bringing of everything into the corpus of the standing-reserve.

The reserve of which we are speaking depends for its existence on that kind of disposing that we have called technological. For such disposing Heidegger employs as central the verb *stellen*. Basically *stellen* means to set or place. By extension it can mean to arrange, set in order, or regulate, to station or post, to furnish (troops), to offer (a proposal or motion), and, in a military context, to engage with the enemy. Thus the verb connotes a placing that is often a supplying and that frequently indicates an assurance or authority that can reach even to combativeness. Heidegger repeatedly uses *stellen*, often in company with verbs compounded from it, to speak of the intent, self-confident actions and effective occurrences that initiate or promote technological processes of every kind.[56]

Such "setting" is always a bringing of something into its proper place and function within the process in question. It is therefore a

---

itself out. Cf. Ch. II, pp. 68f, and n. 50, and p. 89. In all such forward thrusting we glimpse the powerful surging into play that belongs to the happening of Being as such. See EM 55, IM 59, and cf. Ch IX, p. 334.

[55]"The Question Concerning Technology," VA 24, QT 16.

[56]VA 23ff; QT 15ff. For a passage illustrative of Heidegger's employment of *stellen* to portray technological action, cf. this chapter, pp. 230f, where the oft repeated "sets" translates *stellt*.

providing, but a providing of a peculiar kind. It cares nothing for the fullness of the entity that is being fitted into the schema. It cares only for its qualification to serve the use to which it is being given over. And it is unhesitatingly prepared to use up anything with which it has to do, if such using-up will further the purposes whose implementation is underway.[57] Hence, this setting-in-place that furnishes whatever is needed in the undertakings of technology is of a violent cast. This it is that, setting upon nature,[58] commandeers a river, placing it in a subservient relation to a hydroelectric plant and so forcing it to appear only as a supplier of power. This it is that dragoons once fertile land into the service of the mining industry or preempts a forest for the use of manufacturers of paper, so that field and forest can present themselves meaningfully now only as ore-bed or repository of calculable amounts of timber.[59] No less inexorably, at the other extreme, this setting-in-place-as-supply draws persons into the sphere of mass production in every field, from housing construction to education, from the auto assembly line to prefabricated home handicrafts. Increasingly now persons have their *raison d'être* as members of a labor force,[60] as minions of some Department of Human Resources,[61] or as conditioned consumers in thrall to advertising and its patron industries, so that they must constantly struggle to maintain any genuine integrity and self-identity from out of which to meet their world. Heidegger, indeed, sees mere utilization as so exclusively demanded in the world where technology rules, that he can portray as required leaders who will dedicate themselves totally to the thorough-going calculative planning of all "sectors" of activity whatever, and he can confidently foresee the breeding of human beings as they may be needed.[62] In the technological context, he can say, "the need for human material" is, with respect to the all-encompassing organizing of everything that is there

---

[57]Cf. this chapter, pp. 229f.
[58]"The Question Concerning Technology," VA 24, QT 16.
[59]VA 22f; QT 14f.
[60]"The End of Philosophy and the Task of Thinking," SD 64, TB 58.
[61]Cf. "The Question Concerning Technology," VA 26, QT 18.
[62]"Overcoming Metaphysics," VA 93-95, EP 105f. For an explication of relevant aspects of the discussion from which these remarks come, cf. Ch. VIII, pp. 309-321.

underway, in fact to be seen as on a par with "the need for entertaining books and poems," for whose production, indeed, the poet himself is of no more importance than is the bookbinder's stockroom employee.[63]

The "setting" found in modern technology is, then, a commanding-forth that impinges with coercive urgency on everything with which it has to do, leaving everything embattled and largely or wholly unable to present itself as itself. Heidegger brings out this connotation of importunate exaction by specifically defining the bringing to accomplishment of which the verb *stellen* speaks as a challenging or defying (*Herausfordern*).[64] The commanding-forth-into-place-as-supply that is everywhere in play in modern technology is, literally, a demanding out hither. Imposing itself, it draws entities into an enforced self-exposure that is antithetical to what they would be if allowed, in a congenial milieu, to appear in and of themselves, as, in its finding out and displaying of resources *as* resources, it incessantly transpires so as relentlessly to promote (*fördern*) their most expeditious, most economical use.[65]

It is modern man who accomplishes this setting-in-place. In the name of technology he confronts everything with the demand of exploitation, wresting it forth into the domain of technology, that it may serve technological ends. As such, his activity is a perpetual setting-in-order (*Bestellen*), an ever-proliferating arranging and rearranging of countless components of vast processes of production, distribution, and consumption. Ceaselessly he provides for himself—and that means for technology, on which he relies and which he promotes—the needed standing-reserve. He enumerates and inventories, augments and disposes it, ever keeping it ready and available for the appropriate use, through a comportment that Heidegger can call an arming, an all-out combative preparing (*Rüstung*), for thorough-going dominion.[66]

The comportment in question is, wherever and however it shows itself, fundamentally the same. It is, as Heidegger takes the pervasiveness

---

[63]VA 95; EP 105f.
[64]"The Question Concerning Technology," VA 22f, QT 14f.
[65]VA 23; QT 15.
[66]"Overcoming Metaphysics," VA 91f, EP 103. For *Bestellen*, cf. this chapter, p. 259.

of war in our century to show, bent everywhere only on the serving, by
any means and at any cost, of an ever-retreating goal of self-establishment
and self-aggrandizement through an insatiable producing that, driven by
self-absorbed planning and calculating, indiscriminately uses and uses
up.[67] In the lecture that we now have as "The Question Concerning
Technology (*Die Frage nach der Technik*)," as Heidegger began the
lengthy consideration of technology that was the chief concern of the
lecture series in progress, he forcefully made this point by coupling
together undertakings that his hearers could be expected ordinarily to see
as radically different from one another.   With the clear purpose of
riveting attention on the point he was making, he adduced several
activities that would be bound to horrify, precisely so as to identify with
them another that would be widely accepted simply as a given in our day.
With unmistakably brutal intent he said: "Agriculture, now the
'mechanized food industry,' is in its manner of disposingly carrying itself
out (*im Wesen*) the same as the manufacture of corpses in gas chambers
and extermination camps, the same as the blockading and starving-out of
countries, the same as the manufacture of hydrogen bombs.[68]"   If the

---

[67]See "Overcoming Metaphysics," VA 91-97, EP 103-109. For a discussion of the
primary context of thinking in "Overcoming Metaphysics" to which this passage
belongs, which is Heidegger's portrayal of the ruling of Being as the "will to will" that
is disclosed via metaphysics in Nietzsche's "will to power" and is manifested in the
technological, cf. Ch. VIII, pp. 315-319.

[68]This statement appears on page 4 of the original version of the second lecture in
Heidegger's lecture series from 1949, which was delivered under the title "Das Ge-
Stell." (Wolfgang Schirmacher, *Technik und Gelassenheit*, Freiburg, Verlag Karl
Alber, 1983, p. 21, and n. 22.) "Das Ge-Stell" has never been published. The
statement is absent from the published version of the lecture, "Die Frage nach der
Technik." There, instead (VA 21; QT 15), its opening words about agriculture are
used to place the latter in a quite unremarkable series of modern-day activities.
Schirmacher, who quotes directly from "Das Ge-Stell," appears to us quite to
misinterpret Heidegger's original assertion. He takes it to display the meaninglessness
for Heidegger of all evils. This interpretation would be credible only if the horrendous
evils of concentration camps and hydrogen bombs were equated with "the mechanized
food industry" with the latter as the standard. But the equation made by Heidegger is
the opposite of this. It is technologized agriculture that is to be assessed in the light of
the dreadfully extreme activities that are emphatically enumerated.   Heidegger
specifically deplored the kind of superficial viewing that sees everything as equally
without importance, which Schirmacher attributes to him (see "Memorial Address," G

13, DT 45f); although ultimately speaking, thanks to his categorical rejection, in his understanding of the happening of reality, of all moral valuing as possessing any validity, he himself can be seen to be guilty of something that is fundamentally like to it. (Cf. Ch. II, pp. 95f, and Critical Epilogue, pp. 729-735.) But this criticism does not appear to be applicable at this point. Again, Heidegger's own conduct is certainly very much open to the charge that he failed to grasp the significance of the dreadful events, such as the amassing of corpses in gas chambers and extermination camps, that belonged to the recent German past under National Socialist rule. (Cf. Critical Epilogue, pp. 739-744.) But the statement we have just quoted should not be used to evidence that insensitivity. (For such a use see Phillipe Lacoue-Labarthe, *Heidegger, Art, and Politics*, translated by Chris Turner, Oxford, Basil Blackwell, 1990, pp. 34f.) To Heidegger gas chamber and concentration camp belonged to the whole rapaciously wasting mechanism of modern technology. Schirmacher quotes also from the wholly unpublished lecture, "Die Gefahr," that followed "Das Ge-Stell" in the lecture series: "a part of the reserve pertaining to the manufacture of corpses (*Bestandstücke eines Bestandes der Fabrikation von Leichen*)" (Schirmacher, p. 25). No context whatever is provided for these words, but whatever may originally have been intended in them, they can once again certainly be seen to placard the extreme character of technological providing. Heidegger had earlier quite cold-bloodedly spoken of all technological means whatever as necessarily and properly belonging to the executing of the thorough-going planning that was now demanded from out of Being's present ruling in withdrawal. (See, e.g., "Overcoming Metaphysics," VA 94-96, EP 105-107.) And undoubtedly his discerning and his acceptance of that fact sound in the words now in question. He had expounded an understanding of contemporary human comportment that could surely have seen the accumulating of corpses as a standing-reserve to be used in industrial production, to which he clearly alludes, as quite in keeping with the unremitting decimating that was now underway under Being's present mode of dominion. But he was also keenly aware of the uncanniness, with its undoubted element of horror, that belongs to that ruling of Being. Precisely in "Die Frage nach der Technik" he Points out that *Gestell*, the word with which he names Being's present mode of ruling, means skeleton. He then adds that the employment of this word that is now newly required "is equally gruesome (*schaurig*)." (See VA 27, QT 20, and cf. this chapter, p. 260; see also "Memorial Address," G 52, DT 22.) And in the statement now in question that awareness, joined, it would seem, with a genuine moral outrage at a "wrongness" in the ghastly manifestations of the technological that are adduced and also, be it noted, in its manifestation in present-day agriculture, appears—notwithstanding Heidegger's avowed philosophical position—to be what is in play. Later (G 25; DT 54) in a context in which he is speaking of a way of meeting technology that can counter its present mode of ruling, Heidegger speaks of modern agriculture in a manner very different from that found in the statement from "Das Ge-Stell," saying that a meaning can be seen in it. That meaning lies, evidently, in the fact that it is at least a mechanized *food* industry. One cannot but wonder why "Das Ge-Stell" and "Die Gefahr" remained unpublished. They were delivered originally to a select audience. Were the pointed evocations of the immediate German past given in them simply too forceful and too offensive to be allowed general dissemination? In the

character of present-day human conduct as a pursuing of the technological
is to be properly understood, Heidegger unquestionably demands of us
that we recognize that in all technologically motivated activities a like
self-absorbed ruthlessness toward whatever or whoever is being dealt with
invariably rules.

This is what Heidegger would bring us to see as we look out upon
our technological age. And yet this is only a partial vision. Everything
that is now taking place belongs within the unfolding Twofold. If under
the rule of technology modern man coercively exploits and disposes
everything—nature in all its aspects, the products of his own industry,
persons and things of every sort—therewith relentlessly reducing
everything more and more completely to standing-reserve and in that
process heedlessly destroying at will, and if, correspondingly, everything
shows itself more and more pervasively merely in such guise, this can
occur only because the Being of what-is, bringing its disclosive ruling to
fruition through the agency of man, is itself happening compellingly as a
contravening of full appearing that lets man and what-is presence toward
one another only in this discrepant way. The setting in place as supply
that evinces itself so pervasively wherever modern technology is
underway is itself the bringing-to-accomplishment of that peculiar
revealing—that peculiar protective preserving that shows forth
(*Entbergen*)—as which, Heidegger insists, we must view that technology.
And in it the Being of what-is, the ruling Happening that is the wellspring
of every disclosure whatever, is bringing its dominion powerfully into
play.

3

In "The Question Concerning Technology," Heidegger initiates his
lengthy discussion by asking about the "essence" of technology.[69] The
word he himself employs is *Wesen*. In German philosophical usage
Wesen ordinarily parallels comparable uses of essence in English.

---

absence of the texts themselves, we can only wonder.
[69]"The Question Concerning Technology," VA 13, QT 3.

Viewed more generally, the German noun has a wide range of meanings. It can mean existence, being, reality, nature, state or condition, demeanor, way, or again creature or living thing, and, in compounds, arrangement, system, or organization, concerns, matters or affairs. The verb *wesen*, from which it comes, is little used in contemporary German. It is sometimes found in religious contexts, where, used of God, it means to live or work or be creative in persons or things. Behind it lies the Indo-European root *wes*, which meant to remain or dwell, and from that root come in German and English various elements in the composite conjugations of the verbs *sein* and to be.[70] Drawing upon this original meaning and presupposing that coalescence of roots, Heidegger accords to the verb *wesen* a crucial role in his speaking of the happening of Being. Moreover, he early reached behind the ordinary philosophical meaning of the noun *Wesen* to bring forward that verbal meaning as resident within it. "The substantive *Wesen* means originally," he says, "not whatness, *quidditas*, but enduring as presence, presencing and absenting,"[71] and he repeatedly states that *wesen* is the same as *währen*, to endure.[72] It is precisely as this *enduring*—which is itself always an enduring-unto (*Anwähren*) wherein bringing-to-while (*Verweilen*) and whiling (*Weilen*) are in play—that the Being of what-is, unfolding as the Twofold, as presencing (*An-wesen*) governs everything that, maintaining itself on-goingly in its own particularity, presents itself by way of time as the latter is opened out by way of man and lived out by him.[73] It is this ruling happening of Being as provisive enduring that the verb *wesen* names for Heidegger, and for him the noun *Wesen* is decisively determined also by the same dynamic verbal connotations.

We have seen that, whether in its happening vis a vis temporally

---

[70]EM 55; IM 59.
[71]*Das Substantivum "Wesen" bedeutet ursprünglich nicht das Was-sein, die quidditas, sondern das Währen als Gegenwart, An- und Ab-wesen.* EM 55; IM 59.
[72]E.g., "The Question Concerning Technology," VA 38, QT 30, and "Science and Reflection," VA 50, QT 161.
[73]See "Time and Being," SD 12, TB 12, and cf. Ch. I, pp. 39f & 49f, and Ch. II, pp. 72f.

Chapter VI

particular configurings of the manifold of what-is or in its far-ranging
governance as destining upon destining, Being sends itself forth variously
and changefully.[74] In so doing, whether in vast compass or in small, it
rules via a multitude of specificities. Since in every mode and instance of
its coming revealingly upon whatever is so as to permit the latter self-
maintainingly to unconceal itself Being holds sway as a gathering forth
that, precisely *as a gathering*, disposes into particularity,[75] these self-
modifications always govern *groupings* of particulars, whose participants,
in their very particularity, answer to and evince some specific mode of
happening. *Wesen*, whether as verb or as ostensible noun, speaks of this
gathering-disposing-happening in its bringing of itself initiatorially to bear
by way of time.[76]

In some instances of Heidegger's use of *Wesen*, notably for modes
of human comportment, the *Wesen* of poetic composing (*Dichten*) for
example, or the *Wesen* of dwelling (*Wohnen*), the word's connoting of an
on-going happening, an *enduring* that disposingly gathers forth, is partly
hidden. Nevertheless it is fully present. Not only are a wealth of
individual undertakings of the comportment in question gathered under the
latter's name, but the human comportment that is named is always one
that, in accordance with the mode of transpiring fundamental to human
existing, in one way or another carries itself out as a gathering to light of
whatever is so presenting itself as to be meaningfully disposed according
to the accomplishing that is underway.[77] Here we need only think of the
appearances of the heavens and the earth brought to light via poetic
composing or of the "things" that yield their disclosures in intimate
relation with human dwelling.[78]

---

[74]Cf. Ch. I, pp. 47-49. Cf. also Ch. XIII, pp. 495-498.
[75]On the happening of Being as a gathering, cf. Ch. III, pp. 124f.
[76]In his discussion of the proper meaning of *Wesen*, Heidegger points to this
connotation by citing a use by Johann Peter Hebel of an old word, *Weserei*, to mean
"the city hall inasmuch as there the life of the community gathers and village existence
(*Dasein*) is constantly in play, i.e., endures as gathered and disposed (*west*)." Cf. "The
Question Concerning Technology," VA 38, QT 30.
[77]Cf. Ch. II, p. 69.
[78]Cf. Ch. V, pp. 200 & 189-191.

On the other hand, when it is the particulars that comprise the manifold of what-is that are primarily in view, the connotations of *Wesen* as an enduring that disposingly gathers forth are much more immediately evident. Here *Wesen*, in a usage that parallels the noun's usual philosophical employment to mean "essence," speaks of the peculiar mode of enduring, i.e., the peculiar mode of pursuing its course as present, that pertains to a given entity precisely as something akin to other entities all of which take their course in a manner that is discernibly specific to themselves. Each particular entity is disposed from out of the ruling happening of Being into a particular on-going present wherein it subsists within a network of interrelations that is always peculiar to itself.[79] It is ruled from out of a distinctive on-goingness that lets it be surveyed and apprehended unitarily in company with its kindred. Each particular, that is, takes place in accordance with its *Wesen.* Thus, individual trees are ruled throughout in their gathered togetherness by a distinctive mode of on-going and self-presenting, the *Wesen* belonging to the tree;[80] and to individual human beings there pertains a distinctive mode of carrying themselves forward, the *Wesen* belonging to man (*Menschenwesen*).[81]

At this point we must be careful to bear in mind the changefulness intrinsic to the Happening that we name as the Being of what-is. For that changefulness is likewise intrinsic to the enduring that disposingly gathers forth, *Wesen*, as which we are now considering Being in its happening as presencing. When the word *Wesen* is spoken, Heidegger says, concerning house or state, what is meant is "the manner in which house and state hold sway, administer themselves, develop and decay."[82] *Wesen* does not bespeak an enduring that is in any sense a changeless permanence. Rather, it bespeaks a changeful staying-for-a-while. Here

---

[79]Cf. Ch. I, pp. 47f & 49f, and Ch. III, pp. 124f.
[80]"The Question Concerning Technology," VA 13, QT 4.
[81]Cf., e.g., "The Turning," TK 38f, QT 39.
[82]*Schon wenn wir "Hauswesen," "Staatswesen," sagen, meinen wir nicht das Allgemeine einer Gattung, sondern die Weise, wie Haus und Staat walten, sich verwalten, entfalten und verfallen.* "The Question Concerning Technology," VA 38, QT 30.

Wesen names the distinctive manner in which each entity that is ruled by a specific self-manifesting of Being, and that is therewith both gathered into a particular grouping and humanly apprehended as belonging thereto, endures, i.e., presents, even imposes, itself, running its course as something that in one way or another comes to be and passes away by way of some particular on-going present.[83]

The *Wesen* of anything is, then, the gathering, disposing enduring that, happening in specificity via time, gives it to happen as it does. Thinking from this perspective, Heidegger rejects any notion of essence as a universal that pertains generally to a class of particulars, whether intrinsically or as a validly applicable concept. Every such notion is for him a mere static abstraction from the powerfully ruling Happening that *Wesen* should connote.[84] The "essence" of anything, when thus wrongly understood, is nothing other than one more particular of the manifold of what-is. Reflecting this view, Heidegger can say ridiculingly that "That which," as *Wesen*, "pervades every tree, as tree," is in no sense—as for him any conception of essence at all reminiscent of the Platonic prototypical tree must suggest—a tree encounterable among other trees.[85]

The *Wesen* of any entity or of any group of entities, as Heidegger understands it, can in no way be induced from the entity or entities in question. As the specific mode of gathering-disposing enduring that rules therein, it can only be recognized. Such recognition, in the dual sense of discerning and announcing, must take place not through any movement toward conceptualization but simply and immediately through the speaking of language, that primary mode via which, through the medium of man, the Being of what-is accomplishes disclosure.[86] Language happens centrally as naming. On-going in its vouchsafing of itself to speaker after speaker, it names in context after context. Here the ultimate singleness of the Happening that brings itself to pass as the Being of

---

[83]Cf. Ch. I, pp. 49f.
[84]"The Question Concerning Technology," VA 39, QT 31.
[85]VA 13; QT 4.
[86]Cf. Ch. III, pp. 118-121.

whatever is[87] is decisively in play. For a single Happening rules in the vouchsafing of language and in the vouchsafing of that which presents itself and, through language, is brought meaningfully to light. The name names forth. Directly it announces that whose peculiar manner of presenting itself, i.e., of enduring, accords with it, the name in question. Thus "house" names forth, i.e., brings meaningfully to light, that which is, by the on-going ruling of Being, gathered forth and disposed as a house among other houses. As naming, language discloses what-is. As immediate, "naming first names forth what-is *to* its Being *out* of the latter."[88] Accordingly, inasmuch as the Being of whatever is rules after the manner of the specific self-modification of which the word *Wesen* speaks, such naming names forth whatever is to its *Wesen*, i.e., to the manner of self-presenting that belongs to it as something gathered into a grouping and disposed as one among others. And in so doing it names it thus precisely from out of that *Wesen* that rules within it.

Here, then, no securely specifying "essence" can properly be extracted from one or more particulars. Rather, that which "essence" attempts to name, a particular mode of self-presenting, can only be caught sight of as the speaking of language announces it in its manifesting of itself. And no such peculiar self-presenting can be susceptible, as an "essence" in the sense of a universally defining Idea or concept might be supposed to be, of clear and precise delineation. For it remains subject to the variously happening concealedness that ever inheres in the coming to pass of Being, the concealedness that language, indeed, ever evinces in its continual offering, together with trustworthy disclosures, of countless others marked by unclarity, ambiguity, or mere seeming.[89] However, an "essence," a *Wesen* in Heidegger's sense, a particular mode of enduring, can be discerned and named—when genuinely naming language proffers itself—as that enduring instances itself in the peculiar self-presenting belonging to whatever takes place under its sway.

But what of modern technology? Here we have to do with a vast

---

[87]Cf. Ch. I, pp. 27-29.
[88]"The Origin of the Work of Art," UK 84, PLT 73. Cf. Ch. III, pp. 125f.
[89]Cf. Ch. III, pp. 115 & 126f.

complex consisting of a multitude of particulars, which now comprises a great and growing portion of what-is. Within it there rules, as a destining, a particular modification of Being in its self-sending-forth as the directive enduring that gathers forth whatever presences under its dominion. Precisely in the manner in which technology so compellingly impinges on us today, this its *Wesen* is pervasively in play. For Heidegger to describe the technological phenomena of our age is not merely to portray what is happening around us and to us. It is, more importantly, to catch sight genuinely and in a transforming way of the specific manner of Being's happening that, as the gathering-disposing enduring that is accomplishing itself in modern technology, looms as ominously decisive for the present age; for that happening is the very destining now accomplishing Being's historical governance.[90]

Heidegger's name for the ruling enduring according to which everything that lies under the dominion of technology now decisively presents itself is *das Ge-stell*. *Gestell* is a noun that in ordinary German usage means frame, rack, framework, or support and that has a variety of specific uses such as skeleton, bookcase, well curb, bed-stead, or jack. In this word Heidegger finds, on the other hand, the very utterance of the happening of the Being of what now is. In *Gestell* there sounds for him the whole force of that *Stellen*, that commanding forth into place as available supply, that so pervasively and definitively characterizes modern technology. He hyphenates the noun in order to point up the presence of the verbal root. And in so doing he stresses also the force of the prefix *ge-*. *Ge-* as a nominal prefix expresses the gatheredness resident in a collective noun. Heidegger takes it to speak of the active "gathering" that intrinsically informs and binds together the sometimes varied elements that such words name. *Gebirg*, the word meaning mountains or mountain chain, names, he says, the "gathering" that "primordially unfolds the mountains into mountain ranges and courses through them in their folded togetherness." *Gemüt*, a word for one's inner orientation, translatable as mind, disposition, feeling, heart, speaks of the "gathering" in virtue of

---

[90]Cf. "The Question Concerning Technology," VA 13, QT 3f.

which all our varied ways of feeling unfold. Similarly, *Ge-stell* speaks of an in-drawn yet provisive gathering. For it connotes immediately the commanding-into-place-as-supply (*Stellen*) that is in play in modern technology, as that setting-in-place is restrictingly "gathered into itself"; and it therewith shows that constrained setting-in-place as a mode of gathering-to-light that via man disposes and lets endure as present, as it names "the challenging claim (*herausfordernden Anspruch*) that gathers man thither to set in order (*bestellen*) the self-revealing as standing-reserve."[91]

Here we see how immediately for Heidegger the technological reality that we find everywhere about us and that we can up to a point describe and understand for ourselves arises out of the ruling of Being and can disclose that ruling to those with eyes to see. Both the comportment of modern man as the protagonist of technology, who incessantly orders and reorders everything as a reserve ready at hand to serve some projected end, and the ease with which everything in our day lends itself to such ordering manifest it directly. Every exploitative encounter with particular entities that present themselves only as numerable, manipulatable elements in the complex fabric of technology evinces it afresh. The specific mode of presencing, i.e., the distinctive disposing enduring, that makes itself felt in modern technology and is named with the word *Ge-stell* is an enframing summons that gathers everything forth into the spare and rigid structuring of purposeful setting-in-order, and that holds sway immediately in every technologically motivated meeting of man and what-is.

As a summons that brings to accomplishment the revealing of what-is, this *Ge-stell*, Enframing, is nothing other than a happening of Being as primordial showing-saying, i.e., of the disclosing carried to fruition via speaking that, originally and definitively named as *logos*, simultaneously gathers, distinguishes, and sends toward its appointed

---

[91]TK 37, QT 36; VA 29, QT 17. Cf. "Identity and Difference," ID 98f, 34f. *Anspruch*—from *ansprechen*, to speak to, accost, appeal to—means claim, demand, title (to something). In *Ge-stell* as an in-gathered forth-gathering *Stellen*, provisive Happening is clearly discernible. For a parallel portrayal of that Happening when considered as *Ereignis*, cf. Ch. IV, p. 164. For a discussion of the relation to be seen between *Ge-stell* (Enframing) and *Ereignis*, cf. Ch. XII, pp. 484-486.

appearing whatever is.[92] It is a particular mode of the governing summons to which man as man ever responds from out of the constitutive core of his on-going existing.[93]   That summons brings itself to consummation by way of modern man's—technological man's—varied activity, first in language but also in all the spheres of endeavor that language informs and makes possible.   All this activity is, seen from Heidegger's point of view, nothing but our according with that which is addressing itself to us.[94] It is the accomplishing, i.e., the living out by way of time, of a claim that is itself a specific self-imposing of the powerful happening of Being. That claim lies also upon all that is. It summons everything to reveal itself, i.e., to present itself and appear, in a particular guise by way of the open arena that man's existing as openness-for-Being provides;[95] and, by way of the peculiar comportment to which man is now being called. Everywhere everything endures toward us now as material to be ordered according to the dictates of the technology that we so assiduously put into play.

It is precisely as the complex bodying-forth of this disclosive summons that modern technology can be said to be a mode of revealing, a mode of harboring forth, i.e., a mode of protectingly maintaining into self-disclosure.[96] Yet what a strange sort of revealing—indeed of protective revealing—this is! If truly discerned, weird and terrible, it should strike us as nothing less than appalling. Heidegger maintains that his unprecedented application to the manner of happening of technology of the word *Gestell* that can name a skeleton is just as gruesomely eerie (*schaurig*) as is that usage itself.[97] The claim of which the word *Ge-stell* speaks is a *challenging* summons to disclosure, an antagonistic provocation, that, as such, but reveals the horribly wasted, the apallingly destroyed.

In the revealing as which, as the disposing enduring (*Wesen*)

---

[92]Cf. "*Moira*," VA 247, EGT 93. Cf. Ch. III, pp. 124f & 119f.
[93]Cf. Ch. III, pp. 117f & 112f.
[94]Cf. Ch. XII, pp. 499f. Cf. also Ch. II, pp. 107-109.
[95]Cf. Ch. III, pp. 119f, and Ch. II, p. 60.
[96]Cf. Ch. I, p. 24, n. 12.
[97]"The Question Concerning Technology," VA 27, QT 20.

specific to modern technology, that claim brings itself to bear, the gathering that is taking place is an ever more frantic heaping up of that to which all inherent spontaneity is being denied, and the disposing that is underway is a thrusting into place that virtually loses sight of that which is at any given time being disposed. Here the bringing to light that is in play has become not a receptive opening-out of one intrinsically meaningful configuration after another, as concealing encroaches upon each in turn,[98] but an increasingly impetuous rush from configuration to configuration, in which all recognized meaning lies not in what is being gathered and disposed, but in the ever multiplying projected ends in subservience to which the latter is being amassed and set thus restrictedly forth.

In the unlocking and transforming, the storing up and distributing and shunting about that, under the dominion of technology, are practiced upon the energy to be found in nature and ultimately upon the host of particulars that, in the context opened up by the availability of that energy, subsists in the technological realm, it is this profoundly strange, inimical revealing that rules. Those technological ways of dealing with whatever is are manifestations of that revealing, and their incessantly on-going interrelation and proliferation bespeak its most fundamental character. Never does it "simply come to an end," and never does it "stray off into the indeterminate." Its very directing of its own "manifoldly interlocking paths" toward goal upon goal and its securing, through the on-going establishing of the standing-reserve, of the directionality thus continuously put into play constantly open up and maintain those paths, so that the revealing in question, as a protective maintaining that ceaselessly despoils as it brings into minimal disclosure, carries itself endlessly forward, ever structuring and imposing itself anew. The goal-pursuing directing-toward and the exploitative securing through which that directing sustains itself are to be seen, Heidegger says, as the "chief determining tendencies (*Hauptzüge*)" of this peculiar mode of revealing that in modern technology now protectingly maintains into

---

[98]Cf. Ch. IX, pp. 339f & 342f, and Ch. III, pp. 135f & 136f.

disclosure only in challenging-forth and relentlessly commandeering into place as wholly expendable supply everything to which its governance extends.[99]

Here we must come to grips with a revealing that is in truth to an extreme degree a concealing.[100] In it, we have to do with a disclosive happening that maintainingly harbors forth only predaciously, a happening that scarcely lets endure at all. Under the sway of this revealing, self-disclosure is suppressed and the self-maintaining that accomplishes itself via it transpires as an overmastering keeping to self that sunders and withdraws from genuine enduring.

The enframing summons, the *Ge-stell*, that rules now as the initiatory disposing enduring (*Wesen*) in play throughout modern technology remains, like every such *Wesen*, a specific mode of presencing. Presencing is enduring-unto (*An-wesen*). It gathers together. That which it lets genuinely presence sustains itself through accomplishing its self-maintaining as a coming and departing, via time, in compliant interrelation. In presencing in openness, man offers himself thus toward what-is, which offers itself thus toward him;[101] and the particulars of what-is proffer themselves thus, also, toward one another.[102] In the sphere of modern technology, however, man does not make his way in openness vis a vis that which meets him, and nothing is allowed to presence truly as itself in simple harmony with its fellows. Everything is set upon. The responsive concern of man with what-is is constricted into an obsessive gaining of control. Concomitantly, the fullness possible to whatever comes to encounter him is shut away. Each entity is called forth only in severely diminished guise, and each is forced into a configuration that is, in the deepest sense, profoundly inimical to it. Each belongs together with other entities, but only through a kind of interrelating that strips it of intrinsic significance and makes it no more meaningful than a mere cog in a machine.

---

[99]VA 24; QT 16.
[100]Cf. Ch. I, pp. 24-26 & 43f.
[101]Cf. Ch. I, pp. 49f, and Ch. II, p. 71.
[102]Cf. Ch. I, pp. 49f & 54f.

Here presencing has all but withdrawn itself from view. The keeping-to-self as which it happens[103] has come so powerfully to the fore that it evinces now only a skeletal vestige of itself. Under its stark dominance man and what-is meet one another like embattled foes. Man refuses himself to what-is. Instead of entering spontaneously into compliant interrelation with that which he encounters,[104] he isolates himself in ruthless self-assertion, brings everything about him uncompromisingly into subservience, and reduces everything he touches to a mere supply to meet his projected and ever proliferating ends. And, for its part, what-is denies itself to man. Whatever offers itself to him does not do so out of the richness of a complex relational context where uniqueness can come into play;[105] rather, it too isolates itself, precisely as material inherently fitted for man's purposeful manipulation. Its multitudinous components stand before him restricted, defying any sort of discerning apprehending that could selectively gather and dispose them in meaningful relation, and hence directly vulnerable to his single-minded attack. Rigorously subsisting in detachment from one another, so far as inherent meaningful interconnection is concerned, the entities of what-is offer themselves not together but one by one. Cloaked in the apparent simplicity of alikeness, they appear as ready counters in the calculating and ordering that modern man ceaselessly undertakes. And there, having surrendered their genuine individuality in surrendering meaningful interrelatedness, they find their *raison d'être* only in standing by as amassed and available for some projected use. Far from properly *enduring* as themselves—as entities gathering and gathered into significance as belonging to a shifting, on-going manifold that is of concern in and of itself—they simply *persist*, isolated but almost undetectable, as members of the ever malleable standing-reserve, where the only gathering in play is a manipulative assembling and arranging of that which remains unheeded in itself.[106]

---

[103]Cf. Ch. I, p. 24.
[104]Cf, Ch. V, pp. 189f.
[105]Cf. Ch. III, p. 135, and Ch. V, pp. 185 & 190.
[106]On persisting, see Ch. I, pp. 50-52.

Here, in the mode of presencing via which both man and what-is as
standing-reserve (*Bestand*) are being gathered forth under the rule of
modern technology, that urgent thrust toward self-isolating, self-assertive
self-maintaining in "mere permanence (*blosse Beständigkeit*)" that
Heidegger sees as ever informing all particulars of what-is—including
man himself—from out of the very happening of Being as presencing, as a
mounting to excess of the impulse to continuance which alone permits
them to withstand (*bestehen*) at all in a continuance fitting for them
(*fügliche Beständigkeit*) the precariousness of the ever-opening present,[107]
has come so radically to the fore as all but to belie gathering-to-disclosure
as such.

Thus is the protective revealing now underway, paradoxically, a
divisive, stultifying, distorting summons that reveals only through
withholding from genuine self-manifestation whatever it calls forth. At
once fragmenting and setting in tension, the disposing enduring that is
now holding sway in man and in all that is cramps and confines them,
dwarfing their relationship to that of orderer and ordered. Indeed, it even
commands them forth into virtual self-annihilation. For it summons and
directs them to enter into the all-consuming realm of modern technology,
where the very processes of planning and execution constantly devour
ordered and orderer alike, leaving them no genuine significance but that
of means to some further, ever-reposited end.[108]

▨   ▨   ▨

---

[107]"In presencing itself . . . rendering constant rises up. (*Im Anwesen selbst . . . steht
die Beständigung auf.*)" "The Anaximander Fragment," HW 328, EGT 43.. Cf. HW
329ff, EGT 44ff. Cf. also Ch. I, pp. 51f. On the inclusion of man, with respect to this
impulse toward persisting, among the particulars of what-is, cf. Ch. II, p. 85, and n.
101. The self-vitiating self-assertiveness that we find evident under the dominion of
technology is clearly the counterpart of that shown us in the meeting of man and what-
is as early depicted by Heidegger in his portraying of human existing in its transpiring
in inauthenticity. Cf. Ch. II, pp. 85-93. On a want of genuine self-presenting as
constituting actual impermanence (*Bestandlosigkeit*), see Ch. X, p. 370.
[108]Cf. "The Question Concerning Technology," VA 34, QT 27, and Ch. XII, pp.
472f.

Technology concerns us. Its very dominance and rapacity often give us pause. Nevertheless, we are prone to believe that however unwieldy and even destructive technology may be, it is always somehow under our control. We see now how radically Heidegger challenges this assumption. Seen as Heidegger would have us see it, the technology so familiar to us emerges as the supremely determinative phenomenon of our day. Its very ways of proceeding and the vigor and pervasiveness of its impinging upon us bespeak its role as that wherein the Being of what-is is now decisively accomplishing its rule. The harmful character of technology that most alarms us, its awesome power heedlessly to invade and despoil both nature and human life, is nothing less than a manifestation of this governance of Being, as the latter, unfolding and maintaining itself as the Twofold, happens as a directing revealing that suppresses even as it calls forth and masks even as it makes known.

This according of decisive supremacy to technology must surprise us. What a sweeping derogation it is of the other phenomena of our age, most importantly of modern science! After all, is not technology a late-comer on the stage of history? Is it not in fact a kind of derivative of science? And, far from its being supreme, does not the more fundamental, more far-reaching sphere of scientific activity and scientific knowledge actually hold priority over it, offering us a viable and proven standing ground from which to launch our technological enterprises, to assess and to control them? To this, Heidegger would unhesitatingly reply that it is in fact precisely in technology's relation to modern science that the priority belonging to modern technology is to be found most strikingly in play and that, accordingly, we cannot legitimately find in science any superior vantage point from which to survey and cope with technology. Instead of assigning to science a preeminence vis a vis technology, we should, he would say, rather look to it to provide, in its accomplishments and procedures, a singular sphere for insights into the way in which technology, as the vehicle of Being's dominion in our age, itself holds controlling sway.

CHAPTER VII

SCIENCE

The word "science" evokes for us a kind of aura. For more than four centuries science has stood in the forefront of human endeavor in the West, and we ourselves are, for the most part, willing and all but unquestioning heirs of its presuppositions and outlook. When we speak of science, we have in mind an activity that, through disciplined observation and research and often through the use of hypothesis and experiment, attains accurate knowledge concerning all manner of phenomena. Our science has many branches, from physics and biology to anthropology and history. These differ widely in the degree of exactitude that is possible to them in measurement and classification and in the substantiation of data and the anticipation of what is still to be discovered, but all see as their primary task the sheer augmenting of what we know.

To designate such disciplines when seen under this aspect, we use the name "pure science." Pure science, we believe, is disinterested. Without ulterior motive, it probes into reality, laying bear its secrets in one realm after another. For this, its dedicated work awakes in us a sort of awe. For with our forebears we moderns feel, if often dimly, that this dedicated pursuit of discovery, this selfless enriching of the scope of human knowledge, is a cardinal expression and evidence of human dignity and greatness. When we view science under this aspect, we say that it is theoretical.

But beyond this, our science has also a second side. In all the scientific disciplines, repeatedly the knowledge that is gathered is drawn upon for the solving of practical problems and the achieving of results that are of utilitarian worth. Such scientific activity is, we may say, "applied science." Characteristically, it holds for us a lesser place than that accorded to the pure science out of which it springs. Often enough we marvel at and pride ourselves on the achievements that it fathers. But its avowedly practical bent makes us somehow uneasy, subtly depriving applied science in our eyes of the lofty status that pure science enjoys.

For Heidegger, on the other hand, there is no "pure," no disinterested science.[1] To him, science, precisely in its character as "theoretical," never so meets the reality with which it has to do as simply to look out ingenuously upon it and receive from it knowledge directly given. Rather, modern science always approaches reality with a predetermined outlook and predetermined intent.

If we were to retain our usual categories, we would have to say that for Heidegger the attitude prevailing throughout every science is actually that which we ordinarily assign to "applied" science alone. For him, science is forever assiduously looking for a reality that will fit successfully into its own preconceived framework and that will therefore, in the last analysis, be manageable on its own terms and be adequate to serve its own projected ends. Heidegger portrays this character of science by saying that precisely as *theory* it is "an observing (*Betrachten*) that strives after." As such, it is "an entrapping and securing refining or working over (*Bearbeiten*) of the real," which "encroaches uncannily upon it."[2] Far from disinterestedly pursuing discovery for discovery's sake, modern science sets out always toward a specific goal, and, moreover, it does so along self-prescribed paths and in accordance with self-prescribed criteria that it never fails to provide for itself. To Heidegger it is this character, indeed, that distinguishes our science from the science of every previous time and that marks it as belonging to the modern age.[3]

1

Always modern science so proceeds as to discover reality as

---

[1]Cf. "Science and Reflection," VA 56, QT 167.

[2] . . . *das nachstellende und sicherstellende Bearbeiten des Wirklichen* . . . . *eine unheimlich eingreifende Bearbeitung des Wirklichen.* VA 55f; QT 167. In this striving of science we should surely discern the urgent transpiring from out of thrownness of human *Being*-there as such. Cf. Ch. II, pp. 68f, and n. 50, p. 89, & p. 100, and n. 157.

[3]Cf. "The Age of the World Picture," HW 75, QT 122.

something calculable. It looks for sequences of cause and effect that it can follow out; invariably it sees whatever it encounters as possessed of discrete components whose relationships it seeks to discern and measure; in classifying, it reckons on the presence of common characteristics that allow it to categorize and compare; and it confidently expects to find patterns and coherences that will permit it to adduce "laws" on the basis of which it will be able to predict and give an account of phenomena not yet met with.[4]

In comporting itself thus, modern science invariably approaches the reality toward which it looks with a prior knowledge that at once defines and makes possible its work.[5] To assert this is, for Heidegger, to point to the fact that science is, always and everywhere, "mathematical." Its mathematical character does not lie in the fact that science works with numbers, although this is undoubtedly often true. *Ta Mathēmata* means in Greek, Heidegger says, "that which man knows in advance in his observations of whatever is and in his intercourse with things: the corporeality of bodies, the vegetative character of plants, the animality of animals, the humanness of man."[6] Number is but one salient instance of such defining characteristics that are always already known. But because of its prominence, the numerical in time drew the name "mathematical" to itself.[7] In understanding the mathematical character of modern science, then, we must go behind this specialized usage. Our science is mathematical because in coming to reality it always already knows what it is seeing. It has a prior conception of what it will discover, and it necessarily views the reality before it from out of that knowledge.

---

[4]"Science and Reflection," VA 57f, QT 167f; "The Age of the World Picture," HW 73ff, QT 120ff.
[5]"The Age of the World Picture," HW 71, QT 118. For an early portrayal of the sciences as possessed of just this character, see SZ 232, BT 275. There the "circle" that manifests the "interpreting" ever intrinsic to human existing, which, broadly speaking, always happens as a coming-to-know that presupposes a knowing-already, is in question. On that interpreting, cf. Ch. III, pp. 113f.
[6]HW 71f; QT 118. See also FD 58, WT 75.
[7]HW 72; QT 119.

First of all, modern science knows in advance that it is viewing a complex of "objects (*Gegenstände*)," a constellation of discrete particulars of one sort or another that can be grasped as standing over against it in a pattern of causal connections that can be discerned and followed out.[8] But these objects are widely diverse and can be observed under many aspects. Fundamentally cognizant of this, science itself diversifies as the sciences. Each of its branches addresses itself to a particular object-sphere (*Gegenstandsbezirk*), an aspect of reality that it can successfully survey.[9] In so doing, it takes for granted a knowledge of the kind of objects it will encounter and the kind of coherences it will find. And, acting out of that conceptual framework, it opens up to itself its particular object-sphere by rigorously scrutinizing the latter as defined comprehensively and exclusively by that very framework.

Heidegger describes this aspect of the basic approach of science as the projecting (*Entwurf*) of a fixed "ground plan (*Grundriss*)" of the reality with which it has to do.[10] Each science views its specific object-

---

[8]"The Age of the World Picture," HW 71ff, QT 118ff.

[9]HW 76f; QT 123. Elsewhere Heidegger uses *Gegenstandsgebiet*, with a similar meaning. "Science and Reflection," VA 56f, QT 168f, and n. 24. In "On The Essence of Truth" (W 88; BW 132), in a discussion concerned with man's meeting with what-is so as to allow to it disclosure, Heidegger makes the point that what-is, when met in openness "as a whole," is not accessible to the sort of "everyday" comprehending that reckons upon and makes available that with which it has to do. What-is must be seen limitedly, with this specificity or that, in order to be so grasped. It is such grasping that is clearly in question in his subsequent portrayals of modern science.

[10]"The Age of the World Picture," HW 71, QT 118, and note 6. *Entwurf*, a noun ordinarily meaning sketch, outline, or plan, here speaks of the projective prior sketching out of the delineative and hence disclosive groundplan in question. The close relation of *Entwurf* with *Entwerfen*, the "projecting," the casting-forward in a gathering-toward, via which man's existing on-goingly carries itself out, and with *Geworfenheit*, the "thrownness" from out of which that projecting transpires, should here be noted. (Cf. Ch. II, pp. 68f.) Indeed, Heidegger's portrayal of science into which we are now inquiring is throughout closely parallel to his original portrayal of the manner of accomplishing of human existing. From out of a provided basis (here the ground plan), a reaching-toward that gathers-toward in an achieving of self-sustaining is carried out (here via explicating) in the course of an unremitting self-carrying-forward (here on-going activity) that has as its motivating intent the very maintaining, through accomplishment, of the self-carrying-out that is underway (here

sphere as a vast if circumscribed theatre of events whose basic character and manner of interrelation it can stipulate in advance. Such structuring allows the science to portray reality to itself as something that can be clearly grasped. And it allows it, on that basis, confidently to anticipate that events in its object-sphere that have not yet actually been glimpsed will also be comprehensible, since they too will fit into the stipulated ground plan.

Having thus established for itself the ground plan of its object-sphere, a given science is bound in its working to adhere rigorously to it.[11] That science does not, indeed it cannot, question its own presuppositions.[12] Instead, it moves within the boundaries that they prescribe, secure in the assurance that the reality with which it deals will show itself as comprehensible in their light.

Modern physics, for example, takes as its object-sphere what is called "nature," and it understands and views nature in a very specific and circumscribed way. Physics does not simply direct its attention to the world as we see it around us. Rather, it approaches that world with the prior expectation that the latter, considered solely in terms of its corporeality, can readily be brought to exhibit itself and can be understood as a "self-contained system of motion of units of mass related spatio-temporally."[13] For physics, this ground plan that it stipulates is wholly definitive. And the peculiar structuring that constitutes it *is* nature as far as physics is concerned. In all that it does—and this holds for atomic physics and classical physics alike—physics is wholly constrained by the plan of nature that it has projected and ever conforms scrupulously

---

via specialization and institutionalization). Cf. Ch. II, pp. 68f & 73-75. On the structure of Happening here presupposed, cf. Ch. I., p. 33, n. 36.
[11]HW 71; QT 118.
[12]"Science and Reflection," VA 65, QT 176.
[13]"The Age of the World Picture," HW 72, QT 119. For physics space is looked upon as something that can be presupposed as a locus wherein entities are found and events occur. Time has a similar antecedent character, whether it be understood as a series of "nows," as in classical physics, or as a "dimension" of the space-time continuum, as in relativity physics. On Heidegger's understanding of space as first of all opened up by the self-presenting of things, cf. Ch. V, pp. 186f. On his understanding of time as first of all the opening-up to disclosure of the Happening that is named as Being, see Ch. I, pp. 37f.

to the latter's dictates. It performs all of its work with the exactitude demanded by the character of its ground plan and of the axiomatic definitions that follow from it, all of which constantly require the greatest possible precision in measurement and in the determination of data. And it never doubts that this single-minded approach will yield ever new insights into the real structuring of nature itself.[14]

The projecting of a ground plan presupposed in advance and the rigorous adherence to that plan and its requirements, when taken together, are, Heidegger says, "the fundamental event (*Vorgang*)" that always underlies the "procedure (*Vorgehen*)" of modern science. Always they are needed to open up one object-sphere or another as comprehensible. It is their presence at the foundation of science, and not the pursuance of a particular methodology, that is for Heidegger primarily constitutive for modern science as such. For it gives to it its definitive character as "research (*Forschung*)," i.e., as an investigation into a reality that is confidently expected to yield up its secrets in the form that science prescribes.[15] The actual "methodology (*Verfahren*)" of science only arises out of and follows upon the self-constituting event in which the latter projects its determinative plan and accepts the stringent obligation of adherence to it.

The crux of that methodology lies in its character as an "explicating (*Erklären*)" of the actual relationships subsisting among the elements composing an object-sphere that appear within the purview of the ground plan that orients and governs specific work.[16] The plan provides a fixed perspective that, as it were, captures reality and sets it over against the viewer in some kind of predictable pattern. But the reality that is in this way schematized in advance is itself anything but simple and static. No matter where or under what aspect it is grasped, it is filled with complexity and incessantly changing. Science needs, therefore, beyond its rigorous adherence to its projected groundplan, a way of proceeding that can allow it to deal with complexity and with

---

[14]"The Age of the World Picture," HW 72f, QT 119.
[15]HW 72; QT 118.
[16]HW 74; QT 121.

change of every sort by discerning among all the elements in its object-sphere, precisely in their changefulness, interrelations that it can categorize and hence comprehend. Every science assumes in advance a coherence among the elements in its own sphere. Whether it be a matter of forces and motions for physics, or actions and occurrences for historiography, an assumption of likeness is always made. The experimental method that is so characteristic of the physical and biological sciences, and often of the humanistic sciences as well, springs out of that assumption. The phenomena that have once been observed are taken to be like their counterparts that have not yet come under observation. Moreover, it is understood that this is necessarily so. It is axiomatic for the sciences that when the elements in an object-sphere are viewed under some particular aspect that pertains to them in common, their behavior in interrelation, fraught with change though it be, is always governed by some "law," some binding necessity.[17]

Only because coherences can be thus assumed can science rely on experiment to advance its knowledge. Whenever an experiment is performed some law rooted in assumed likenesss of character, behavior, and manner of interrelation is presupposed as its basis. The experimenter believes himself able to provide conditions, identifiable in dependable detail, that will permit some particular sequence of events to be followed out to a conclusion foreseen in advance. He takes for granted the existence of patterns of necessary connections on the basis of which he can plan his experiment, calculate its needed course, and anticipate the results it will achieve.[18]

In the historical sciences experiment in this sense is not possible, for the subject matter that they have in view is always something that is already past. Nevertheless, fundamentally the same scientific approach is in play in their work also. There "source criticism," Heidegger says, parallels experiment in other sciences. A ground plan of history is projected that takes history to be a system of events, centered in human

---

[17]HW 73f; QT 120.
[18]See HW 74f, QT 121f.

action, related causally by way of a temporal context. Out of the perspective that is thus provided, relevant source materials are gathered and assessed, collated and interpreted, and, as such work proceeds, the existence of coherences among the phenomena under investigation by means of those sources is once again presupposed. History is seen as an object-sphere made up of varied events among which likenesses can be discerned. Comparisons are made. Calculations regarding those events in their interrelation are undertaken. Prediction of the course of future events may even be attempted on the basis of patterns detected in the past.[19]

In each of its forms of procedure vis a vis its subject matter, modern science, starting from a premise that posits beforehand the explicability of the reality with which it is concerned, is able indeed to explicate that reality—complex and varied and rife with movement though it be—solely from out of the knowledge that it itself has gained and is continually gaining. Thus, constantly, science after science, each diligently pursuing its observing of its chosen object-sphere, through explication clarifies for itself more and more of the ever-changing reality with which it has to do, advancing steadily over against that reality and bringing what is not yet known into the horizon of the presupposed framework by means of which so many observed phenomena have already been made intelligible and secured in some patterned configuration.[20]

This clarifying explication (*Erklärung*) that is wrought out through the methodology of modern science is constitutive for science itself. Science unquestioningly expects to accomplish it and relies confidently on its having been accomplished. And yet it is an explication that remains always provisional. Each science finds itself able to undertake only limited observation of its object-sphere. Only some of the available phenomena that constitute the latter at any particular time are brought under its scrutiny. These exhibit coherences among themselves in keeping with the projected ground plan of the science, and they thus

[19]HW 76; QT 122f.
[20]HW 74; QT 120f.

permit the positing of "laws" in accordance with the prior expectation that just such coherences will be found. But the evidence that they provide remains incomplete. It does not suffice either to establish absolutely the validity of the laws in question for *all* the phenomena with respect to which they would be applicable or to insure that those laws exhaust the range of the interrelations that are in play among the elements constituting the object-sphere. Therefore every science, even as it explains to itself the reality with which it has to do, is constantly faced with the need to seek still further and to verify through additional evidence the very principles of interpretation that it has just been successfully employing. Science remains aware that the laws that it propounds, at least insofar as its own knowledge of them is concerned, are in fact not laws in an absolute sense but are rather what it calls "hypotheses." They are established bases for its work, fixed beforehand in accordance with known data and properly accepted as deserving of credence.[21] But at the same time they are always in need of verification and always being tested, whether through experiment or through the interpretation of further sources, even as they are being applied. Thus the explication of reality that science accomplishes has always a twofold character. "It accounts for an unknown by means of a known," Heidegger says, "and at the same time it verifies that known by means of that unknown."[22]

Through the application of this method the extent of the reality that has been brought into the horizon of scientific knowledge, i.e., the range of that which has become secured as intelligible in the light of one ground plan and one set of hypotheses or another, steadily grows. It is just this increase in reliably secured knowledge that modern science as research is bent on achieving. And the way in which scientific work structures itself in carrying itself forward is determined specifically by that goal.

---

[21]HW 75; QT 121f.
[22]HW 74; QT 121. We should here recall the parallel previously adduced between the structure visible in the transpiring of modern science and that belonging to human existing (*Dasein*) (see this chapter, p. 270, n. 10), and should note the very close parallel between this scientific explication and the interpreting, a viewing on the basis of an already having laid hold of in a seeing, that characterizes that existing. Cf. Ch. III, pp. 113f, and this chapter, p. 269, n. 5.

It is fundamental to the character of modern science that it is what Heidegger calls "on-going activity *(Betrieb)*."[23] As such activity, science never proceeds in a random spirit, working at this and that, investigating here and there.[24] Instead, all its working and all its investigations are directed toward the solidifying of its position in order that it may reach out further and further to draw more and more of reality into the scope of its comprehension. Constantly it advances, but not in a linearly simple way. New findings do not so follow those already made as merely to augment their number. Rather they throw light upon them, now confirming the interpretation accorded to them, now demanding the revision of that interpretation. As this happens, earlier discoveries and hypotheses are taken up into those that succeed them.[25] Science, we might say, builds itself forward.

In order to proceed in this way, assimilating the results of its research and adapting itself on the basis of them, modern science has throughout its history presupposed and required some kind of stable community of knowledge and of contributory work. In our day this "institutional" character of science has come dramatically to the fore. The many branches of science, each intently pursuing its research into its particular object-sphere, have been entrenching themselves more and more firmly, first in universities but also in the many research organizations that exist under the auspices of governments, industries, and private foundations, which themselves depend so extensively on those universities, even as they supplement their work.

Through this burgeoning of the institutionalization of science, the latter's effectiveness has continually increased.[26] Through it, science has become increasingly specialized. Constantly it proliferates into disciplines and subdisciplines. This tendency to specialization, Heidegger insists, should not be viewed, as it so often is, as nothing but a necessary evil that increasingly afflicts the sciences in consequence of the very number and

---

[23]HW 77; QT 124, and n. 10.
[24]HW 90; QT 138f.
[25]HW 76; QT 124.
[26]HW 77-79; QT 124-126.

complexity of the results of their research. It is, rather, a direct expression of the character of modern science as such.[27] From its inception the latter has circumscribed and apportioned the reality with which it has concerned itself. It has invariably directed its gaze toward one object-sphere or another, striving with the appropriate approach to render intelligible the components of each. Now science has gone far toward that goal. An abundant reservoir of results and tested procedures exists. On its basis particular sciences can readily individualize themselves still further. When this happens, ever greater circumscription of object-spheres is achieved, perspectives are narrowed, and the possibility of securely grasping the reality in question within the confines of the newly specialized sciences grows apace.[28]

For science in the highly institutionalized and specialized form that it has now assumed rich opportunities lie open. Results of research can be readily exchanged and reconfirmed; joint projects can be undertaken; particular methodologies can be borrowed or combined across disciplinary lines; and the knowledge and talents of research workers can be readily shared.[29] Thus, far from losing themselves in fragmentation as a result of their specialization, the sciences are through its means actually establishing for themselves the solidarity and unity appropriate to them.[30]

The goal of modern science is the formulating of a methodology that can, on the basis of a projected plan of reality, bring that reality securely under its sway. Now science, ever more inclusive in scope, even as it becomes ever more incisive in its attack, is drawing markedly nearer to that goal. More and more, under the dominance of science as self-adaptive, on-going activity that continually grows in accuracy and power, "the projective tracing-out (*Entwurf*) of an object-sphere" is forever being "built into whatever is."[31] Increasingly, indeed, in our age all manner of entities *are*, for science, and also for those outside the immediate confines

[27]HW 76f; QT 123.
[28]HW 79; QT 126.
[29]HW 77f; QT 124.
[30]HW 79; QT 125f.
[31]HW 77; QT 124.

of scientific work, only as seen by the research that, through its complex carrying-out of itself, has thus so largely succeeded in bringing them to stand before it in a secured, surveyable configuration.

2

Only in our day, when science in its forward growth has attained this formidable pitch of organization and precision in its work, has it begun to display clearly the character that Heidegger sees as most intrinsic to it, a character, namely, that is inherently technological in cast.

In science, incessantly, ever newly projected inquiry and observation are being carried out, and through that on-going activity the reality[32] that is under investigation is being ever more thoroughly ordered and categorized in accordance with the presuppositions from out of which science carries itself forward. Here the fundamentally "technological" character of modern science begins to be discernible. Under the dominance of that science, the particulars of the reality that is brought under observation are systematically emptied of their significance as the particulars that they are, even as they are gathered to light as evidence so as to further the investigative advance of science itself.[33] In this connection, the words of the twentieth century physicist Max Planck are striking: "That is real which can be measured.[34]" That is real, that is to say, which can be assessed and grasped by means of some already established standard that allows whatever is being considered to be met with not in itself but only insofar as it shows itself in terms of its commensurability with whatever else the particular "measure" in question is intended to identify and bring to view.[35] And just this measurability of the measurable "real" assures the latter's availability for the amassing of a

[32]Modern science as described above is "the theory of the real." "Science and Reflection," VA 46, QT 157. See this chapter, p. 268. For the meaning of "real," see VA 49ff, QT 160ff. See also this chapter, p. 282.

[33]On just such comportment as characteristic of technology, cf. Ch. VI, pp. 242-245 & 259-261.

[34]"Science and Reflection," VA 58, QT 169.

[35]For Heidegger's understanding of the full meaning of "measure," see chapter V, pp. 195-218, especially pp. 200f & 216f.

body of reliable knowledge on whose basis subsequent measuring and explication can proceed.

This attitude toward the "real" is clearly characteristic of the physical sciences, but it is definitive for the biological and humanistic sciences as well. It is not only electrons or hydrogen atoms or supernovae that have their importance for science not as individuals but as examples of a class whose behavior is studied, measured, and used as the basis for further research. Plants and animals in biology and human beings in psychology have their decisive importance for their sciences too solely in virtue of their membership in some particular class. Anthropology assiduously studies particular cultures, but ultimately in its work the novel is always fitted into some structure of classification and so made available to aid in future investigations. Even historiography, which takes as its object-sphere the vast panorama of human events seemingly so diverse, constantly undertakes to classify those events. It does not view them in their uniqueness. Rather, it sees them as examples of trends or outcomes. They are for it, finally, *comparable* phenomena whose precise character falls away into insignificance. Even "the great," Heidegger points out, becomes for this science merely "the exceptional." It is thus fitted into the ruling conceptual framework and hence rendered incapable of impeding the progress of research.[36]

More and more, calculating is becoming dominant in every science.[37] Increasingly the information that is handled in all the sciences and that is exchanged among them is stripped of extraneous details regarding the phenomena it concerns. Everywhere the computer is assuming an increasingly central place in scientific research, and data processing is becoming all-important. Heidegger points to the fact that all the sciences are now being drawn into the domain occupied by the science that has always been the normative one, mathematical physics. Biology, he says, is transforming itself into biophysics, anthropology into anthropophysics. Today cybernetics, that statistical study that aims at

---

[36]"The Age of the World Picture," HW 75f, QT 122f.
[37]On calculating as central to the carrying forward of technology, cf. Ch. VI, pp. 241f.

controlling the flow of information in particular systems, is, he says, the place where all the sciences meet.[38]

In all this the science of our time stands squarely on technological ground.[39] Increasingly the elements composing the object-spheres of the various sciences are being reduced to a state very like that of the contents of the standing-reserve that is continually being set in order under the rule of modern technology.[40] For increasingly those elements are losing even their standing as objects—even as such standing is regularly being forfeited by whatever is met with specifically in the domain of technology[41]—and are being accorded significance in their sciences only in virtue of their relationships with one another, in systematizations in which their actual importance lies simply in their availability for further research, i.e., for a further ordering of reality in keeping with some projected scientific ground plan. The modern atomic physicist, Heidegger says, still dreams of "being able to write one single fundamental equation from which the properties of all elementary particles, and therewith the behavior of all matter whatever, follow."[42] In his own science, viewing subject and viewed object have already disappeared, and the relation between them is alone taken into account.[43] Were the hope of the "one single fundamental equation" ever to be realized, only the *relations* subsisting among particulars would be of any importance at all. And nothing whatever but standing-reserve would remain.[44]

But is science, then, indistinguishable ultimately from technology?

---

[38]"A Heidegger Seminar on Hegel's *Differenzschrift*," translation by William Lovitt, *The Southwestern Journal of Philosophy*, Vol. XI, No. 3, Fall 1980, p. 30. For the complete German rendering of the passage, which, like the English, is a translation from the French (*Questions IV*, Paris, Gallimard, 1976, p. 258), see VS 51. Heidegger can remark that physics, important though its present role must be acknowledged to be, "remains double-entry accounting." See "The Nature of Language," US 210, WL 103.

[39]Cf. Ch. VI, pp. 141f & 235-239.

[40]Cf. Ch. VI, pp. 244-249.

[41]Cf. Ch. VI, pp. 244f.

[42]"Science and Reflection," VA 61, QT 172.

[43]VA 61; QT 173.

[44]Cf. VA 58f, QT 170.

Heidegger's answer to this question would be yes. The very destining of Being that rules in modern technology[45] rules in modern science as well. As that destining comes more and more overtly into play, the comportment of science and of those who further its work presses forward ever more intently into the realm where technology is already so formidably carrying itself out in its assigned manner.[46] Nevertheless, the distinction that we make between science and technology is not simply illusory. There is even a precedence that belongs to science vis a vis technology and an inherent interrelation between the two.[47]

Modern science is a forerunner. Machine technology has arisen on the ground prepared through it. This is not to say at all that science has fathered technology as something derivative from itself. Rather, Heidegger says, science has arisen in our age in order to accomplish its given task in the service of the technology that, chronologically speaking, succeeded it by more than a century.[48] The destining of Being that rules in both is the sole determinant of their interrelation.

Always the Being of what-is, unfolding as the Twofold, in its governance sends itself forth changingly, on-goingly, by way of time.[49] Today it holds sway as a destining, Enframing (*Ge-stell*), that is a challenging summons that, through the taking place of technology, threatens to obliterate even as it reveals.[50] There, constantly, the intensity of its depredations grows. The degree to which it directly manifests itself is changing, and it has changed hitherto. Modern science is the harbinger of the destining that is, as it were, only now bringing its

---

[45]Cf. Ch. VI, pp. 257-259.
[46]"The Age of the World Picture," HW 78, QT 125.
[47]Here and elsewhere in this study the reader should be careful rigorously to separate "relate" from its usual meaning. Here "relate," in contexts where the Being of what-is is in any way under consideration, does not speak at all of a connecting of inherently discrete elements, one of which is to be subordinately referred to the other. The relating that is now in question is a *holding-toward-one-another* of participants that, while they are distinct from one another, are ultimately one. For we are using "relate" to speak of the happening of the single Twofold. See further, Ch. IV, p. 145, n. 10.
[48]"The Question Concerning Technology," VA 29, QT 22.
[49]Cf. Ch. I, pp. 257-259, and Ch. XIII, pp. 495-498.
[50]Cf. Ch. VI, pp. 257-259 & 260f.

manner of ruling fully to bear in the unfolding of the manifold of what-is.[51]

Modern science arose as an observation intent on so structuring the components of a field of objects as to be able to scan them, fix them in intelligibility—through procedures directed at their assumed measurability—and further its own forward progress on the basis of its knowledge of them. In so doing, it both responded to the manner of disclosure ruling in the reality it met with and brought that disclosure to completion. If science, in its looking, saw before it expanses of commensurable objects among which it confidently expected to find patterns of coherence and causal connections, this took place, ultimately, because the presencing that ruled in that reality had the character of objectness (*Gegenständigkeit*), i.e., of a letting-stand comprehendedly over-against in discreteness, wherein the interrelating that united the particulars of the presencing manifold of what-is was happening as a connecting of initiating act and consequent result.[52]

The "real" in German is *das Wirkliche*. This adjectival substantive derives from the verb *wirken*, to work, whose fundamental meaning is, Heidegger says, "to bring hither and forth."[53] For him, when we speak of reality (*Wirklichkeit*), we are speaking of the unfolding Twofold, the Being of whatever is. For reality, when adequately thought, means "that which, brought forth hither into presencing, lies before," and it means equally "the presencing consummated in itself of what brings itself forth."[54] Hence, when modern science observes the real, it becomes a participant in the particular manner of "working" holding sway in the self-presentation of that which encounters its scrutiny. That working is nothing other than the discrepant bringing forth into presencing that is the peremptory summons to a relentless setting-in-order that Heidegger names with his pregnant word *Ge-stell*.[55]

---

[51]Cf. "The Question Concerning Technology," VA 29f, QT 22.

[52]"Science and Reflection," VA 51ff, QT 161ff; VA 58, QT 169. On the coming to dominance of causality as the mode of relating found to characterize reality in its particularized happening, cf. Ch. XI, pp. 409 & 410f.

[53]VA 49; QT 160. On working, see further Ch. IX, pp. 347f & 352f.

[54]VA 49; QT 160. On the dual-unitary happening of the Being of what-is, the Twofold, see Ch. I, pp. 27-33.

[55]Cf. Ch. VI, pp. 258f.

Heidegger does not use *Ge-stell* itself when speaking of the engagement of science with the object-spheres whose components are its concern. But its presence lies close at hand. Repeatedly it resounds through words in which he speaks of science as an observing of the real. Thus the real is, when there precisely defined, "what-presences-that-sets-itself-out-hither (*das sich herausstellende Anwesende*)."[56]   It is as possessed of this character that what presences so manifests itself in our age that "it brings its presencing to a stand in objectness." So doing, it sets itself in position (*stellt sich*) for the kind of observation with which science advances upon it. For its part, science aims, in its observing, directly at that objectness and thereby challenges forth the real to appear in the light of its projected ground plan and in that light alone. In so doing science "sets upon (*stellt*) the real" in a manner that is intended so to situate (*stellen*) the latter that it will therewith exhibit itself (*sich darstellt*) as a surveyable network of interacting causes. Science strives in this way to achieve a securing (*Sicherstellen*) of the real confronted in its objectness, by making sure of it as something certain, something that can be comprehended and traced out. Hence it compartmentalizes the real and sets it fast (*stellt . . . fest*) in one object-sphere or another. It sets it before itself through a representing (*Vorstellen*) that is always actually an entrapping (*Nachstellen*) of the real in some conceptual framework prepared beforehand that, delimiting and structuring in advance, can alone serve to place any object-sphere in association (*beistellen*) with a branch of science. It is precisely in this way that every science accomplishes what Heidegger calls its *refining* of the real, its working over and adjusting of the latter through its own entrapping observation. Each science looks attentively toward the real, striving to grasp and fix it for itself. This it achieves, often to an impressive extent. Yet it can see the reality that falls to its scrutiny only as an object-sphere. It cannot encompass and come to grips with (*umstellen*) the fullness of the manner of presencing belonging to that reality, since the objectness in which the latter is always already offering itself is but one way in which, in the full

---

[56]VA 56; QT 167. Cf. VA 50, QT 161, and especially n. 10.

sense of its happening, it displays itself (*sich* . . . *herausstellt*), and it is objectness alone that awaits the scientific gaze. Science is ineluctably required to direct its attention toward that objectness, for only through the responsive revising of the real into an object-sphere that each science accomplishes can that objectness come to manifestation. Only by way of such "refining," which is ever in progress, "does objectness first define and determine itself," for only there is it "expressly set forth (*eigens erstellt*)."[57]

This telling and repeated use by Heidegger of *stellen* and of verbs formed from it strikingly bespeaks the ruling in the phenomenon of modern science of the imperious summons to ceaseless derogating ordering that now governs in modern technology. As in technology, so in modern science as such, a purposeful ordering of the particulars of the manifold of what-is is continually accomplished, and through it those particulars are brought into a type of relationship with one another—that of the measureable, linked via cause and effect—that necessarily precludes their showing themselves as themselves and restricts their recognizable significance to that of comparable useful phenomena that are always to be seen ultimately as links in a chain leading toward some result. Presencing in this guise, those particulars have their standing in the scientific milieu, as in the technological, because of their availability to serve some end beyond themselves.[58] Here they matter, finally, because they are at hand

---

[57]The discussions using verbs composed of *stellen* that have here been gathered together are to be found in "Science and Reflection," VA 56, 61f, 67; QT 167, 173f, 178. Several of the verbs, notably *sich herausstellen, nachstellen,* and *sicherstellen* occur with some frequency elsewhere in that essay. On science as refining observing, cf. this chapter, p. 269.

[58]Cf. Ch. VI, pp. 259 & 262f. It is interesting to note that in his discussion of science in general Heidegger does not use the verb *bestellen*, which figures so centrally in his discussion of the ordering of the standing-reserve that takes place under the rule of technology. Cf. e.g., Ch. VI, p. 259. There *bestellen* appears only in the context in which Heidegger remarks, parenthetically (VA 61; QT 173), that in contemporary atomic physics the relation between subject and object becomes a standing-reserve to be set in order (*wird ein zu bestellendes Bestand*). It therefore appears that he reserves the word *bestellen* as a *terminus technicus* to be used only in specifically technological contexts. But see "The Question Concerning Technology," VA 29, QT 21, where the application of *Bestellung* is specifically extended to include the ordering accomplished in science.

to confirm a projected ground plan, to help establish the lawfulness governing an object-sphere, or to undergird the further investigations of science as research. The manifold of what-is reveals itself for science precisely in this way, in accordance with an initiatory happening of Being as a presencing that, as objectness, withdraws that manifold into a standing-over-against that brings particulars to interrelate in commensurability via causality.

Once again it is man who is called upon to bring this ordering to fulfillment. As scientist, he is summoned to order the self-revealing, not now as standing-reserve, but rather as areas of objects whose coherence must be disclosed. Again a "challenging claim" imperatively "gathers" him to his appointed place,[59] setting him over against what-is, in order that, equipped with an eye for its objectness, he may work over and adjust whatever presences to his observation, in accordance with a projected ground plan that is always itself nothing other than an immediate putting into play of the objectness that is the determining goal of his every attitude and procedure.[60] A demand lies upon him that through his scientific undertakings object-sphere after object-sphere of what-is should be surveyed and explicated with respect to its movements and interrelations, to the end that what-is as a whole may stand forth increasingly in "lawfulness" and intelligibility.

Here we come upon the ultimate basis of Heidegger's contention that there is no pure science that is disinterested, that is, there is no science that is "without a goal (*zweckfrei*)."[61] Every science, as science, sets out with a specific if unavowed intention that is in fact constitutive for it. In its observing, modern science always lays hold upon the real in such a way as to render it congruent with presuppositions already held.

---

[59]For the identification of this "challenging claim" and of the role to which, as ruling in modern technology, it summons man, see "The Question Concerning Technology," VA 27, QT 19. See also Ch. VI, pp. 259f.

[60]See "Science and Reflection," VA 57, QT 169. The ubiquity among the sciences of this concern with objectness that arose as definitive of science as such is vividly shown by the fact that through elaborate procedures atomic physics, also, diligently strives to make visible, even if very indirectly, the "objects" with which it has to do. See VA 61f, QT 173.

[61]VA 56; QT 167.

Invariably it strives to set the real forth before itself coherently in objectness. Indeed, it can do no other. For the goal that shapes its manner of proceeding is given in the destining of Being that calls it forth. Science, in the last analysis, so comports itself as to serve that destining and hence also to serve the technology in which the latter brings its ruling to consummation.

In the rise and the ever more complex growth of modern science, the initiatory disposing enduring that now lets technology presence as it does[62] has unremittingly prepared the ground for technology as such and hence also for its own arrival into full dominion. How readily a revealing of the particulars of what-is in terms of an objectness that serves primarily to disclose structural coherences could pass over into a revealing of those particulars in terms of an availability that snuffs out individuality and leaves in evidence only the statistical interrelatedness characteristic of the standing-reserve! So easy, indeed, was this transition that science, for all its characteristic orientation toward objectness, has been able, often without even a sense of disjunction, to let its definite concern with objects as such slip from it in favor of an even more engrossing preoccupation only with calculable, statistically statable interrelations as they are being fitted into some preordained process. With the inception of the stance and methodology that characterize modern scientific research as such, the manifold of what-is could be ever more inclusively approached and grasped as something well assured, something that could be reckoned out and reckoned upon by way of a controlling conceptualizing that accomplished the defining determination of the way in which that manifold would presence, the way in which it would *be*, in this technological age.

This, science as theory has long been achieving. But in this role science does not directly lay hold of the reality with which it has to do *in its concreteness*. True, it handles certain particulars of what-is in experiments or in the assembling of evidential materials. But the genuinely significant ordering that it brings about takes place in the realm

---

[62]Cf. Ch. VI, pp. 257f.

of ground plan and viable hypothesis, discernible law and reliable prediction. Our so-called "applied science," while maintaining close touch with the realm of theoretical research, moves toward a direct dealing with particulars. It subsists, as we are so keenly aware, on ground somewhere between science proper and technology. But it is only in technology itself that a full involvement with the multitude of the concrete particulars of what-is takes place. Technology literally lays hands upon them. It harnesses rivers and distributes electricity, mines coal and stockpiles it, grows timber and manufactures paper, recruits persons and assigns them to posts. It deals immediately with every sort of entity, continually procuring and fabricating, amassing and allocating, utilizing and consuming, in pursuance of its multitudinous undertakings. In all this, technology presupposes the accomplishments of theoretical science. Indeed, it lives out of them, taking for granted in its own planning the laws and coherences that science has disclosed and relying, with respect to countless entities, on a predictability of behavior that science has established. But the ordering of what-is that technology itself carries out is more complete than that of science, precisely in being more direct. Technology brings to fulfillment the work that science as such begins, just because it does *apply* that work *in concreto* and totally captures and secures the particulars of what-is in the full immediacy of their standing at hand. It is the particulars themselves, and not the particulars as represented in any conceptual configuration, that are gathered by technology into the standing-reserve. No matter that in it their individual identity is submerged; they stand there *as themselves*, ready and available, under the dominion of the inimical revealing that only there brings its power to despoil fully into play.

Almost paradoxically, it is the most abstract of modern sciences, mathematical physics, that has played the most salient role in preparing for technology's very concrete work.[63] In the first place, physics, with its fundamental concern to exhibit reality as a coherence of motions among units of mass viewed in spatio-temporal relation, has always

---

[63] "The Question Concerning Technology," VA 29, QT 22.

grasped and displayed nature as a "surveyable network of forces,"[64] a portrayal that has become even more thorough-going with the identification of mass and energy in twentieth-century physics.[65] In so doing, physics provides to technology that assessment of nature as a vast storehouse of reserves of available energy (*Energiebestand*) that is fundamental to the latter's happening.[66] The demanding summons that so rules in technology as to command forth what-is into the standing-reserve does so ultimately through revealing nature precisely in the guise that physics has already discerned. Secondly, the exactitude of measurement that mathematical physics pioneered is absolutely indispensable to technology in the execution of its mandate to order everything as standing-reserve.[67] Often technology employs physics directly,[68] not least because of the specificity and inclusiveness of the data that it can supply. Here we need only think of the military and industrial utilization of knowledge from physics about atomic fission or of the dependence of the aeronautic and now the aerospace industry on information concerning velocity, gravity, or friction.

The relationship between physics and technology is in some respects reciprocal. The research of modern physics can proceed and its precise determinations can be gathered only through the use of the sophisticated apparatus that technology provides. This becomes increasingly the case, and indeed not for the science of physics alone. But the priority belonging to an activity that draws upon a contributory undertaking remains with technology.[69] The scientific achievements that the latter makes possible take place ultimately for the sake of technological advance. Above all, technology relies on the precision and accuracy with which those scientific achievements are attained. And it itself, indeed of necessity, adopts the same approach in its handling of that

---

[64]VA 29; QT 21.
[65]VA 31; QT 23
[66]Cf. VA 29, QT 21. Cf. also Ch. VI, pp. 245f.
[67]Cf. VA 30f, QT 23.
[68]VA 29; QT 21.
[69]VA 21f; QT 13f.

with which it has to do. Continually and increasingly it applies the rigorous exactness of measurement and calculation of which physics is the exemplar *par excellence*.

In physics, the quintessential science among the sciences, with its immediate, intrinsic reliance on mathematics, the destining ruling in the modern age has called forth for technology, in advance, the method for dealing with reality—a directing toward it of a measuring and calculating that can be relentlessly precise—that could alone enable technology at once to aim its attention directly at the minimal, i.e., the statistical, relation subsisting among the particulars of what-is, to lay sure hold upon those particulars through its planning, and to set them continually in order under the aegis of but one criterion, their availability for subsequent use. And it is, at the present, through just this thirst for exactness that all the sciences are being transformed into forms of physics and being driven ever more inexorably overtly into the sphere of technology itself.

Beyond this preeminent role of physics, the destining now governing in modern technology has prepared for its own full dominion via the latter's course by mandating for modern science as such a procedure that, in every one of its fundamental aspects, prefigures and opens out upon the more direct and concrete approach and methods of technology in its encountering of what-is.[70] Thus, science always sets out on its course with the projecting of a ground plan of some object-sphere or other that offers in advance a conceptual framework via which the reality under scrutiny is then brought into view in objectness and rendered intelligible in that guise. Correspondingly, technology characteristically undertakes its work with the projecting of some plan that allows it to see and evaluate the materials under consideration with regard to their appropriateness, availability, and employability in pursuance of the specific end that it has in view. Science adheres rigorously to its ground plan, never questioning it and never looking beyond the sketching-out of reality that that plan achieves. Technology intently follows through with its calculations and its planning, never doubting the viability and

---

[70]For discussion of the intent and methods of technology, see Ch. VI, pp. 228-232 & 241-250. For detailed consideration of those of science, cf. this chapter, pp. 268-279.

superiority of its manner of operating and never glimpsing anything in any other light. The methodology of science, as an explicating of the real, uses the phenomena that it discloses in their coherences to carry itself forward to further discoveries. In so doing, it brings those phenomena into an interrelation that is basically one of utility. Those already known serve to suggest or predict what is yet to be found, while those that are newly discovered serve to confirm or establish what has already been made known. And for science the final importance of all phenomena lies in just these functions. The methodology characteristic of technology aims at so ordering everything as to have it immediately at hand for technology's use. Everything is continually assembled into an available, utilizable reserve, where entities bear significant relation to one another and to the technology that assigns them their place only insofar as they are being fitted into some preordained process. Those entities do not, like the phenomena of science, foreshadow or corroborate one another or throw light on the order in which they stand. Rather, they stand ready, supplied by and supplying the exigencies of technology and allowing the latter to press forward to further ordering that will eventuate in a still greater availability of everything for use. Science carries itself out as "on-going activity." As such it concentrates its gaze more and more, by dividing itself into specialty after specialty, while at the same time it makes the extent of its surveying ever more inclusive by institutionalizing itself so that its disciplines can draw upon one another, build themselves up in concert, and constantly extend their mastery over the real. Technology ceaselessly proliferates. New needs are perceived or posited. Techniques and production processes multiply, spawning ever more sophisticated successors for themselves. At the same time technology entrenches itself ever more firmly. New avenues of communication are opened up, information and know-how are exchanged, particular methods and materials are more and more widely used, standardization grows, projects and processes play into one another to an increasing extent, and technology, therewith consolidating itself even as it ceaselessly divides and subdivides into new undertakings, continually extends the range and strength of its control over whatever comes under its sway.

In each of these instances, science, as harbinger of the manner of holding sway that is bringing itself to fulfillment in the modern age, displays in advance, if in a guise peculiar to itself, the very characteristics that are definitive for technology. Small wonder, then, that now that technology is invincibly underway, science is eagerly transforming itself into cybernetics and so moving steadily into the sphere of technology as such. Science has played its part. It has prepared in thought and attitude and action for the ascendancy of technology. Now objectness has withdrawn, and objects as such have disappeared. The revealing presently holding sway as decisive is a revealing via which only calculable relationships stand forth and the interrelated is but standing-reserve placed in some needed order to serve some intended end. Science as science has been superseded, or, more accurately, the day has come when science can display its true character and appear as technology itself.

▦   ▦   ▦

At this radical juncture every depiction of science as preeminent over technology and every hope that science may, from that preeminence, be able to control and direct technology's course must, from Heidegger's perspective, be engulfed in futility. And yet science is but one phenomenon of our modern age. What of the rest? Surely our hopes of gaining mastery over technology do not and need not lie with science alone. Above all, in face of technology's often menacing presence we are prone to believe that it will be possible so to effect a transformation in our fundamental outlook on life and on all that fills our world that through new attitudes, or, as we say, through the establishing of new priorities, we shall succeed in relegating technology to its proper place and function and shall be able consistently to use it wisely in pursuance of goals chosen wholly from beyond its own sphere. Is there not some new philosophy, some new and compelling orientation of mind, that can provide for us the vantage ground vis a vis technology that we so often crave? Here once again Heidegger's answer would be no. Under no circumstances can philosophy serve us successfully in the combating of technology. For philosophy is not an autonomous activity able to find and assume new postures for itself. Quite to the contrary, philosophy, and with it every kindred phenomenon of our age, has long been taking place and must take place still under the rule of the directing destining that governs so inexorably in technology itself. If we look to philosophy, therefore, we shall find but a further witness to the peculiar character and pervasiveness of that rule.

CHAPTER VIII

PHILOSOPHY

When we say philosophy, the word connotes for us first of all an approach to an understanding of the world that is consistent with itself and that on the basis of that consistency succeeds in giving a coherent explanation of diverse phenomena, indeed often of all phenomena, that come into its ken. Beyond this basic definition we usually have in mind two quite disparate conceptions of what philosophy is. On the one hand, philosophy is for us an academic discipline. In it the interpretations of reality that are characteristic of various thinkers are expounded. When these are drawn from the past, they regularly strike us as irrelevant and outdated. They are like quaint pictures, highly interesting perhaps, but not speaking to us from our own perspectives or meeting our own concerns. The work of more contemporary philosophers—those of the last two hundred years—engages us more directly. There we find viewpoints and insights that often appeal strongly to us and that we take to ourselves, incorporating them into or making them the basis of our own fundamental outlook. There too, we find interpretations that repel us. Against these we may react vehemently in opposition, so that our own views are in fact shaped and given cogency by those interpretations and by the threat or adverse pressure that we feel from them.

On the other hand, there is current among us another, more popular, meaning for the word philosophy. We can speak of the "business philosophy" or the "educational philosophy" or even the "sporting philosophy" of individuals or groups. Indeed we often take it for granted that "everyone has a philosophy," i.e, a way of looking at life and hence a way of living that is consciously held, although not necessarily well thought through, and is adhered to with considerable constancy.

A philosophy can, then, be for us a doctrine or system of doctrines or, at the other extreme, something like a mere temper of mind. In every case, we are apt to attribute to it a highly individual character. To us, great philosophers and common people alike find or work out their philosophies, and it is always the right and privilege of every person to evaluate any given philosophy, to align himself with it, to borrow from it for his own self-orientation, or to reject it out of hand.

For Heidegger such an understanding of philosophy is wholly mistaken. For to him philosophy is by no means primarily a human enterprise wherein, through the centuries, individuals have attempted to think out meanings and discover structures as they looked out upon their world.[1] And it is certainly not an enterprise open to all and sundry, where the "rightness" of a philosophy or a philosophical insight can be decided by any interested person[2] and a particular philosophical stance can be accepted or rejected at will. Philosophy is, rather, a thinking that pursues its way in accordance with the dictates of the Being of what-is. It brings to utterance in accordance with the various "stampings" as which the latter, ruling determinatively as the presencing of what presences, manifests itself from age to age.[3]

For Heidegger, philosophy is metaphysics. It concerns itself with what-is as such, as the latter is considered in relation to its ground as which—concealing both its self-differentiation vis a vis what-is and the happening of the two as a unitary Twofold and showing itself as though participant in one simple structuring—Being itself rules.[4] Indeed, in accounting for and establishing that ground by thinking and saying it forth, metaphysics, at any given time, "grounds an age," for it takes place as the very unconcealment of what-is.[5] However varied its pronouncements and discussions may be, it is always "the truth of what-is as such in its entirety."[6] Happening thus, time after time, via its words it

---

[1]Cf. "The Word of Nietzsche: 'God Is Dead,'" HW 193f, QT 54.
[2]Cf. "The End of Philosophy and the Task of Thinking," SD 62, TB 56.
[3]Cf. Ch. III, pp. 131f.
[4]"The End of Philosophy and the Task of Thinking," SD 62, TB 56. "Overcoming Metaphysics," VA 74, EP 87. Cf. Ch. III, pp. 130f. On the manner of happening of the Being of what-is that is here hidden from view, cf. Ch. I, pp. 27-33.
[5]"The Age of the World Picture," HW 69, QT 115.
[6]. . . die Wahrheit des Seienden als solchen im Ganzen. "The Word of Nietzsche:

speaks forth Being according to the latter's mode of self-unconcealing and therewith brings that self-unconcealing immediately and constitutively into play.

1

The metaphysics of the modern age—the decisive manifestation of the latter's philosophical thinking—has precisely this grounding, revelatory character. From its outset, which takes place with the thinking of Descartes, it brings what-is as a whole into a new mode of disclosure through its speaking. It sets it forth as something that *is* on a new basis, and one peculiar to the present age. Therewith it grounds that age and all its phenomena in a way that is, so far as overt understanding is concerned, radically new.

In his philosophy generally and above all in his *ego cogito (ergo) sum*,[7] Descartes gave utterance to and hence brought determinatively into play a radical transformation in the stance of man vis a vis all other components of reality and in the way in which those components in their turn were brought into relation to man. Behind this sudden enunciation of a grounding metaphysics lay changes in man's attitude within his world that had been gradually taking place. Under the hegemony of the Roman church, in the later Middle Ages, men had found themselves quite at home in a world whose stable hierarchical structure was assumed. God, understood as first cause, was seen as creator and sustainer of the whole of creation. Created beings were viewed as participant in an ordered structure of levels of being, and that structure found its guarantee in the fact that its pattern and prototype, to which the orders of creation corresponded, lay in the mind of God.[8] Men could know this scheme and know their place within it. The written word of scripture and of church doctrine stood as a locus of certainty. Assent to it and to the practice that it enjoined offered believers the possibility of an inner secureness, born of

---

'God Is Dead,'" HW 193, QT 54. On truth as unconcealment, cf. Ch. I, p. 38, and n. 49.
[7]Heidegger makes *ergo* parenthetical. "The Age of the World Picture," HW 100, QT 149.
[8]Cf. HW 93, QT 141.

rightness of outlook and life, with regard to this world and the next.[9]

At the end of the Middle Ages, however, the stability that had prevailed began to give way. The authority of the church weakened. The certainty offered in its doctrine no longer bound men or sufficed them in either their religious or intellectual questioning. And these two spheres of inquiry rapidly fell asunder. In the sphere of faith, men asked in a new way concerning their salvation. They asked, that is, Heidegger says, concerning their assured continuance (*Beständigkeit*).[10] How might they be certain of that continuance? This question was the question of justification (*Rechtfertigung*), i.e., of a rightness before God that could ensure the security that was sought.[11]

Both here and, more importantly, in a wider and increasingly secular context of thought and inquiry men began to assume as their proper prerogative a freedom to decide for themselves what standard should hold for them, what should be authoritative.[12] Yet even as men claimed that freedom, they remained determined by their former ways. Precisely as they rejected the certainty of salvation proffered by the church, they yet craved some other certainty that could undergird for them a comparable assurance regarding their own unassailable continuance, i.e., regarding their very being. They required a new sort of knowledge in the midst of their world that could confer rightness upon them and so render them secure.[13]

It was out of this age in transition that Descartes spoke. His thought and his words offered that which consciously men were imperatively seeking. In them there came overtly to light a new basis

---

[9]Cf. "The Word of Nietzsche, 'God Is Dead,'" HW 226, QT 90, and "The Age of the World Picture," HW 75, 83; QT 122, 130.

[10]This rendering of *Beständigkeit* draws on the fact that it can mean continuance, persistency, stability, constancy, steadfastness. In the context in which the present thought appears, a connotation of sureness is clearly intended by Heidegger for the continuance in question. For a related use of *Beständigkeit* by Heidegger, cf. Ch. I, p. 55.

[11]"The Word of Nietzsche: 'God Is Dead,'" HW 226, QT 90.

[12]"The Age of the World Picture," HW 99, QT 148.

[13]"The Word of Nietzsche, 'God Is Dead,'" HW 224-226, QT 88-90; "The Age of the World Picture," HW 99f, QT 148f.

upon which men could achieve a grounding assurance and make bold to assess their world.

For Descartes the mathematical, i.e., that whose knowability and certainty are intrinsic to itself, presented itself as the standard for his thinking. Accordingly, he himself set out on a reflective quest in search of something that he would be unable to doubt.[14]   In his second *Meditation*, he finds that he can doubt the existence of everything that surrounds him, even his own body. Only one thing he cannot doubt: himself as thinking. Even if he were to have to acknowledge himself to be deceived about everything else, yet he would be deceived *as one thinking*. Thus, for Descartes, the awareness voiced in his "I think (*ego cogito*)" imposes itself as indubitable, as certain, as something self-evident that can in no way be gainsaid. And his unassailable assertion itself, expressive as it is of a self-containedly assured knowledge, has for him the desired character of the mathematically certain.[15]   From this certainty that belongs to Descartes' thinking there follows, moreover, an immediate correlary likewise undubitable.   The certainty in question bespeaks the sure existence of that which is certain. Hence Descartes can with like assurance immediately extend his statement and say, "I think, therefore I am (*ego cogito* [*ergo*] *sum*)."[16]

With this statement, Descartes arrived at and set forth precisely that which his age required. There men were seeking something certain that could render them secure but that, far from making them subservient to any external authority, would, through the assurance that it provided, free them to be themselves. This needed "something" Descartes brought to light. The indubitably certain and secure was indeed nothing external. Descartes had discovered it as intrinsically his own, and by implication his words held for all his fellows.[17]   Sure continuance, being, was guaranteed in the certainty of thinking as a self-evident reality and was validated with this seal of self-evidence. This "thinking" could offer to

---

[14]FD 80; WT 103f.
[15]FD 80; WT 104. On the "mathematical," see also Ch. VII, p. 69.
[16]"The Age of the World Picture," HW 100, QT 149.
[17]HW 101; QT 151.

men the security, the adjudging of themselves as right, for which they so eagerly longed.

The thinking that is here in question is of a peculiar character. It is not at all a directly responsive apprehending of something reaching the thinker from beyond himself. It is, rather, Descartes' grasping, in a cognitive act, of himself as engaged in cognition. In saying, "I think," Descartes does not *simply* think, as though something were offering itself to be thought about. Rather, he sets himself, as thinking, before his own inquiring mental gaze. Such thinking is "representing *(Vorstellen)*," a setting-forth-before. It is, in this crucially important case, the human self's presenting of itself as something viewed to itself. The certainty and security that Descartes achieves derive solely from this mode of thinking, undertaken in this radical way. Without it they could not arise.[18] Thus, the new sure foundation that Descartes discovers involves the positing of a new manner of comportment for man as knowing, which obtrudes itself as of fundamental validity.

This "thinking" of which Descartes speaks, and which he undertakes to have in view *as* undertaken, is not at all to be identified as a contentless state of mind. It is not something abstracted from ordinary human awareness. Rather, it is that awareness itself. As such, it encompasses all actions and passions, all modes of relating that the self knows.[19] It is intrinsically possessed of content that reaches it from beyond itself. That content too is seen in the catching-sight-of-self that prompts Descartes' assertion, "I think." It is himself as thinking something specific that he sets before himself. And hence the content of his thinking also receives the imprimatur of certainty and reality that he accords to his thinking self in immediately adding, "therefore I am." But like that thinking itself, any content of thinking now *is* for Descartes not directly—in itself and as itself—but only as set forth and grasped in a reflexive act of knowing, wherein precisely those criteria that govern the latter are in play. With his acceptance of his act of thinking as the

---

[18]"The Word of Nietzsche, 'God is Dead,'" HW 224f, QT 88f; "The Age of the World Picture," HW 99f, QT 149f.
[19]"The Age of the World Picture," HW 100, QT 150.

indubitable and certain that implies its own being, Descartes confines his own horizon of reality to the arena of that thinking. There, in representing to himself himself as thinking, he discovers what *is*; but having put into play his criteria of clarity and self-evidence in this way, he can, in seeking to discover anything that is, in no way turn away from this mode of representing thinking or go beyond it. In it alone can certainty be experienced. Through it alone as possessed of that certainty can the being of anything be posited and affirmed.[20]

Representing, as a particular kind of thinking, does not and cannot partake of untrammeled openness. For it is informed by intent.[21] Descartes, in turning his mental gaze upon himself as thinking, is already seeking the self-evident. Within his knowing lies a willing,[22] which evinces itself in that seeing. In advance, he brings criteria to that which he sees. And it is the meeting of those criteria that satisfies his willed seeking. It is the clarity and distinctness in which his own thinking confronts him that renders its reality incontrovertible for him.

The volitional character belonging to Descartes' self-apprehending, which inheres in his coming with criteria to his task, is in fact subtly contained in the very verb that he uses, *cogito*, which we translate as "(I) think." The Latin verb *cogitare* is a contraction of *coagitare*, a verb comprised of a form of *agere*, to drive, and the prefix *co-*, together. Heidegger, drawing on this basic meaning, points out that Descartes' thinking, spoken of in *cogito* understood as *cogitatio*, is a driving together.[23] Everything, the self as thinking and every element of its content, is gathered before the self in the latter's act of self-scrutiny. This gathering has a forceful character. The self does not simply subsist in on-

---

[20]HW 100; QT 150.
[21]Cf. Ch. XI, pp. 417f, and Ch. XIV, pp. 544f.
[22]Cf. "The Age of the World Picture," HW 100, QT 150, and "The Word of Nietzsche, 'God Is Dead,'" HW 225, QT 88.
[23]HW 100; QT 150. In considering this volitional character of Descartes' representing thinking we should be reminded of the self-casting-forward from out of an impetus given that is ever intrinsic to human *Being*-there as, from out of thrownness, it carries itself forward. Cf. Ch II, pp. 68f, and n. 50, p. 89, and p. 100, and n. 157.

going openness. Rather, the self deliberately withdraws to itself. In so doing, it asserts itself. It reaches toward itself, masters itself, and brings itself to a stand, fixing itself before its own predisposed, intentional gaze.[24] This forceful bringing to a stand is the self's "objectifying" of itself and of all the content of its awareness, for an object (*Gegenstand*) is precisely that which stands over against (*das Gegen-ständige*).[25] And this making-stand-over-against-as-object is the hallmark of the representing, the setting-forth-before (*Vor-stellen*) that Descartes (although no word precisely equivalent to *Gegenstand*, with its connotation of standing-over-against, was employed by him)[26] is actually designating with his word *cogito*.

In such will-fraught objectifying thinking, what Heidegger calls a calculating (*Rechnen*), an evaluative adjudging, always takes place. Representing thinking always assesses from its own prior perspective and in the light of its preexisting assumptions whatever it sets before itself. It counts on that with which it reckons in its seeing to be amenable to such assessment, even as Descartes unquestioningly counts on that toward which his thinking looks to exhibit itself with the clarity and distinctness that he has chosen to posit as definitive of the indubitable. This thinking seeks to discover the certain and secure. And only by presupposing such

[24]HW 100f; QT 150f. The structure according to which representing thinking takes place is closely similar to the structure that Heidegger describes as belonging to the transpiring of human existing as such. From a given stance already assumed, which is embodied in the criteria with which this thinking is undertaken and which parallels the thrownness of human existing, that thinking reaches away from itself and gathers itself and its content toward itself so as to establish and validate itself as itself, even as man's Being-there-as-openness, as Being-in-the-world, casting itself forward, draws what it meets as world toward itself in a movement that allows it, via the opening-up of time, to carry itself onward as itself. Cf. Ch II, pp. 68f & 73-75. On the fundamental identity of this structure with that characterizing the happening of Being, cf. Ch. I, p. 33, n. 36.
[25]HW 100; QT 150.
[26]The word *Gegenstand*, which is used by Heidegger in his consideration of Descartes, was introduced by Kant. Descartes uses *objectum*, ultimately from *ob-*, facing or against, and *jicere*, to throw. Heidegger points out (FD 81f, 105f) that this use of *objectum* to denote that which has its reality through being thrown up opposite the thinking subject, via the latter's representing thinking, radically transforms the word's meaning, since *objectum* formerly denoted something imagined, something which would not be found actually to exist.

calculability and by proceeding in accordance with it can it guarantee to itself in advance the finding of something lastingly certain, as it sets that with which it has to do fixedly over against itself as object, in the light of its own predetermined seeing.[27]

Descartes was by no means the first thinker to endeavor by conceptualizing, by representing to himself (*Vorstellen*), to discern and keep in view something stable in the reality he encountered.[28]   But Descartes was the first to seek and find the stable and secure by looking solely upon himself as representing, in his very act of representing, and so to acknowledge both certainty and Being only as these were vouched for by that representing itself.   In this, Descartes put into play a "subjectivity" that was new.   The reality of the subject, the *subiectum*, was not new as such.   It was known to the Greeks, Heidegger says, under the name *hypokeimenon*, which means "that which lies forth before (*das Vorliegende*)." As an island or a mountain looms up before a traveler, so a "subject" in this sense presents itself from out of itself in its immediacy. It encounters the apprehending of the human *subiectum* that encounters it.[29]   But now, with Descartes, this meaning of subject is with one stroke swept away.   Nothing so lies forth before him as to receive from him an acknowledgement of its intrinsic being.   Instead, anything that *is*, *is* only as it appears when—from out of his determinative intending—it is set fast over against him in his representing thinking.   Everything seen as resident in that thinking is object.   Descartes himself, who sees, remains subject, albeit in a peculiar sense.   All else has exchanged its subjectness for objectness.   Descartes, who, with prior intent, looks at his thinking and its content, alone goes forth directly from out of himself to an encounter. And only in respect to himself does he affirm being that is in any way independent and intrinsic.   For he encounters only that which he has set

---

[27]On the reckoning (*Rechnen*) definitive of such calculative representing-thinking, cf. Ch. XI, pp. 417f. See also "Science and Reflection," VA 58, QT 170.

[28]See Ch. X, pp. 370-372.

[29]See the "Seminaire du Thor, 1969," *Questions IV*, Paris, Gallimard, 1976, p. 260. For the German translation, by Curd Ochwadt, see VS 65f. Cf. "The Word of Nietzsche: 'God Is Dead,'" QT 68, n. 9.

before himself as object: himself as thinking and all that stands as content in that thinking. Finding indubitability, certainty, in that at which he looks, he finds concomitantly, ultimately and most importantly, his own being, not, however, solely as the one who, self-objectified, is observed in thinking, but rather as the one who, at once objectified and objectifying, is assured that he *is* by the very secureness of his representing of himself to himself.[30]

In all this, man as subject has risen up into sudden, unparalleled preeminence in relation to all that is.[31] Descartes speaks for himself. But in the most fundamental sense, in his *ego cogito ergo sum* he speaks decisively for Western humanity as such. His words bring to overt utterance what Heidegger can call the "insurrection" of man in the modern age.[32] Implicit in the certainty inhering in his *ego cogito ergo sum*—the certainty of the self set before itself in its own representing thinking—there rules a hitherto unknown "I-ness *(Ichheit)*." In it a way of being human via self-establishment through self-consciousness that is new and is now determinative is brought into play,[33] and a new foundation, accessible and convincing, is laid for all human thinking and doing. With it the age in transition becomes decisively the *new*, i.e., the modern, age.[34] Now, in the comportment of individual and group alike,[35] the new subjectivity made manifest in Descartes' certain establishing of himself through representing thinking begins everywhere to hold dominion. Transpiring self-intendingly as self-establishing and self-authenticating, that subjectivity assumes as its right the establishing and authenticating of all else as well. Always it subsists in preoccupation with itself, bent on the knowing that is its self-validation.

In speaking of this self-conscious knowing, this modern "consciousness," Heidegger can use the word *Ge-wissen*, whose common

---

[30]HW 100f; QT 150f.
[31]Cf. "The Age of the World Picture," HW 98f, QT 147f.
[32]"The Word of Nietzsche: 'God Is Dead,'" HW 236, QT 100.
[33]"Overcoming Metaphysics," VA 86, EP 98f.
[34]The German expression that Heidegger continually uses, which is regularly translated as the modern age, is *die Neuzeit*, literally, "the new age [time]."
[35]"Overcoming Metaphysics," VA 86, EP 98f.

meaning is "conscience." He hyphenates the word, thereby letting it show what he takes to be its proper meaning: a gathering of knowing. This is precisely Descartes' self-concerned knowing, which gathers everything together into his self-conscious gaze.[36] This new subjectivity of modern man as subject is *new* as self-consciousness.

In this new mode of being subject, modern man begins and grounds an assault on all that is. Happening intentionally as reflexive self-setting-forth-over-against, his subjectivity becomes, whether he knows it or not, a prevailing. It renders secure that which it gathers before its purposeful knowing; but precisely in thus bestowing secureness it despoils. It encompasses everything within itself. Now "everything that is, is therefore either the object of the subject or the subject of the subject."[37] Everything, including the subject itself as grasped in its self-knowing, is mastered into object. Everything is brought under control and so prevented from being, i.e., from sustainedly presenting itself, simply as and of itself.

And with this the objectifying subject itself is in turn self-bereft. Precisely in its quest for self-certainty, it exalts itself in lonely and erosive self-assertion. Setting itself before itself, it blocks its own spontaneous and direct going-forth, thus confining itself within the subjectivity that it has brought to light. Therewith it deprives itself of any meaningful encounter from beyond itself. No correlative *subiectum* can meet it in immediacy. Everything must submit to the objectifying in representing thinking through which this peculiar "subject" makes itself secure. Hence, everything falls away into the subordinate. Everything becomes something to be dealt with, not something to be met and acknowledged as

---

[36]See "The Word of Nietzsche: 'God Is Dead,'" HW 224, QT 88. The original meaning of the verb *wissen* is, as Heidegger points out, "to see." Thus the German words built on *wissen* that speak of consciousness—*Gewissen* (conscience), *Bewusstsein* (consciousness)—allude implicitly to seeing. So too do words like *Gewiss* (certain) and *Gewissheit* (certainty). Hence in Heidegger's discussion of Descartes and of the Cartesian legacy, the stress on seeing is far stronger and more pervasive than it can be in a discussion pursued in English. On the prefix *ge-* as connoting a gathering together, see Ch. VI, pp. 258f.
[37]HW 236; QT 100.

and for itself.

Thus in the new age does modern Cartesian man, entering into a mode of existence shaped by self-assertive subjectivity, arrogate to himself sweeping dominion. He affixes the seal of certainty. He accords being to whatever *is* as set before himself. He rises up into an unremitting control centered in himself and extending to all that he surveys. Fundamentally *he rules*. And yet his ruling is, after all, not his own. In the new interrelating[38] of man and what-is that is now underway, a new manner of happening of the Being of what-is has come radically to the fore. The Being of whatever is, is now ruling at once and unitarily in the seemingly disparate guises of self-assertive subjectness and objectness.[39] Precisely as it proclaims itself in Descartes' *cogito ergo sum*, just so does it hold sway concomitantly in every aspect of the complex, reciprocal relating between man and what-is that is there embraced and initiated.

Happening as self-maintaining self-unconcealing, the Being of that which is happens determinatively now as a self-securing self-intending. Hiddenly accomplishing its sway newly via Descartes' philosophical thinking and speaking, it begins to rule decisively as on-striving will to will (*Wille zum Willen*),[40] i.e., as self-assertiveness that maintains itself

---

[38]Here and elsewhere in this study the reader should be careful rigorously to separate "relate" from its usual meaning. Here "relate," in contexts where the Being of what-is is in any way under consideration, does not speak at all of a connecting of inherently discrete elements, one of which is to be subordinately referred to the other. The relating that is now in question is a *holding-toward-one-another* of participants that, while they are distinct from one another, are ultimately one. For we are using "relate" to speak of the happening of the single Twofold. See further, Ch IV, p. 145, n. 10.

[39]On Being's happening as such subjectness (*Subjektität*), see "The Word of Nietzsche, 'God is Dead,'" HW 236, QT 100. On its happening as objectness, see Ch. VII, p. 282.

[40]On this concealed modern ruling of Being as "will to will," see "Overcoming Metaphysics," VA 78, EP 91. In the phrase "will to will," the second "will" is nominal in the German and not verbal as the English phrase suggests; although certainly the noun is intended to speak of what we may perhaps call a very active engaging, a carrying forward of willing. The phrase is wholly parallel to Nietzsche's "will to power," where power is named as always self-accomplishingly in play. See this chapter, p. 302. In this will to will the structure intrinsic to Happening, a going-forth that is itself a gathering-toward (cf. Ch. I, p. 33, n. 36, and this chapter, p. 310,

by relating itself back to itself. On the one hand, it holds sway in the self-assertive subjectness of modern man with its reliance on will-fraught, reflexive representing thinking. On the other, it rules also, if more hiddenly, in the happening of what-is as object. There too an assertive self-maintaining by way of a relating back to self takes place. Whatever now is, is as represented, as set in place, by man as representing subject. To achieve this manner of being, whatever is offers itself in such a way as to appear in representing as object. It offers itself in the fixity, the assertiveness among its fellows, of standing-over-against.[41] In so doing, it distances itself from itself and thus arrives into on-going self-presenting, i.e., into *Being*, only by way of the subjective representing that as such allows it to *be* only as standing in relation to its concealed self.[42]

This new mode of happening of the Being of all that is makes its dominion evident in the shutting away of man and what-is from open meeting with one another that marks the modern age. If modern man, in finding in self-consciousness the sure ground of certainty, prevails, in so doing, over whatever is—even over himself—and therewith despoils as he makes secure, this comportment bears witness in the last analysis to the fact that, in a new self-sending-forth of Being itself, the pull of self-concealing has gained in dominance and has laid arresting, wasting hold on everything that is.[43] And here again, in a particular guise, that isolative impulse toward self-assertion in unassailed persisting that comes

---

n. 63) is very evidently to be discerned. In it, also, the forcefulness intrinsic to Being, as which that Happening is named, comes clearly into view. On that forcefulness, cf. Ch. I, p. 51, and n. 100.

[41]On self-assertiveness as belonging intrinsically to the manifold of what-is in its happening in particularity, cf. Ch. I, pp. 50-52.

[42]Cf. "The Word of Nietzsche, 'God Is Dead,'" HW 218, QT 79f. The parallel between the self-assertiveness predicated of modern man in his subjectivity and the inauthentic mode of existing, also carried out in self-alienating self-assertion, that for Heidegger characterizes man as man should here be noted. Cf. Ch. II, pp. 85f & 89-91. On Heidegger's depicting of the happening of the Twofold as such as marked, in the very structure pertaining to it, by the alienation-from-self that is here shown to us as specifically in play via Being's happening as subjectness and objectness, cf. Ch. I, pp. 45f, and n. 77. On the fundamental structure in question, cf. Ch. I, p. 33, n. 36.

[43]Cf. Ch I, pp. 44-46 and Ch. VI, pp. 259-264. On Being's happening first of all as concealing, see Ch. I, pp. 24-26.

upon all particulars of what-is, including man, from out of Being, in the latter's ruling as the presencing of whatever presences,[44] must surely be seen to be in play.

2

One destining rules in the modern age.[45]   In metaphysics it is possible to glimpse that destining in its most direct disclosure. For there the Being of what-is, happening as primordial showing saying, embodies itself in words that at once contain, display, and bring to bear the enframing summons, *Ge-stell*, that is the self-sending-forth as which Being now governs.[46] No thinker after Descartes seriously questioned his fundamental stance or the rightness of his seminal words that spoke it forth. Those words imposed themselves on subsequent thinking as authoritative. For in them the newly ascendent destining of Being was manifesting itself in formative power. Descartes' work is, as it were, the fulcrum on which the whole of reality is swung over onto a new basis. If at virtually the same historical moment we find modern science emerging as research, i.e., as a self-authenticating knowing that establishes itself ever more securely through a purposeful encroaching upon the real that objectifies the latter in what is, structurally, a representing thinking, and, grasping it in accordance with a preconceived plan, reckons on it as calculable in advance and so fits it to take its place within the sphere of certainty, in order that it may support further investigation,[47] this can happen, Heidegger says, only when the certainty vouchsafed in representing thinking is accepted as itself disclosive of Being. Precisely

---

[44] Cf. Ch. I, pp. 50-52, and Ch. II, pp. 85f, and n. 101. For another evidencing of this fundamental tendency in the happening of the Twofold, see Ch. I, pp. 45f. Cf. also Ch. IV, pp. 163-166. The forcefulness discernible in this self-assertion again bespeaks the manner of happening intrinsic to Being as such. Cf. Ch. I, pp. 51f, and n. 100.

[45] Cf. Ch. VI, pp. 257-259, and Ch. VII, pp. 280f.

[46] Cf. Ch. III, pp. 119 & 130. For a discussion of the self-utterance of the happening of Being in words, albeit those spoken not in the domain of metaphysics but in that of Heidegger's own thinking, cf. Ch. XV, pp. 602-616. On *Ge-stell*, cf. Ch. VI, pp. 258-264.

[47] Cf. Ch. VII, pp. 268-275.

this acceptance is first given utterance and the determinative defining at once of certainty and of Being in terms of representing (*Vorstellen*) first comes decisively into play in the metaphysics of Descartes.[48] Again, if modern machine technology arises out of the work of science, which prepares its way,[49] and then presses forward in imperious self-assurance into every realm of human life, invincibly certain of its manner of looking at and dealing with reality, and if it allows and can allow all manner of entities to count only as factors calculable in advance for a ceaseless, ever-proliferating planning that continually orders and reorders everything so as to set it securely in place to serve some projected end,[50] this too happens immediately on the ground that Descartes has provided. For even where the object thus disappears and the representing that objectifies gives way to a mere calculating intent that never catches sight of anything in such a way as to accord to it a particular identity but rather sees every entity only as something to be set in place as standing-reserve,[51] the fundamental structuring of man's relating to whatever is that Descartes brought to light is still determinative. Man still strives toward whatever he encounters, in the self-assertiveness that distinguishes him as subject in the modern age; and whatever is still comes to appearance only as mastered and set in place by him. For in the utterances of Descartes, the enframing summons that is the initiatory manner of enduring that rules in modern technology and modern science grounded its dominion over the modern age.[52] And every phenomenon of that age must somehow stand

---

[48]See "The Age of the World Picture," HW 80, QT 127. In testimony to the identity of modern metaphysics and science that is here in question, in a remark regarding "the descendants of metaphysics" Heidegger can immediately explain, "i.e., of physics in the broadest sense, which includes the physics of life and of man, biology and psychology." "Overcoming Metaphysics," VA 86, EP 99.

[49]Cf. Ch. VII, p. 281.

[50]Cf. Ch. VI, pp. 241-252.

[51]Cf. Ch. VI, pp. 241-244.

[52]Cf. Ch. VI, pp. 257-259, and Ch. VII, pp. 280f. Well before he named the disposing enduring holding sway in modern technology with his word *Ge-stell*, Heidegger stated clearly the relationship, arising out of that sway as such, that he saw in our age between technology and metaphysics: "The name 'technology' is here so understood after the manner of disposing enduring (*wesentlich verstanden*) that in its meaning it covers itself congruently (*sich . . . deckt*) with the rubric: completed

revealed in the light of his words.

Philosophy after Descartes continued to make known, in various specific formulations, the manner of happening of what-is that he first disclosed for the modern age. Every thinker took as his point of departure the assumption that Descartes had established, namely, that the sphere of the subjective—where knowing was the self-assertive grasping in reflexive representation of the known as object—was the sole sphere in which the validity of the real was to be sought. Thus Leibniz, for example, specifically brings forward "will" as definitive of the Being of whatever is. The Being of all beings whatever is to be understood as force (*vis*).[53] It has the character of *nisus* (striving, endeavor, inclination).[54] At the same time, those beings or monads, infinite in number, that constitute the universe all partake to a greater or lesser degree of the character of mind. Each is appetitive and percipient. Each strives to maintain itself as subsisting over against all others, through representing itself to itself by a positing to itself of the universe as it perceives it. This representation to itself—its perception—brings to accomplishment the forceful striving that most fundamentally constitutes the monad, thereby allowing it to maintain itself uniquely as itself.[55]

Kant, like Descartes, takes his stand within individual subjectivity, although unlike Descartes his thinking centers explicitly not on the individual thinking subject but on "consciousness in general."[56] He is concerned with that subjectivity as transcendental, i.e., as itself possessed

---

(*vollendete*) metaphysics. It contains the recollection of *technē*, which is a grounding condition (*Grundbedingung*) of the unfolding of the disposing enduring (*Wesensentfaltung*) belonging to metaphysics in general. ("Overcoming Metaphysics," VA 80f, EP 93.) On *technē* see Ch. IX, pp. 332-354. On the rise of metaphysics in the milieu in which the disposing accomplished as *technē* was originally centrally determinative, see Ch. X (pp. 367-402). On the "completion" of metaphysics, see this chapter, pp. 324f. On the meaning of *Wesen*, cf. Ch. VI, pp. 252-257.
[53]"The Word of Nietzsche, 'God Is Dead,'" HW 226, QT 90.
[54]HW 211; QT 72.
[55]See HW 226, 211; QT 90, 72. Certainly here the parallel with Heidegger's particulars of what-is, each of which strives to persist in and of itself, is strikingly evident. Cf. Ch. I, pp. 50-52.
[56]"Overcoming Metaphysics," VA 86, EP 98.

inherently of the forms and structures of mental activity by which it molds into knowability the content engendered within it by whatever meets it from beyond itself. The unstructured wealth of percepts that is comprised of sense data is synthesized into stable objects (the word *Gegenstand* was introduced into philosophical thinking by Kant) and presented to itself by the knowing subject as a unified world. This achievement of unifying through objectification evinces and confirms the intrinsic unity of the knowing subject that is presupposed. That subject cannot know itself and hence know that unity directly, for it knows itself only as the object of its own knowing. It is this, its self-consciousness as representing objectification, that discloses to the knowing subject its own inherent character and that in so doing makes it certain of itself and hence secure. And all Being of whatever *is* in the domain of the transcendental subject is grounded in "the transcendental making possible of the objectivity of the object."[57]

The self's carrying forward of itself via this structuring of knowing is, by Kant, understood by way of "the concept of practical reason as pure will."[58] Informing itself via rational consideration, the will presents to itself the "ought" as the principle of its own forward movement; although, for Kant, within this defining structure, the will may or may not act in accordance with that principle.[59] Here Heidegger discerns a disclosure of the "will to will" as which Being has now begun to hold sway, for that will is, so far as the disposing enduring underway within it is concerned, characterized by "limitlessness (*Ziel-losigkeit*)." Nothing impinging upon it from beyond itself sets any bounds to its self-carrying out. Rather, its goals are provided by itself to itself. And just this manner of ruling begins to be evident in Kant's delineation.[60]

---

[57] "The End of Philosophy and the Task of Thinking," SD 62, TB 56. See also "Overcoming Metaphysics," VA 75, EP 88; "The Word of Nietzsche: 'God Is Dead,'" HW 226, QT 90; and Heidegger's detailed discussion in *What Is A Thing?* FD 105ff, WT 134ff.

[58] "Overcoming Metaphysics," VA 88, EP 101.

[59] See Immanuel Kant, *Fundamental Principles of the Metaphysics of Morals*, translated by Thomas K. Abbott, New York, The Liberal Arts Press, 1949, pp. 24ff.

[60] "Overcoming Metaphysics," VA 89, EP 101. On *Ziel-losigkeit*, see this chapter, p.

Hegel too bases his thinking directly on Descartes' disclosure of the *subiectum* as self-conscious knowing subject. But his vision is yet more expansive. The "subject" in question is now neither the individual nor consciousness in general as instanced in individuals but rather the Absolute. As such it is absolute Idea. It is a thoroughly structured plan, immanent in all that is as the latter happens in specificity by way of history. It attains to consciousness of itself and therewith to the fulfilling of itself as itself through the thinking—the thinking of metaphysics—that grasps it in this its self-concretizing and mediates it to itself in a dialectic that mirrors its own self-movement. Here thinking, in the full sense, lets be that which is. Indeed, with respect to the absolute subject, thinking and Being are identical. Hegel makes no place for a dichotomy of subject and object where the "object" would be taken in any sense to derive ultimately from beyond the sphere of consciousness belonging to the knowing subject. But he nevertheless retains as wholly fundamental that structure of a self-accomplishing, self-authenticating turning of the self back upon itself, in self-knowing, that is the very structure of objectifying representing thinking as such.[61] As direct heir of Descartes, his is the "metaphysics of absolute knowing as the spirit that is will."[62]

The ultimate spokesman for the destining of Being, the manner of happening of all that is, ruling in the modern age, is Nietzsche. In his thinking, the grounding disclosure provided through Descartes is brought to completion. Nietzsche's thinking is, Heidegger says, the metaphysics of the will to power.[63] For to him "will to power" denotes that which, as

---

321, n. 107. On the phrase "will to will," which is itself expressive of this limitlessness, see this chapter, p. 304, n. 40.

[61]See "The Onto-Theo-Logical Constitution of Metaphysics," ID 123-125, 56-58. On the structural similarity between such representing thinking and the manner of transpiring of human existing as such, with its evincing of the structure of the happening of Being, cf. this chapter, p. 300, n. 24.

[62]. . . *des absoluten Wissens als des Geistes des Willens.* "Overcoming Metaphysics," VA 76, EP 89. *Geist* is a word of rich connotation. It means spirit in the sense of mind or intellect. The genitive governing "will" has here been overtly rendered as objective, so as to show clearly that the thought expressed is that of spirit or mind carrying itself out after the manner of intention, will.

[63]Cf. "The Word of Nietzsche: 'God Is Dead,'" HW 215, QT 76. Nietzsche himself

the very "reality of the real," is determinatively ruling in everything that is.[64]

That will to power, the determinative "stamping" in which the direct ruling of the will to will as which Being hiddenly rules in the modern age is to be discerned,[65] is a striving forward, self-directed and unconditional, that is ever establishing itself in power precisely in order that it may advance to greater power.[66] As such striving, although it thrusts ever forward as though toward something still beyond its reach, this will carries itself out as an unremitting exercising of dominion that in fact never looks beyond what it itself already possesses within itself. As ever-intending striving, it has ever already consummated itself; for in transpiring as it does intrinsically as consciousness it renders real as object and already holds before itself in general concept, as something "expressly known and just as expressly consciously set forth to itself, that for which it strives."[67] Happening thus via objectifying consciousness inherently as a will to mastery,[68] it intently wills itself in the fullness both of its forward thrusting and of the conscious content as which it knows

---

rejected and sought to overturn traditional metaphysics, but Heidegger asserts that in this very overturning Nietzsche still stands squarely within the metaphysical tradition. See HW 214, QT 75. We should note that in the discussion now in progress the structure visible in representing thinking, which parallels the structuring that Heidegger finds for human existing as such and for the happening of Being manifested therein (see this chapter, p. 300, n. 24, and Ch. I, p. 33, n. 36), is shown us with peculiar clarity as pertaining in Nietzsche's thinking at once to the manner in which Being is itself now disclosing itself—as will to power—and to everything that is, i.e., to whatever is and to man in particular.

[64]HW 223; QT 86.

[65]"The disposing enduring belonging to the will to power lets itself be comprehended only from out of the will to will. (*Das Wesen des Willens zur Macht lässt sich erst aus dem Willen zur Willen begreifen.*)" "Overcoming Metaphysics," VA 82, EP 95. On the will to will see this chapter, pp. 304f, and n. 40.

[66]"The Word of Nietzsche, 'God is Dead,'" HW 216-219, QT 77-80. The parallel between this structure found for the will to power and the structure that Heidegger portrays for the self-carrying forward of science and of technology must immediately strike us. Cf. Ch. VII, pp. 272-276, and especially n. 10, and Ch. VI, pp. 228f, and n. 12.

[67]"Overcoming Metaphysics," VA 88, EP 100.

[68]"The Word of Nietzsche, 'God is Dead,'" HW 216, QT 77.

itself and in accordance with which that very on-thrusting carries itself forward.[69] So transpiring, in its self-strengthening advance the will to will, happening thus as will to power, continually turns back in self-scrutiny toward itself so as to secure itself in the level of power achieved, in order that it may at the same time surpass itself and attain to further power.[70] This reflexive turning-back upon itself is what is meant, Heidegger says, by Nietzsche's phrase "the eternal return of the same." For Nietzsche "the way in which that which is, in its entirety—whose *essentia* is the will to power—exists, its *existentia*," i.e, the manner as which it maintains itself as itself, is precisely this *returning*.[71]

This movement the will to power accomplishes through its character as a self-aware knowing that is a seeing. Always it has something in view, something presupposed at which that seeing aims and by which it directs its course. Nietzsche designates this intrinsic aspect of the will's self-accomplishing "value-positing."[72] Here value means both the determining orientation, the point of view that directs the seeing in question, and that at which that seeing aims, that which is in view for it.[73] Such valuing inherently involves a ranking, a presupposing of what Nietzsche can call "a numerical and mensural scale.[74]"

Only as value-positing can and does the will to power maintain itself as itself. Ceaselessly it gathers to itself that which it sees and makes it secure for itself as something securely known. Having thus established and secured itself as knower, it enables itself to go beyond itself, to direct its gaze toward the yet unknown, and to order the latter in its turn,

---

[69]"Overcoming Metaphysics," VA 82, EP 95.

[70]"The Word of Nietzsche, 'God is Dead,'" HW 220f, QT 83f.

[71]HW 219; QT 81f, and n. 21.

[72]HW 219; QT 80f.

[73]HW 210; QT 71f. In his discussion Heidegger uses *Gesichtspunkt* (point-of-view) and gives to it this double connotation.

[74]HW 210; QT 71. We have noted that the structure that Heidegger finds for representing thinking is clearly apparent in Nietzsche's will to power (cf. this chapter, p. 310, n. 63). Nietzsche's positing of a "mensural scale" offers an interesting, if oblique, illustration of Heidegger's contention that such thinking is inherently calculative in character. Cf. this chapter, pp. 300f.

drawing it into its own ever-expanding dominion.[75]

In accordance with this manner of happening of the will to power, Nietzsche sees two values as of primary importance to the will as value-positing. The first of these, the value that orients the will and is in view for it in its establishing and preserving of itself in each level of power attained, is "truth" understood as the certainty and secureness of that which the will already claims as known.[76] The second is what Nietzsche calls "art." At once an assumed stance and a goal in view, art is here an initiating ordering that opens up new perspectives into the as yet unpossessed, by conferring knowability and hence making possible the will's advance to greater mastery, greater power.[77]

Although Nietzsche sees this purposeful will to mastery and to ever greater power as informing everything that is, it is only as happening in man, where it partakes fully of self-knowledge, that it can come to conscious, disciplined accomplishing of itself. Nietzsche finds this to have occurred only rarely, in outstanding individuals—Caesar, Napoleon, Goethe. Ordinarily history displays something quite different. One may see in the barbarian the spontaneous, unself-conscious wielding of power. In man as shaped by platonized Greek culture and by the ethos of Christendom, one finds, on the other hand, a self-consciousness that abandons and fears spontaneity. Here man, subservient to norms imposed from beyond him, turns his strength inward upon himself, succumbing to guilt and to the inhibiting dictates of consciousness as conscience. Gaining in sensitivity, he simultaneously surrenders his capacity for bold self-initiated action. His is but a slave mentality.

Man of this latter sort is for Nietzsche "man up to now."[78] This is man, Heidegger says, as unwilling and unable consciously to actualize the will to power.[79] In contrast to him, Nietzsche speaks of overman

---

[75]Cf. HW 221, 238f; QT 83f, 102f.
[76]HW 221f; QT 84f.
[77]HW 222f; QT 85f. Cf. "Overcoming Metaphysics," VA 87, EP 99.
[78]For the characterization by Heidegger of Nietzsche's man up to now, see HW 232-234, QT 97f.
[79]HW 234; QT 98.

(*Übermensch*). This man, who is now only foreseen, would be man who had mounted beyond man up to now into a new mode of being man.[80] In him forceful spontaneity and a self-consciousness marked by sensitivity and profundity of awareness would be fused into a new humanness. This overman would be single-mindedly directed solely from within himself. Self-controlled and possessed of the magnanimity of strength, he would venture to rely exclusively upon the integrity, the self-contained, self-validating unity, of his own willing and perspective. In a spontaneity grounded in self-certainty, he would press resolutely forward, valuing everything solely from out of his own unquestioned orientation and confidently bent on controlling everything in accordance with the norms and goals that it would be his sole prerogative to provide.[81]

Heidegger stresses the fact that the word "overman" should not be given an individual connotation.[82] Nietzsche's overman is man as such. This is the humanity that Nietzsche sees as demanded and that, if men prove equal to the conscious effort of will required, may in the future come to be.[83]

Here, in Nietzsche's thinking, the subjectness brought to light by Descartes receives extreme statement. Man as subject who, moved by inherent intentionality with its accepted goal of achieving certainty, securely established himself and established his world by consciously laying hold of himself and of that world in his objectifying act of thinking, has now, with Nietzsche's overman, become man as subject who, made secure through his willed knowing of himself as embodying the imperious, self-authenticating will to power, in utter confidence lays hold on everything, in accordance with his own immediate, unerring assessment of it, and sets it fast in a configuration determined and validated solely via his own authoritative consciousness.[84]

---

[80]Cf. HW 232ff, QT 96ff.
[81]For Heidegger's discussion of overman as living out directly the happening of Being as will to power, see HW 232-234, QT 96-98.
[82]HW 232; QT 96.
[83]HW 234; QT 98.
[84]Cf. HW 232ff, QT 96ff, and this chapter, pp. 302f.

With Nietzsche, self-assertive *will* that is a ceaseless striving toward ever greater power is brought decisively forward. It centers in man, the self-conscious subject. All else that is falls by right within his grasp. Itself informed by the same will to power, it may resist his determined assault upon it, but the right of command and disposal is his. Whatever is can *be* only by submitting, in its own assertion of itself over against him, to the purposeful, evaluative scrutiny and controlling activity with which he advances against it.[85]

Here the insistent setting-over-against that is intrinsic to the subjectiveness ruling in the modern age comes to sharply focused manifestation. For Nietzsche his overman is man of the future, who may exist, but only isolatedly, here and now. But beyond this, Nietzsche also vividly portrays the insurrection of modern man in the subjectness of the will to power that is already decisively underway. In the words, "God is dead," Nietzsche boldly proclaims the dissolution of the suprasensory world of norms and values that has for centuries governed the life of Western man; and, Heidegger avers, nothing less than the uprising of man to assume a place of ascendancy, as opposed to the place left empty, is announced in those words as well.[86] In image after telling image Nietzsche depicts man as having annihilated that which is as that which, having its own independent reality, *is* in and of itself. Modern man has swallowed up what *is* in this sense. In the power of his new subjectivity he has turned it into object. He has transformed its direct appearing into the appearing of the object-in-view that is always subservient to an intentional, evaluative seeing that eyes it as something that exists only within the framework of the viewer's self-aggrandizing, purposeful forward-striving.[87]

The killing of God of which Nietzsche dares to speak—the abolishing of all externally provided norms that is implicit in man's imperious taking evaluative charge vis a vis whatever is—has also, to Heidegger's mind, implications that extend far beyond those of which

---

[85]Cf. HW 234, QT 98.
[86]HW 234f; QT 99f.
[87]HW 241f; QT 106f.

Nietzsche himself is aware. For with Nietzsche's putting forward, in intended contradistinction to metaphysics but in fact in unguessed consummating of the latter,[88] of value positing as definitive for the self-accomplishing of the will to power that centers itself determiningly in man, Being has itself been transformed. It has been degraded into a value.[89] Now Being has become something posited and validated by man and is no longer even considered as belonging to that which *is* in and of itself. It is indeed Nietzsche himself, Heidegger says, who, if unwittingly, overtly accomplishes this transforming of Being. For it is he who, for all his attempted annihilation of all former values and valuing, himself brings to light value-positing as something intrinsic to reality as such. In his metaphysics of the will to power, he carries to most explicit and trenchant statement the relationship in over-mastery and estrangement now existing between man and what-is. And precisely in so doing, through his elevating of value-positing to a fundamental principle, he brings to nothing the Being of what-is as in any sense the independently encounterable appearing of that which appears from out of itself.[90]

Here, where Being has let itself be definitively named as a mere value, so that its absence has been directly affirmed, nothing less than a profound, all-pervading negating has, Heidegger insists, come to dominion. With the work of Nietzsche, when the supremacy of man vis a vis all that is has been boldly and decisively set forth, when certainty has become a regnant value and self-securing has been proclaimed as everywhere centrally in play, it is in fact this absence, this want of Being in its withdrawnness, that is, with respect to the will to will in its self-affirming self-assuredness, for Heidegger ultimately determinative in the on-going happening that is in question. The will to will that manifests itself as will to power is precisely a holding sway of Being as Nothing.[91] This will, as the Being-masking self-manifesting ground of what now is, is a ruling instantiation of Being to which Being as provisive Happening—

---

[88]HW 200; QT 61. See also HW 238-247, QT 102-112. And cf. this chapter, pp. 324f.
[89]HW 238; QT 102.
[90]HW 238; QT 103.
[91]HW 239; QT 104.

as presencing that lets presence, appearing that lets appear—is lacking. Therefore neither that which, in its on-happening centeredly via human consciousness, the ruling will gathers to light as set before its self-concerned knowing, nor the will itself that thus via consciousness carries itself ever forward partakes of true manifestness. The certainty, the "truth," that this will via its ruling supposedly sights upon and affirms as secure in its affirming of itself is in fact deviant error. It is not an unconcealing or safeguarding of Being.[92] It is, rather, a delusive assumption of Being where no Being is genuinely proffering itself. Being, ruling now hiddenly in a reality bereft, in concealing its absence so holds sway as to deceive. In the self-securing and sureness characteristic of the will to will, it is only an on-going bringing into play of Beinglessness that is underway. Heidegger, portraying the ultimacy of that which he has in view, can give to this understanding extreme statement. He writes that for the will to will, whose willing volunteer is modern man, "all truth becomes that realm of error (*Irrtum*) which it requires so that it can set securely before itself its delusion regarding this: that the will to will can will nothing other than the empty nothing over against which it asserts itself without being able to know its own completed nullity."[93]

Here man himself and all that meets him as what-is via his intent-fraught representing thinking, and even the ruling happening as which Being, in its withholding of itself, ubiquitously holds sway, take place as it were in a void. Self-assertive self-securing is everywhere determinative, but it has to do, finally, with nothing but a radical dearth of self-presenting that can only serve radically to undercut its very accomplishing of itself. Under the dominion of the will to will that brings to bear the withdrawnness of Being, nothing proffers itself in the simplicity, the singularity, that permits a world genuinely to disclose itself.[94] Rather, "everything of the world here manifests itself only after

---

[92]For this meaning of truth, see Ch. I, p. 38, and n. 49, and p. 54, n. 109.
[93]"Overcoming Metaphysics," VA 72f, EP 86.
[94]On the disclosing of world via the unique, see Ch. III, pp. 135-137, and Ch. V, pp. 173-175 & 183.

the manner of a monotonous uniformity. Via this will's ruling, instead of the disposing enduring (*Wesen*) that lets anything whatever be as itself, an absence of genuine disposing, a non-ruling of disclosive enduring (*Unwesen*), which is all but completely removed from true manifesting, is in play,[95] and only an "unworld (*Unwelt*)" is disclosed.[96] Accordingly, all the accomplishing of setting-fast that is underway in keeping with the dominion of the will to will as will to power that has been brought to dominance now via Nietzsche's metaphysics can be only an exercise in futility. Nothing that presents itself as itself can be met with, and no genuine securing whatever can be achieved.

Thanks to this emptiness, nothing gained ever satisfies that will and, marked as the latter is by a striving to make itself secure, everything it can reach can appear to it only as rightful material for the further fueling of its own self-securing self-aggrandizement. Impelled by the very emptiness of its achieving, however hidden from it the latter's vacuity may be, this will to will manifests itself in a far-flung, complex "struggle for power," in a "planetary mode of thinking" and in the calculative comportment that exploits and orders, as it reaches out avidly planet-wide to claim everything for itself.[97]

Heidegger portrays the will in question as wrought upon also even more inwardly by the lack within itself that its ruling as a happening of Being in withdrawnness entails. This will is, in the interest of its self-securing, wholly calculative. As such, it takes account of and counts on everything toward which it looks as established by itself as what it knows. Knownness pervades its carrying forward of itself. Accordingly, it is constantly pursuing and augmenting its calculative knowing. It drives forward, ever forward, ceaselessly devising fresh means, searching out fresh grounds, revising its aims and pressing them to an extreme, always in a manner that assesses them and reckons them up in the service of its self-sustaining. Yet precisely here, since in its happening as a self-

---

[95]"Overcoming Metaphysics," VA 89, EP 101. For an elucidation of the meaning of *Wesen* see Ch. VI, pp. 252-257.

[96]VA 97; EP 108.

[97]See VA 90, 77, 83, & 80f; EP 102, 90, 95, & 93.

manifesting of the ruling of Being that is bereft of Being "delusion and
deception" lie at the root of its transpiring, the achievings of this will
continually fail, once again, to suffice it. Unable to know the nullity
intrinsic to all its self-carrying-forward—and indeed to all that with which
it has to do—it is ever still unsatisfied and, for all its reliance on certainty,
is never assured. Consequently, its inquiring is a veritable inquisition. It
mistrusts itself and, belying its show of self-secureness, is reserved, never
committing itself directly and unqualifiedly to what has been achieved but
continually concerned, in its quest for the sure basis that eludes it, with
"nothing other than the securing of itself as power itself."[98]

With Nietzsche, the destining of Being that began newly to disclose
itself in the metaphysics of Descartes brings itself to a fresh grounding
disclosure that is, as such, a radical concealing of itself. Here man,
brought to understand himself and all with which he has to do, and
likewise to pursue his role as man, solely from out of Being's self-
masking instantiation as will to power, discerns nothing that imposes itself
determinatively upon him, let alone the ruling immediacy of the deeply

---

[98]VA 88f, EP 100f. *Der Wille zum Willen ist die höchste und unbedingte Bewusstheit
der rechnenden Selbstsicherung des Rechnens . . . . Daher gehört zu ihm das
allseitige, ständige, unbedingte Ausforschen der Mittel, Gründe, Hemmnisse, das
verrechnende Wechseln und Ausspielen der Ziele, die Täuschung und das Manöver, das
Inquisitorische, demzufolge der Wille zum Willen gegen sich selbst noch misstrauisch
und hinterhältig ist und auf nichts anderes bedacht bleibt als auf die Sicherung seiner
als der Macht selbst.* The published English translation of this passage contains a
seriously misleading phrase. It speaks of "the miscalculating exchange and plotting of
goals" as belonging to the will to will. Here a form of *verrechnen* (to reckon up,
charge to an account) has evidently been mistaken for *sich verrechnen* (to
miscalculate). *Ausspielen*, which follows and is translated accordantly, does not mean
"to plot" but "to carry to term." This rendering introduces into a passage in which the
deceivedness that intrinsically belongs to the will but is hidden from it is in question a
superficial mistakenness, which is in fact not borne out by Heidegger's depictions
elsewhere of the self-confident, if vain, comportment that is being portrayed. The
same mistranslation of *verrechnen* appears again (VA 94; EP 104), where the matter in
question is clearly the accomplishing of all-inclusive, total calculation and *not* the
accomplishing of total miscalculation. See VA 73, EP 85f; VA 98, EP 109. Cf. also,
e.g., (of technology) Ch. VI, pp. 230f & 246f, and (of science) Ch. VII, pp. 272-278.
On the uncertainty inhering in supposed certainty that is here in question, see also VA
88, EP 100.

concealed governance of Being. Instead, man now stands alone. As the supremely preeminent center for the actualization of the will to will discerned as the will to power, he knows himself and knows himself only as unquestionably entitled to claim lordship over all that he surveys and as possessed, likewise, of the needed ability and power to gather everything unhesitatingly and ever more sweepingly into the sphere of his command. Here modern technological man comes clearly into view, and the metaphysical basis for the thorough-going pursuance of that comportment that decisively marks him in his age is both made manifest and brought into play. In the ruling of the will to will now brought overtly to bear, the technological is definitive.[99]   Under its now all-pervasive sway modern man, in his planning and ordering and stockpiling of everything into a standing-reserve, by means of a continual setting of goals in whose light everything met with is to be rigorously assessed and made use of, has launched a calculated assault upon everything that is.[100]   So drastic and thorough-going, indeed, is that assault that the undertakings through which man now moves to make himself lord over everything basic to his intended governance can be identified by Heidegger as nothing less than an arming, a military mobilization (*Rüstung*),[101] and Heidegger can portray as demanded an all-encompassing, total reckoning up and utilization of every raw material, including "the raw material man," through the efforts of leaders especially capable of and dedicated to just that unsparing mobilization for conquest.[102]   And the metaphysics of the will to power, with its unequivocal insistence on the will's need to establish within its grasp a secure and stable reserve that will assure the possibility of its advance to ever greater power[103] and on its need to do so through a positing of values in accordance with which whatever it gathers to itself can be so assessed as properly to serve this its self-enhancement,

---

[99]"Overcoming Metaphysics," VA 80, EP 92f.
[100]Cf. Ch. VI, pp. 241-252, and Ch. VII, pp. 283f.
[101]"Overcoming Metaphysics," VA 91f, EP 103.
[102]VA 93-95; EP 105f.   The English translation of this passage contains a seriously misleading use of "miscalculate" for "reckon up."   On this point see this chapter, p. 319, n. 98.
[103]"The Word of Nietzsche, 'God is Dead,'" HW 221, 238f; QT 84, 102f.

boldly asserts the appropriateness, not to say the necessity, of just such human dominion over all the earth.[104]

At the same time, this grounding metaphysics concealedly and unknowingly brings man into a chimerical, profoundly empty pursuit. The technology that he indefatigably carries into play endlessly consumes all manner of materials, man included, for the sake of further production. Yet that which it thus claims and consumes and that which it provides is alike devoid of genuine self-presenting. Whatever is, is suspended now in emptiness. It takes place by way of the withdrawnness of Being. The emptiness in question can in no way be filled up by an amassing of what-is, certainly not as thus bereft. Accordingly, no goal of accumulation and use is anything but a vain invention, invincibly ephemeral, ever worthy only to be surpassed. Technology is both insatiable and endlessly resourceful in providing "substitute-things" in an endeavor to meet its perceived, ever on-going needs. And yet, for all its ceaseless producing and providing, that technology is, when rightly viewed in reference to the withdrawnness of Being, simply "the organization of lack."[105]

Via technology the will to will, the "restless," uncertainty-haunted "securing of the order of ordering (*der Ordnung des Ordnens*)," bringing itself into play via modern man, whom it has claimed into its sway, pursues its course and "can be its 'self' as the 'subject',"—the establishing, self-establishing provider—"of everything."[106] But since such ordering (*Ordnen*) is accomplished by the self-securing will that makes its way only in unguessed nullity and relates itself only to itself in its concern for that which it has provided to itself, it remains ultimately vain. It is, as wrought out through technology, finally no more than a securing concerned with limitless and hence ultimately aimless doing, i.e., with a doing that is without discernible, attainable limit or termination and is therefore, ultimately speaking, quite without a goal.[107]

---

[104]HW 233f, 241f; QT 98, 107. Cf. also ZS 58, QB 59.

[105]"Overcoming Metaphysics," VA 95, EP 106f.

[106]VA 96; EP 107.

[107]VA 95; EP 107. *Da aber die Leere des Seins, zumal wenn sie als solche nicht erfahren werden kann, niemals durch die Fülle des Seienden auszufüllen ist, bleibt nur,*

Chapter VIII

At this juncture the destining of Being, Enframing (*Ge-stell*), which is the disposing enduring holding sway in modern technology—that skeletalizing power that despoils whatever it lets endure and frenziedly proliferates in its insatiable ruling[108]—is shown clearly to us in the guise of all-consuming, self-engrossed will. Precisely from out of its self-withdrawal, Being rules. Hidden in itself, hidden, indeed, from the very metaphysical thinking that discerns its mascarade as "will" and proclaims the latter to be the ultimately governing ground of whatever is, this self-sending-forth of Being governs concealedly after the manner of the ever self-aggrandizing will as which it lets itself be descried and foundingly spoken forth. This will, bringing itself into play now via the comportment of modern "technological" man, in its headlong outrunning of the capacity of what-is to slake its craving for self-securing, "forces the impossible upon the possible." The profoundly deceptive doing (*Machenschaft*) that continually arranges (*einrichtet*) this importunate compulsion and unremittingly sustains its dominion "arises," Heidegger

---

um ihr zu entgehen, die unausgesetzte Einrichtung des Seienden auf die ständige Möglichkeit des Ordnens als der Form der Sicherung des ziellosen Tuns. Ziellos is formed from the noun *Ziel*, goal, aim, end, term. Its predominant meaning here clearly cannot be "aimless" in the ordinary sense of the word. The technological has always its goal (cf. Ch. VI, pp. 228f and "Overcoming Metaphysics," VA 97, EP 108), and the revealing as which it takes place neither "simply comes to an end" nor "strays off into the indeterminate." (Cf. Ch. VI, pp. 261f.) Heidegger is here speaking of an inherently limitless activity which, because of the withdrawnness of Being vis a vis which it takes place, can, in a nullity-fraught transpiring whose character is hidden from it, only plunge forward above an abyss. And yet precisely as thus without limit this activity should be thought of as "aimless" in a deeper sense. For while it continually acts purposefully in pursuit of its ultimate goal, i.e., the secure setting in order of everything with which it has to do, there is no genuine goal, no affirming achievement arrived at, that can possibly suffice it. It is an activity that is invincibly pointless, that always comes to nothing. It cannot arrive at any genuine disclosure of anything. Or, more properly, no true self-disclosure can reach it so as arrestingly to impose itself as itself and hence to give pause, as a meaningful goal, to the interminable activity in question. (Cf. Ch. IX, pp. 348f.) With respect to the coupling of limitlessness and nullity we should remember Heidegger's view that it is only through a limiting, a bringing of something into its proper bounds, that that something can be permitted to be whatever it is. (See Ch. V, p. 187f, and n. 69; Ch. IX, pp. 347f & 352f; Ch. X, p. 383; and Ch. XI, pp. 436-438.)
[108]Cf. Ch. VI, pp. 260-262.

says, "from the disposing enduring belonging to technology," where the word technology is intended as equivalent to the concept of metaphysics as the latter displays itself in Nietzsche's thinking, i.e., "of metaphysics consummating itself (*der sich vollendenden Metaphysik*)."[109]

This coming together of modern technology and modern metaphysical thinking into what is at bottom one event evinces with peculiar force the unity and pervasiveness with which the destining of Being now ruling in modern technology at once exercises and extends its sway.   Philosophy as metaphysics has indeed no independent basis. Rather, as the relation between Descartes' thinking and the establishment of science and between Nietzsche's thinking and the establishment of technology shows, metaphysics serves precisely to undergird the coming of modern technology into its own.

Heidegger can unhesitatingly allude to the initiatory enduring that disposes and gathers forth (*Wesen*) holding sway in modern technology as "identical   with   the   disposing-enduring   (*Wesen*)   of   modern metaphysics."[110] To him, clearly, in the modern age metaphysics no less than science has served to prepare the way for the emergence and increasing entrenchment of technology. Indeed, metaphysics can be seen to be the more fundamental of these two.   Modern science, as research, arose on the philosophical ground provided by Descartes, and its development continued to be based in and sustained by subsequent philosophical thinking.   Thus Spinoza offered a portrayal of reality in which a wholly self-contained and inclusive network of causal interrelations underlay all motion and all change; Leibniz set forth force as most fundamentally constitutive of everything real; and, perhaps most importantly, Kant first named and delineated the object and objectification and decisively established the sphere of the empirical, the experienceable, as the valid realm for the comprehending of reality via reliable structures of knowing innate to the human mind.

---

[109]"Overcoming Metaphysics," VA 99, EP 110.
[110]"The Age of the World Picture," HW 69, QT 116. See also "Overcoming Metaphysics," VA 80f, EP 93.   For a detailed discussion of *Wesen* cf. Ch. VI, pp. 252-257.

324   *Chapter VIII*

Metaphysics, science, technology—these three seemingly
distinctive human activities constitute in fact for Heidegger one complex
on-going event that is the out-working simultaneously in multifaceted
ways of one ever intensifying dominion, namely that of Being's self-
sending-forth as the estranging, impelling yet transfixing summons,
Enframing, that rules inexorably in our age. Modern technology is now
clearly ascendant. Through the work of Nietzsche, it has now been
grounded firmly upon its needed basis. Metaphysics has completed its
work. Accordingly, Heidegger can speak of the "end" of philosophy as
having come. Here, he insists, the primary meaning of "end *(Ende)*" is
neither terminal point nor dissolution, but fulfillment, a gathering into the
most extreme possibility.[111] And yet at the same time this fulfillment is
itself a demise. Via the on-going grounding disclosing wrought out
through philosophical thinking, nothing less than "nihilism" has come
now to hold disposing dominion. In ruling increasingly as an
unconcealing determined by withdrawnness, Being has come to hold sway
as negation; not, certainly, as simple negation, but rather as a deceptively
concealed yet wholly powerful happening that rules through a negating of
itself.[112] Accordingly, in Nietzsche's metaphysics of the will to power,
nullity has engulfed metaphysics itself and has come to pervade
everything that metaphysics undergirds. Thus Heidegger can speak of the
overcoming, the conquest *(Überwindung)*, of metaphysics and can say that
the latter "has entered into its negating-ending *(Ver-endung)*." It is
defunct as genuinely on-going and is only as having been.[113] In this
overcoming, Being has been subdued *(überwunden)* from out of that very
enowning Bringing-to-pass, itself at once a dispossessing, that gives it into
its every disclosure *(Er-eignis)*,[114] so that it shows itself now only as a
value captive to the will. Metaphysics as a mode of disclosure has now
no role. It makes manifest now only "the submersion *(Untergang)* of the

---

[111]"The End of Philosophy and the Task of Thinking," SD 63, TB 57.
[112]See "Overcoming Metaphysics," VA 91, EP 103. See also, "The Word of
Nietzsche, 'God is Dead,'" HW 239f, QT 104f.
[113]"Overcoming Metaphysics," VA 71, EP 85.
[114]VA 71; EP 84f. On *Ereignis*, cf. Ch. IV, pp. 148-151 & 163-166.

truth," the safeguarding unconcealment, "of what-is," as which Being might rule; and that submersion is metaphysics' own completion (*Vollendung*).[115] With Nietzsche, this paradoxical fulfillment has been accomplished in the sphere of metaphysical philosophy itself. With him, the most explicit statement of the ground needed by this technologically determined age has been given. But thanks to the radical self-withdrawl of Being from self-affirming ruling, that ground is in fact a groundlessness.[116] In face of it the technological, accomplishing itself via illimitable willing while yet blind to its abandonment by Being and its governance by the latter in puissant self-negation, self-confidently pursues what can therefore be only its negating, ravaging course.

Beyond Nietzsche, Heidegger says, philosophy will be able only to restate and recast what in metaphysics has been already said.[117] In so doing, it will exhibit a tendency already strongly in evidence in movements like positivism or the philosophy of language analysis. Determined by the subjectivism of the age, it will, that is, as what Heidegger calls "philosophical anthropology," ever more persistently evaluate and interpret everything that is from the point of view of man and only in relation to him.[118] Phenomenology and existentialism have for Heidegger this same epigonic character.[119] Philosophy can but live out its fulfillment. It cannot surpass it.

In a wider and no less fundamental sense, philosophy—and indeed by way of the grounding that it itself provides—has been brought to completion in the rise of modern science and in the technology for which that science has prepared the way and into which it is now beginning to merge.[120] The sciences are the offspring of philosophy. And so too is

---

[115]VA 72; EP 86.
[116]On such groundlessness, see further, Ch. XI, pp. 449-456.
[117]"The End of Philosophy and the Task of Thinking," SD 63, TB 57.
[118]"The Age of the World Picture," HW 86, QT 133. Cf. "Overcoming Metaphysics," VA 86f, EP 99.
[119]See "The End of Philosophy and the Task of Thinking," SD 69f, TB 62f. Cf. "Letter on Humanism," W 145ff, BW 193ff, which is in its entirety written against the subjectivism of existentialism, in particular that of Jean-Paul Sartre.
[120]Cf. Ch. VII, pp. 279-281.

technology, even in its most extreme manifestations. The metaphysical thinking that we trace in the modern age, with its ultimate reliance on a presupposed perspective and on the setting fast in view, for the sake of the viewing subject's own self-establishment and self-securing, of that which is to be validated and rendered corroborative by such a judging securing, has its required fulfillment in the cybernetics, the processing of information for utilization, that, brought into play in technology as the method for dealing with reality most appropriate for it, now stands ready to engulf and thus technologize the sciences themselves and hence looms as the definitive mode of human activity now extant in our time.[121]

At this juncture we find ourselves confronted with precisely the same sort of impasse that we encountered when considering modern science. Again Heidegger's last word is technology. And the philosophical thinking that seems to us so varied and so firmly rooted in individual insight and predilection is presented to us as but one factor, albeit a central one, in the arriving upon the historical scene of just those technological attitudes and procedures that we would like to mitigate and control. Heidegger's picture is unsparing. And, quite surprisingly, he has preempted "new" as a designation for that which has already taken place. In philosophy as in science, he tells us, every avenue of seemingly fresh discovery and insight only leads us deeper into the realm where technology rules. From Heidegger's standpoint this must be so. In our persistent hopefulness for the future, he would surely say, we merely exemplify the subjectivism that is characteristic of our age. We center everything upon ourselves, believing that somehow we, through our exercise of reason or wit or audacious perspicacity, shall be able to find a new grounding for our lives and transform the fundamental character of our age.

Heidegger, on the contrary, keeps his gaze steadily fixed on the

---

[121]"The End of Philosophy and the Task of Thinking," SD 63f, TB 57f. Cf. Ch VII, p. 279.

happening of the Being of that which is. Everywhere he discerns it. It alone genuinely initiates that which comes to be. And that happening is at once spontaneous and uncontrollable *and* possessed of an inner consistency, a drift that leaves its traces in a patterning of whatever is and that can be glimpsed when the latter is seen in a long perspective that lets change and interrelationships appear. It is just this patterning that Heidegger shows us in his depicting of the modern age. There everything, when probed by his gaze, displays the same lineaments, the same fundamental manner of happening. A single basic structuring of reality in the relating of man and what-is is everywhere pointed out, even as that structuring is shown to grow ever more explicit and all-consuming with the advance of the age.

But, beyond this, Being is historical precisely as a Happening that, out of a unique and unitary self-initiating, transforms itself in an interrelated succession of self-sendingsforth.[122] Hence, the emergence of this new manner of transpiring of reality can be nothing arbitrary or suddenly prepared. Rather, it is a climactic occurrence whose origins lie distant from itself. If Heidegger is sure that no basis for the transforming of our age can be found within the confines of those endeavors that characterize it, that assurance does not rest primarily in his scrutiny of the age itself, crucial though that scrutiny be. Rather, it rests in his prior, wider discerning of the manner in which the Being of what-is brings itself to pass and in his discovery of it in the long sweep of that history as which Being unfolds itself as the Twofold by way of the openness lived out by man as time. If we are to understand thoroughly Heidegger's accordance of invincible supremacy to the technological in our age, we must, with him, turn our attention to the historical origins of our technology as these are seen with respect to his one touchstone for thinking, the determinative happening of the Being of what-is.

---

[122]Cf. Ch. I, pp. 47-49, and Ch. XIII, pp. 496-498.

PART THREE

Antecedence

CHAPTER IX

TECHNE

When Heidegger turns his attention to origins, he characteristically concerns himself with that which is *Greek*. For us the word "Greek" is bound immediately to evoke the thought of a specific people who, at a particular time in history, achieved remarkable heights in what we think of as cultural attainments, notably in philosophy, mathematics, and the arts. We are heirs of the Greeks. And we readily recognize that much in our culture has, through a process of successive transmission, reached us from them. Heidegger is as aware as we of the presence in history of the Greek people. But in keeping with the particular understanding of history and of time that marks his thinking, his fundamental way of viewing that people differs strikingly from our own.[1] Heidegger is not seriously concerned to look back through linear time so as to discern and depict the life and thought of a people that can be primarily defined and judged in terms of its chronological distance from us here and now. Rather, he is concerned to discover and to show what manner of relationship[2] with the Being of what-is was vouchsafed to the Greeks. This and this alone is, for him, to consider the Greeks in genuinely *historical* perspective, for it is to view them with respect to the self-sending-forth, the destining, of Being that governs in the specifically historical manner of existing into which, through it, they are summoned and sent forth.[3]

---

[1]On the meaning of history for Heidegger, cf. Ch. I, pp. 46f.

[2]Here and elsewhere in this study the reader should be careful rigorously to separate "relation" from its usual meaning. Here "relate," in contexts where the Being of what-is is in any way under consideration, does not speak at all of a connecting of inherently discrete elements, one of which is to be subordinately referred to the other. The relating that is now in question is a *holding-toward-one another* of participants that, while they are distinct from one another, are ultimately one. For we are using "relate" to speak of the happening of the single Twofold. See further, Ch. IV, p. 145, n. 10.

[3]On a people as historical, see Ch. III, p. 137. On Being's self-sending-forth as destining see Ch. I, pp. 47-49.

For Heidegger, the distance from the Greeks that is truly significant for us is of another order than that which we characteristically measure out and understand by means of linear time. It is, instead, a qualitative distance, a deviance (*Irre*) in the manner of transpiring of human existing, that arises from and evinces the divergence of one manner of Being's self-sending-forth from another.[4] For him the word "Greek" does indeed speak—and far more inclusively and profoundly than we ordinarily guess—of the origin of that which now characterizes us and our age. But it does so, when truly heard, not, as we suppose, with respect to discoveries or habits of mind or creative accomplishments that we at once admire from afar and invoke as decisive influences and factors shaping us and our lives. "Greek" speaks, rather, of an origin with respect to the very happening of Being by way of time that has unfolded itself as the self-interrelating Twofold[5] in what has taken place as the history of the West. Ultimately understood, the adjective "Greek" presents to us, Heidegger can say, "the earliness of the destining as which Being itself clears, opens itself lightingly, in what-is and lays rightful and demanding claim to a determinative manner of enduring belonging to man that, as taking its character from destining, has its historical course in the way in which it is set free from Being although never separated from it."[6]

1

The origin of the modern technology whose particular manner of happening we have been examining and tracing out in its pervasiveness

---

[4]"The Anaximander Fragment," HW 311, EGT 26. Cf. Ch. I, pp. 48f. On the happening of what-is into deviant divergence (*Irre*) from Being by Being itself, which is here in question, cf. Ch. I, pp. 45f.

[5]On the Twofold, see Ch. I, pp. 27-33.

[6] . . . *griechisch ist die Frühe des Geschickes, als welches das Sein selbst sich im Seienden lichtet und ein Wesen des Menschen in seinen Anspruch nimmt, das als geschickliches darin seinen Geschicksgang hat, wie es im Sein gewahrt und wie es aus ihm entlassen, aber gleichwohl nie von ihm gretrennt wird.* "The Anaximander Fragment," HW 310, EGT 25. On the character of this relating as considered with respect to Being's happening-away of what-is, whose manifold includes man, cf. Ch. I, p. 53, and Ch. II, p. 108, and n. 190. On the meaning of *Wesen*, cf. Ch. VI, pp 252-257.

lies in what the Greeks called *technē*. Like "technology," the word *technē* speaks for Heidegger of a phenomenon central to its age and one by way of which the specific manner of Being's ruling in that age brings itself to fulfillment.

Originally *technē* denoted for the Greeks, Heidegger says, a masterful knowing (*Wissen*) marked by thoroughness and perspicacity that was, as such, an assured ability to accomplish.[7] As such knowing, *technē* is a skilled governance that informs every sort of intentional providing and that does so via an immediate human responsiveness vis a vis the happening Twofold. It is "the discerningly knowing having-of-disposal over free planning and ordering (*Einrichten*)—over planning and ordering, that is, which are directly open toward self-proffering reality[8]—and over a concomitant "mastering of that which is fitly provided and instituted which is similarly free."[9] *Technē* as this provisive discernment is an enabling of that with which it concerns itself. It so takes place as to engender, to raise up, i.e., it takes place as a knowing and able "bringing-forth."[10]

---

[7]N I (German) 192; (English) 164. We should here remember that the original meaning of the verb *Wissen* is "to see." Cf. Ch. VIII, p. 303, n. 36.

[8]See Ch. II, pp. 105-107.

[9] . . . *das wissende Verfügen über das freie Planen und Einrichten und Beherrschen von Einrichtungen.* EM 13; IM 13f. The wealth of meaning in the verb *einrichten* and the noun *Einrichtung* extends far beyond the denotative capacity of a single translation, even a complex one such as that just given. *Einrichten* means to arrange or order, to put right, to adapt, to contrive, to organize, or to install. *Einrichtung* can mean arrangement, adjustment, contrivance, equipment, furnishings, household appointments, establishment, or institution. Behind both words lies the verb *richten*, to direct, to order, to judge, within which inheres the original root meaning, to move in a straight line. (*The American Heritage Dictionary of the English Language*, Boston, Houghton-Mifflin Co., 1971, p. 1536) We have tried to capture something of the inclusiveness of Heidegger's meaning, not, however, without sacrificing the rich specificity that his words contain. We should here recall Heidegger's use of *Einrichtung* to identify modern technology in its instrumental character (Ch. VI, p. 227) and also the centrality of planning as definitive of that technology (Ch. VI, p. 228)—usages that show the kinship of *technē* and technology that we have now under consideration.

[10]*Die Technē ist Erzeugen, Erbauen, als wissendes Hervor-bringen.* EM 13; IM 13. The parallel between this portrayal of *technē* and Heidegger's defining description of the intention-fraught concern (*Besorgen*) belonging to human existing as a receptive

In the work, *Introduction to Metaphysics*, at whose outset the above characterization of *techne* appears, in a discussion that often closely parallels that in *Being and Time*, Heidegger presents this knowing, *technē*, as that which definitively characterizes the human existing, the *Being*-there-as-openness (*Dasein*), that is peculiarly Greek.[11] The openness to Being's ruling that Greek man lives out is determinatively constituted by it. That is, the openness that he, in his relational existing, offers for the disclosive happening of the Being of what-is as the unfolding Twofold always ultimately takes place via the happening of just such discerning knowing.

The Greeks experienced the unfolding Twofold, in its happening as juxtaposed to them, primarily as *physis*, i.e., as a reality that, in whatever centerings and configurations, arose abruptly and commandingly from out of itself.[12] It burst forth into manifestation, sometimes clearly, sometimes deceptively, maintained itself powerfully and lastingly in an ever nuanced appearing, and of itself departed, falling decisively into concealment. Here Being as self-unconcealing self-concealing, happening self-maintainingly as an imperious, eruptive disclosing, held sway as a powerful appearing that now gathered together into the stability of a genuine self-maintaining that manifold as which it bodies itself forth and now brought the components of the unfolding manifold together and to light in the elusiveness of a sham lasting on.[13]

Greek man was called upon directly and out of his own self-

---

out-reaching that frees the ready-to-hand to its intended taking place should be noted. Cf. Ch II, pp. 64-68. So too should the parallel between *technē* and dwelling. Cf. this chapter, pp. 355-359. On the identification of *technē* as *Hervorbringen* see this chapter, pp. 345-348.

[11]Cf. EM 130, IM 142. For a detailed discussion of human existing (*Dasein*) as such see Ch. II, pp. 57-110. Heidegger's portrayal of *technē* often reminds one forcibly of his discussions of *Dasein*'s "authentic" transpiring. (See especially Ch. II, pp. 94-103 & 105-109.) Its "inauthentic" transpiring is rarely noticed.

[12]Physis *meint das aufgehende Walten und das von ihm durchwaltete Währen.* EM 11f; IM 12. Cf. EM 47, 75ff; IM 51, 83ff.

[13]Cf. EM 76, IM 85, and "On the Essence of Truth," W 89ff, BW 132ff. Cf. Ch. I, p. 44. See also "The Way to Language," US 252f, WL 122. On Being's happening fundamentally as self-maintaining self-unconcealing self-concealing see Ch. I, pp. 24-26.

maintainedness to confront and to administer (*verwalten*) the mighty self-disclosive ruling that he encountered in encountering *physis*.[14]   This he did via *technē*, i.e., via the assured and able knowing that could genuinely discriminate and discern and hence genuinely gather and let appear. Through *technē*, seeing out beyond on the basis of what had already been seen and known,[15] he could exert mastery over what met him, so encountering it as to make it manifest at once as itself, i.e., as something standing forth in delimitation, and also as something defined and determined in interrelation with that which environed it.[16]

Technē displayed itself first of all in the language through which the Greeks, naming, called forth into appearing that which offered itself to them.[17]   And it likewise wrought itself out in every arena of constitutive endeavor that language, in its foundational role, undergirded and opened up.[18]   In poetry as preeminent and in all the arts, in rigorous, inquiring thinking, in the deeds of devotion that made manifest the nearness of a god, in the founding of states, in crafts, in the skills of mariner and hunter and ploughman—everywhere the intrepid knowing called *technē* was brought decisively into play.[19]   Not all attained to it.

---

[14]EM 131ff; IM 144ff.

[15]"The Anaximander Fragment," HW 318f, EGT 33f. On the seeing on the basis of a having seen, constitutive of *Dasein*, that is instanced in this seeing, cf. Ch. II, pp. 73f, and Ch. III, pp. 113f.

[16]EM 122; IM 133. Cf. Ch. III, p. 135.

[17]On man's prescient knowing as a finding of his way to language, see EM 120, IM 131, and EM 131, IM 143. On naming, see EM 131, IM 144. Cf. Ch. III, pp. 125f. The present discussion is based primarily on a long and complex passage in EM 112ff, IM 123ff, in which Heidegger explicates the meaning of *technē*, first by way of the words of a chorus from Sophocles' *Antigone* and then through the elucidating of words of Parmenides. In the course of his discussion Heidegger shows us that poet and thinker disclose the same character for the relation between man and Being. Because of the nature of the discussion that our own presupposes, we have frequently found it peculiarly difficult, as in the present instance, to provide notes that do more than give frustratingly inadequate glimpses into Heidegger's argument and into the interconnection of themes that is built forward in it. We urge the reader to turn to the Heideggerian presentation itself and to read it, indeed, in the context of the entire work in which it stands.

[18]See "The Turning," TK 40, QT 40f. See also Ch III, pp. 129-138.

[19]EM 47; IM 51. See "The Question Concerning Technology," VA 20, QT 13, and

Most contented themselves with the comfortable sphere of the customary and with a random and superficial seeing that could not with any assurance detect and transcend the obscure or the spurious and so indiscriminately brought forth that which came into view to an appearing that left it jumbled together and devoid of an identity that was clear. But always there were those who, genuinely achieving *technē*, cut incisively into the awesome reality presenting itself to them, abjured entanglement in the obfuscation of mere semblance, however alluring, and dedicated themselves to the achieving of unclouded disclosure, the clear manifesting of Being.[20] Submitting themselves steadfastly to the rule of *technē*, they brought to fulfillment the existing, the *Being*-there-as-openness, to which Greek man was summoned, for they brought to consummation the self-unconcealing of Being in that which, as some insightfully prepared delineate work, via their disclosive knowing, came to appear.[21]

Where Greek man thus achieved the openness-for-Being incumbent upon him, he brought to bear an apprehending (*Vernehmen*) that brought Being to a stand, i.e., to a happening as disclosive enduring, through a preemptive taking charge (*Übernahme*) in the face of that which bore in upon him from beyond himself.[22] Apart from the knowing that *technē* designates—as that knowing accomplished itself in the putting into play of language that indeed underlay all human doing[23]—this that came upon man from without could have met him only as something dark, shut-away, concealed, where all could but crowd in upon him in a sheer pressing-toward (*Andrang*) that could not be opposed or stemmed.[24] But via *technē*, focused thus in language, man was able boldly to meet and lay hold of that which thrust itself upon him. Even as he broke free of the obscuring pull of complacence and mere appearance, so he could, precisely in so doing and in a movement provisive in fact of the latter's

---

EM 118-120, IM 129-132. See also "The Origin of the Work of Art," UK 68f, PLT 61f.

[20]Cf. EM 127f, IM 140, and EM 80f, IM 89f.

[21]EM 130; IM 142.

[22]EM 105ff, IM 116; EM 128f, IM 141.

[23]"The Turning," TK 40, QT 40f.

[24]EM 131; IM 144.

very possibility of transpiring, break into the closedness of the environing, impinging reality as such—whether the realm of the self-contained sea or earth or the realm of the latent powers constitutive of himself—and through the bringing of something particular to stand forth in and of itself, could wrest from that unmanifestness a happening of Being.[25] He had power definitively to distinguish between this and that, to separate and set apart. And just in so doing he could discriminatingly select and gather into relation that which came under the dominion of his knowing. In thus bringing that which he was encountering into comprehensibility and discernibility, he brought it into a lastingness that disclosed it and let it disclose itself as a world.[26]

The counterpart of *Andrang*, the opaque, unresting on-thrust of the concealed, is *Anwesen*, the presencing, the restrained on-coming that at once particularizes and keeps in relation and therewith lets that which tarries as present shine out, identifiable and governable, in the distinctiveness arising from interrelation that unconceals.[27] Where, among the Greeks, we find genuinely juxtaposed the imperious self-disclosive arising of *physis* and the human apprehending that, as *techne̅*, knowingly grasps and gathers it via time—the constitutive determinant of man's *Being*-there—so as to let it linger in disclosure, there we find Being directly accomplishing itself as presencing. For the Greeks, Heidegger says, "that which is, is that which arises and opens itself, which, as what presences, comes upon man as the one who presences, i.e., comes upon the one who himself opens himself to that which presences, in that he apprehends it."[28]

Physis and *techne̅* belong together, for these words name, as

---

[25]EM 117ff; IM 129ff.

[26]EM 47; IM 51. On the meaning of "world" that must be taken to be fundamentally in question here, cf. Ch. II, pp. 61-63. On the crucial role in the transpiring of human existing as openness-for-Being of the power to differentiate that is here instanced, cf. Ch. II, pp. 77f.

[27]Cf. Ch. I, pp. 24 & 49f.

[28]*Das Seiende ist das Aufgehende und Sichöffnende, was als das Anwesende über den Menschen als den Anwesenden kommt, d.h. über den, der sich selber dem Anwesenden öffnet, indem er es vernimmt.* "The Age of the World Picture," HW 83, QT 131.

originally spoken by the Greeks, the Being of what-is and the disclosive apprehending that answers to it, the apprehending to which Being happens man away from itself in order that, ruled still by Being, he may let what-is come to appearance, may let it *be*.[29]

At the beginning of the happening of this belonging together, which was indeed its happening as *Greek*, Being was holding sway as *logos*, as the selecting gathering-together that let lie distinguishedly and interrelatedly forth before.[30]   Precisely as *physis*, Being ruled as a gathering-together[31] that was, as such, a setting-apart, a fruitful contending (*polemos*, *Auseinandersetzung*), that let the distinguished, *as distinguished*, be gathered unifyingly into interplay together.[32] Everywhere this distinguishing-gathering ruled as a fitly disposing ordaining (*dikē*, *Fug*), in the manifest and the merely seeming no less than in the clearly disclosed[33]—all instances of the one Happening named definingly as the Being of what-is.   Intrinsically ruled by Being in its happening thus as *logos*, *physis*, that which met man gathered itself together toward him and toward the appearing that he alone could bring to fulfillment; and man, for his part, likewise ruled in his distinguishedness by Being as *logos* and himself carrying the latter into play,[34] gathered himself together against the encroaching of all vitiating, dispersive tendencies bearing upon him and therewith gathered himself toward that which met him, drawing it unifiedly toward himself in an intentness of self-assured apprehending that could alone take disposal over it and so let it genuinely appear.[35]

---

[29]Cf. EM 125-127, IM 136-139. Here we should bear in mind the belonging together of thinking and Being as affirmed by Parmenides (cf. Ch. III, pp. 129f), for *technē* includes in itself thinking.

[30]Cf. Ch. III, pp. 124f.

[31]EM 124f; IM 136f.

[32]EM 47; IM 51.

[33]EM 126f; IM 139f.

[34]EM 130f; IM 143.

[35]EM 128f; IM 141. Cf. EM 109f, IM 121. On the lineaments of the structural parallel obtaining between Being's happening as the Being of what-is and human existing as disclosive of that happening, see Ch. I, p. 33, n. 36. In Heidegger's portrayal of *technē* as a forth-going discerning knowing on whose already given basis a

Heidegger's first depiction of the encountering brought to pass by Being in order that it might among the Greeks manifest and maintain itself by way of the openness that man was called upon to provide is given in terms of an immediate and demanding confrontation in power. In that confrontation, ruling and counter-ruling meet. The manifold of that which is, as *physis*, arises and bursts upon man as that which, in the sovereign power of Being, overwhelms and subjugates (*das Überwältigende*); and man, launching himself determinedly toward this onslaught via *techne̅*, his prescient, skillful knowing, in the answering to Being peculiarly vouchsafed to him, through a movement that takes place as violence against Being's happening as concealing and dissembling, masters (*bewältigt*) that manifold, disposes it, lets it appear.[36] This confronting is a conflict (*Kampf*), not in the ordinary sense of a struggle between foes but in the sense of a decisive counterposing in radical self-accomplishment, that can alone fulfill the belonging-together of Being and man and permit the genuine achieving of Being's self-manifestation.[37]

Every human enterprise in which *techne̅*, as such violent countering, is truly brought to bear is a deed of daring on man's part.[38] He who commits himself to put into play this knowing ventures out into the uncharted. He cleaves open the reality that crowds in upon him, imposing upon and tracing into it the separating-relating configurings that his masterful seeing there discerns. Thus he wrests forth genuine disclosure into full appearing.[39] In his knowing, and from out of it, he rises superior to the Being of whatever is.[40] He it is who speaks the poetic word or formulates the grounding thought or accomplishes the work of intellect or craft or art, where, in whatever he thus concretely provides, Being as unconcealing decisively happens. Through what he

---

gathering-together that gathers forth takes place in intense commitment to *techne̅*'s own self-carrying out, the constitutive manner of transpiring of human existing as Being-in-the-world is clearly to be seen. Cf. Ch. II, pp. 68f & 73-75.
[36] EM 122; IM 134. Cf. EM 114ff, IM 125ff.
[37] EM 115; IM 126f. Cf. EM 47, IM 51.
[38] EM 123; IM 135.
[39] EM 123; IM 135. Cf. EM 117, IM 128.
[40] EM 121f; IM 133f.

has wrought, the otherwise inscrutable reality into whose midst he has entered may find itself suddenly illumined, gathered unitedly into one complex of distinctive elements, and so disclosed and sustained as a meaningful world.[41] Such venturing is only for the few. It requires of man an intense concentration toward the pregnant encounter in which he must engage.[42]

Heidegger designates the apprehending that thus single-mindedly directs itself toward the accomplishing of unconcealing "Ent-scheidung."[43] *Entscheidung* ordinarily means decision. But Heidegger is not here using the word in an ordinary sense, to denote a familiar psychological event, e.g., a passing of judgment or a choice made through an exercise of the will, where will is considered as one human faculty among others. Rather, *Ent-scheidung* is, as its orthography indicates, a severing (*Scheiden*) that, as such, is a going forth or away.[44] It clearly parallels the urgently transpiring self-opening-up (*Entschlossenheit*)—itself not something *willed*, but the closedness, the delimitedness, that is alone genuine going-forth—that characterizes man's *Being*-there-as-openness as such in the staunchly borne precariousness of its authentic happening.[45] And it should undoubtedly remind us also of the uniting-sundering (*Unter-Schied*) as which pure Happening brings itself to pass as the Being of what-is, by way of which, as that sundering brings itself into play via the transpiring of human existing, every specific distinguishing, as a bringing to consummation of disclosure, is brought to pass.[46] Whoever, attaining to the fundamental decisiveness of self-opening that *Ent-scheidung* names, accomplishes *technē* and brings unconcealing to happen, must be understood to live out with intensity and directness human existing as

---

[41]For the meaning of "world" that is here in the foreground, see Ch. III, pp. 136f.
[42]Cf. EM 132, IM 145f.
[43]EM 128; IM 140f.
[44]On the prefix *ent-*, cf. Ch I, p. 24, n. 12.
[45]Cf. "The Origin of the Work of Art," UK 76, PLT 67. See Ch. II, pp. 99-101, and n. 156. On "will" as used by Heidegger with reference to the self-carrying-out of human existing (*Dasein*) as openness-for-Being that is Being-in-the-world, cf. Ch. II, p. 72.
[46]Cf. Ch. I, pp. 27 & 29-33, and Ch. II, pp. 77f.

openness-for-Being as such, in its peculiar happening as Greek.[47]

In himself, in his doing and in whatever it may be that he provides, such a one grounds and establishes with his lordly achieving of unconcealing the very *Being*-there of the community from among which and on behalf of which he undertakes his skilled confronting of Being.[48] Yet his is a lonely venture. He necessarily leaves behind the familiar and accepted with which his fellows are content. Presupposing it as something known, he departs radically from it, so as to enter into his disclosive contending both with it and with the as yet unmanifested reality on-coming in power. He is the strange, the uncanny, the *unheimlich* one, who dares to cut himself off from the established, from home, in the hope of wresting-forth and bringing to a stand a wholly new happening of Being as appearing.[49]

In such a one, even as in that which he addresses in his knowing, Being rules. In that meeting, where *logos* as accomplished by man can be named as *technē*,[50] man powerfully achieves. Yet in this, ultimately, it is Being that, via *technē*, achieves, through man, its self-maintaining happening as unconcealing. Ruled by Being, kept in hand by it, man is yet happened away out of Being, handed over to happen of himself,[51] that that self-maintaining may be brought to pass through him. In his self-collectedness in puissant knowing, the one who, accomplishing *technē*, undertakes to bring genuinely to light the Being of what-is fulfills the claim of Being that remains masked from others. To him that claim is a necessity, a distressing need (*Not*)[52] wherein his banishment from home, from all that is familiar, and his intent contending with the imperious self-offering of the reality with which he has to do fuse as an all-demanding

[47]EM 133; IM 145f. Cf. EM 110, IM 122.
[48]Cf. EM 117, IM 128.
[49]EM 115f; IM 127f. One should here keep in mind Heidegger's portrayal of the uncanniness of human *Dasein* in its authentic transpiring as Being-in-the-world as involving an awareness of being without abode, which is juxtaposed to an everyday, inauthentic mode of existing with its pervasive being-at-ease and complacency. See Ch. II, pp. 94f & 87-90.
[50]Cf. EM 129f, IM 142.
[51]See Ch. I, p. 53, and Ch II, pp. 108f, and n. 190.
[52]EM 124, 130; IM 136, 142.

dedication to his undertaking, whose achieving is his achieving of the role
required of him from out of Being. It is just in so happening as this
distressful need that man's existing ceaselessly happens as the openness
via which Being maintains itself through an unresting, disclosive
embodying and reembodying of itself as the Being of what-is.

But man, although empowered thus to contend toward the genuine
disclosure of Being, *as* man *in Being*, is yet but man. Therefore he is not
and cannot be, from within himself, a match for Being. And Being
happens as self-unconcealing self-*concealing*.[53] He who ventures into
radical relation with the Being of what-is, so as to accomplish
unconcealing, exposes himself radically, precisely in so doing, to the
concealing that is always surmounted therein. Everywhere he is beset by
concealing. His self-confident departure toward the attainment of
unconcealing necessarily always rests back on a discerning to which the
clear manifesting of the Being of what-is has not yet been vouchsafed.[54]
And invariably Being as concealing, in the indomitable power of its
happening as *physis*, sooner or later rises superior to any openness of
unconcealing that he is permitted to provide.

A man who gives himself unreservedly to the accomplishing of
technē may achieve a revelatory deed that establishes his community in a
world, bringing the elements of that world to appear clearly, gathered and
disposed with respect to one another, particularized and cohering as a
meaningful whole[55] and permitting genuine enduring to happen. The
taking-place of Greek existing as *historical*, i.e., as sent and determined
from out of Being, and as disclosive of the latter, is grounded always and
carried forward in the intensely dedicated, power-fraught answering of
logos to *logos* that happens in just such deeds. But the man who through
his skillful knowing attempts a mastering of Being may also fail. The
ever primary dominion of the Being of what-is, in its happening as
concealing, may thwart his venture, refuse the happening of unconcealing
to his work, and, in its insurmountable intransigency, vouchsafe to him

[53]Cf. Ch. I, pp. 24-26 & 43f.
[54]Cf. Ch. III, p. 115.
[55]Cf. Ch. III, pp. 136f.

not disclosive mastery but fruitless defeat.[56]   In undertaking in full immediacy the intent, receptive apprehending proper to his existing, Greek man, in presencing directly toward what directly presences toward him, must, in that extreme openness that answers to the sovereign self-opening of Being, suffer in acquiescent, resolute self-launching-forth an onslaught that, for all his skilled and knowledgeable seeing, he cannot finally foresee. Greek man, on-goingly living out the openness-for-Being that Being, in happening him away to his existing, provides for itself in the midst of what-is, is summoned to gather into unconcealment and there preserve that which, via his apprehending, "opens itself in its openness." And precisely in so doing "he must remain exposed, laid open (*alētheuein*), to all its sundering confusions."[57]

The more truly a man accomplishes the existing incumbent upon him, the greater is the risk that in his pursuance of it he may be shattered against the sheer, ungovernable happening of the Being of what-is. And even when he triumphs on behalf of unconcealing, with some masterful stroke of disclosure, his grounding accomplishment and the world that it gathers and brings to clear disclosure last only for a season. Concealing encroaches upon unconcealing, and the brightness of full appearing sinks away in time into mere semblance. Mere opinion replaces clear-sighted knowing, the familiar becomes the norm, as Being conceals and dissembles in happening what-is, and a world once genuinely discernible in Being is misapprehended and misunderstood.[58] Only a new audacious deed in some constitutive sphere of human knowing can let genuine appearing again take place.

In a discussion of Sophocles' *Oedipus Rex*, Heidegger, speaking of the all-consuming, reckless knowing named elsewhere in the lengthy surrounding presentation as *technē*, identifies that knowing as "the fundamental passion (*Leidenschaft*)" of the existing that is specifically Greek. As such a passion, that knowing is a readiness to dare and to

---

[56]See EM 123f; IM 135. Cf. also EM 134f, IM 148.
[57]"The Age of the World Picture," HW 84, QT 131.
[58]EM 121; IM 132f.

undergo, "for the sake of the unveiling of Being."[59] It is a "struggle," a radically illumining setting-sharply-in-juxtaposition that gathers disclosively together, which is engaged in over nothing less than Being itself.[60] Wherever Greek *Being*-there-as-openness was lived out to its uttermost, man, with all the power at his command, set himself boldly over against Being in order that he might, claiming the vantage ground proper to him, wholly and unflinchingly surrender himself to the very sovereignty of Being, i.e., in order that he might offer himself as the open arena via which, precisely in that extremity of interrelation, Being might openly happen as unconcealing. To be himself as man meant, through his knowing so to receive and withstand the complex reality that pressed upon him as to respond with word or thought or deed or fashioned work that could impose itself in transparent clarity, disclosing itself incontestably as itself and as something incontrovertibly in Being. He who achieved this bent his powers single-mindedly toward the bringing-forth of something that would allow the members of the community from whose midst he went forth in the lonely venture of resolute dedication to come to genuine seeing and to orient themselves toward and within a world that had come suddenly to light through what he had brought to be. He might succeed. Or he might fail. He himself might even, like Oedipus, pay for his passion genuinely to disclose Being by being himself overwhelmed by what he unconcealed. The risk was awesome. In its portrayal we can surely discern the shadow of Heidegger's "being-toward-death."[61] Greek *Being*-there-as openness, seen purely, shows us nothing other than man's *Being*-there-as-openness (*Dasein*) as such. It shows us man as possessed of an existing that demands of him that he be himself, in the intense centeredness of authentic individuality in community,[62] precisely in

---

[59]Heidegger's statement, one phrase of which has in this sentence been very freely rendered to bring out its proper connotation, runs in full: . . . *wir müssen in Oedipus jene Gestalt des griechischen Daseins begreifen, in der sich dessen Grundleidenschaft ins Weiteste und Wildeste vorwagt, die Leidenschaft der Seinsenthüllung, d.h., des Kampfes um das Sein selbst.* EM 81; IM 90.

[60]EM 80f; IM 89f.

[61]See Ch. II, pp. 97-101. Cf. EM 121, IM 133.

[62]Cf. Ch. II, pp. 101-105.

abandoning himself to the imperious rule of Being; that he remain
steadfastly open to the happening of Being in all its inscrutability; and that
means that, under Being's governance, he direct himself of himself, in his
disciplined seeing, passionately toward the most extreme possibility that
can befall him in his undertaking of the knowing that is *technē*: the
possibility of his own destruction.

2

The word *technē* is ordinarily translated as "art" or "craft," an
interpretation that narrows the original richness of meaning that
Heidegger early found for it. These meanings correctly mirror those most
immediately suggested by ancient Greek texts. In them, as significances
put forward for *technē*, the thought of a skilled knowing is clearly
present. Yet how muted it seems in comparison with Heidegger's
urgently vivid portrayal of *technē*, born of a more subtle probing, that we
have just followed out. From Heidegger's point of view this contrast is
possessed of an inner logic and cannot be surmounted by a simple choice
between disparate interpretations. For in it is manifest at a central point
and in the complexity belonging to an historical course the happening of
the distinctively "Greek" relating of Being and man as that relating arose
and pursued its way among the people we know as Greek.[63]

In Heidegger's later writings a meaning seemingly close to that
ordinarily given to *technē*—a meaning presented early by him[64] but not
elaborated—comes to the fore. He still insists on *technē*'s fundamental
denoting of an experienced and skillful knowing and on the wide range of
the word's application to a variety of human pursuits.[65] But he uses as a
decisive characterization for *technē* the verb *hervorbringen*,[66] which

[63]See Ch. X, pp. 394-398. Heidegger does not intend that the thought of an eruptive,
power-fraught taking place should ever be absent from the word *technē*. See the quite
late allusion to *technē* in "The Question Concerning Technology," VA 19, QT 10f,
referred to on p. 347 below.
[64]EM 13; IM 13. See this chapter, p. 333, and n. 10.
[65]"The Question Concerning Technology," VA 21, QT 13.
[66]VA 19; QT 10.

speaks not of knowing *per se* but of the achieving in which the knowing in
question eventuates. *Technē* is ordinarily derived from the Indo-European
root *teks*, to fabricate or weave.[67] Heidegger, on the other hand, derives
it from *tek*, the root of the verb *tikto*, to bring into the world, to bear or
beget, to produce (of the vegetative). Accordingly, he equates the import
of both noun and verb with that of the German verb *hervorbringen*, which
can mean to bring forth, to yield or produce, to beget or generate, to
utter, and to elicit.[68]

As a word used to express the meaning of *technē*, *hervorbringen*
could easily be misinterpreted as alluding to "producing" as we
customarily understand it, i.e., as a making that is initiated and carried
out by the "producer" himself and that is decisively under his control.
Hervorbringen does indeed speak, for Heidegger, of a "producing," but
one of a quite different order. In translating a sentence from Plato's
Symposium, Heidegger uses the word in parallel with *poiēsis*, the Greek
noun from the verb *poiein*, to do or to make: "Every occasion for
whatever passes over and goes forward into presencing from that which is
not presencing is *poiēsis*, is bringing-forth (*Her-vor-bringen*)."[69] The
originating of the peculiar "making," the peculiar "doing," here in
question does not lie with any particular "maker." For Heidegger is
using *Hervorbringen*—here, as often, hyphenated as *Her-vor-bringen* and
hence properly translatable as bringing-forth-hither[70]—as a primary way
of speaking of the providing happening as which Being brings whatever is
forth and hither into the disclosive nearness of presencing.[71] True human
producing, as a making, is seen as under the governance of Being so
understood. As such, it is a pro-ducing in the literal sense of a leading

[67]*The American Heritage Dictionary*, p. 1545, citing Julius Pokorny's
Indogermanisches Etymologisches Wörterbuch, Berlin, 1959, p. 1058.
[68]Cf. "Building, Dwelling, Thinking," VA 159, PLT 159.
[69]*he gar toi ek tou me onton eis to on ionti hotoioun aitia pasa esti poiēsis.* VA 19;
QT 10. "All creation or passage of non-being into being is poetry or making, . . . "
(*Symposium*, 205b, Jowett translation). Cf. "Building, Dwelling, Thinking," VA 160,
PLT 159.
[70]See "The Question Concerning Technology," VA 19, QT 10, and n. 9.
[71]VA 18-20; QT 10-12.

forth.

It is this leading forth, this bringing-forth-hither into presencing—a "making" understood as a disclosing, *poiēsis*—of which *technē* speaks.[72] As with Heidegger's identification of *technē* as a bringing into play of *logos*, so here, the answering of *technē*, understood as a bringing-forth, to the happening of Being as the bringing-forth that accomplishes all disclosing of what-is is direct and of the closest possible sort.

The bringing-forth, the disclosive doing, *poiēsis*, that is here in question is itself *physis*, the irruptive "arising from out of itself," that directly brings to presence and to which *technē*, interposing to bring to presence through itself, immediately corresponds.[73] *Physis*, the happening Twofold wherein Being rules as *logos*, is, Heidegger says, *thesis*, i.e., "from out of itself to lay something forth, to set it hither, to bring it hither and forth, that is, into presencing."[74] From the Indo-European root *dhe* found in *thesis* come also German *tun* and English "do." The *thesis* as which *physis* happens is, in the primal sense of the word, *doing*. As such, it is working, *Wirken*,[75] not as an effecting, but as this disclosive bringing hither and forth into presencing.[76]

Like *logos*, *thesis* speaks of the particularity toward which Being happens whatever is. It is the very placing that lets lie forth before. As central in speaking of *thesis*, Heidegger uses the verb *stellen*, which we have found to be so crucial in his portrayal of the happening of modern technology.[77] In a Greek context, *stellen* and its compounds display for him the meaning of *thesis* as naming the original placing, the original bringing into specificity and hence into unconcealment, as which bringing-forth-hither accomplishes itself. Here the "setting" in question

---

[72]We should here note the parallel between this depiction of *technē* and the portrayal that Heidegger gives in *Being and Time* of human intending concern (*Besorgen*) as fulfilling the intending self-presentation of the ready-to-hand (*das Zuhandene*). Cf. Ch. II, pp. 64-68.

[73]"The Question Concerning Technology," VA 19, QT 10.

[74]"Science and Reflection," VA 49, QT 159.

[75]VA 49; QT 160.

[76]VA 50; QT 160f.

[77]Cf. Ch. VI, pp. 247-249.

is a disclosive "setting-hither and presenting (*Her- und Dar-stellen*)."[78] It happens as a "bringing-*hither* into the unconcealed, *forth* into what presences."[79] As such, it is a bringing into specificity. It involves a "fixing" i.e., a "setting-fast (*Feststellen*)," which, through the bringing of any entity into delineateness, that is, into the boundedness of the immediate detailing that shapes and concretely presents it, frees that entity to presence—even to impose itself as thus incisively lifted forth into specificity—as itself.[80]

The configuring as which that delineating carries itself out as such disclosive specifying is, Heidegger says, the meaning in this context of *Gestalt*—structuring, form. And the placing that as bringing-forth-hither gathers itself into play in such particulars can be named *Ge-stell*.[81] Here this word, familiar to us from such a disparate context,[82] speaks of the letting arise, the setting determinately hither, as which "working" in the Greek sense happens. Thus, it names the gathering placing "as which a work originatingly disposes itself and endures (*west*)," insofar as that work offers itself in the immediacy and full specificity of its delineate presence.[83]

In presenting *techne* from the new perspective offered by *Her-vor-bringen*, Heidegger has not changed his fundamental understanding of *techne* itself. The Greek who genuinely accomplishes *techne* fulfills through his producing—his intent and able working—the ruling working as which Being encounters and governs him. In a confrontation in which perceptive, eager skill accords with the urgent presentation to it of that which is called for at a particular juncture, *logos* answers to *logos*, and bringing-forth to bringing-forth.[84]

In a masterful bringing-forth that is at bottom a responsive

---

[78]"The Question Concerning Technology," VA 29, QT 21.
[79]"The Origin of the Work of Art," UK 95f, PLT 82.
[80]UK 95f; PLT 82f.
[81]UK 98; PLT 84.
[82]See Ch. VI, pp. 258f.
[83]"The Origin of the Work of Art," UK 71f, PLT 64. On these connotations of *Wesen*, see Ch. VI, pp. 253f, and n. 76.
[84]Cf. "The Question Concerning Technology," VA 17-19, QT 8-10.

obedience, the doer of *techne̅* brings to appearance something whose unrealized lineaments he and he only can antecedently discern. He places before his fellows in striking immediacy something that—whether great or small—is altogether appropriate, altogether fitting and needful.[85] His working discloses something that imposes itself as incontrovertibly right and that therewith brings everything that gathers about it to be seen in its light. Through his achieving he sets powerfully forth something that is at once his work and a thing worked from beyond himself. It appears through his working, but by way of that working its own impending Being is in truth sending it forth, placing it, setting it determinatively hither, so that by its very standing-in-the-midst in the power of its presence it may, in a suddenly illumined present, proclaim and fulfill the happening of Being as unconcealing. Art, Heidegger says, fulfills this role of *techne̅* in the highest degree. It "most immediately brings Being, i.e., the appearing that stands there in itself, to a stand in something presencing (in the work)."[86]

But *techne̅* as such, wherever and however as a far-seeing provisive knowing it comes fruitfully into play, has ever that same character of an accomplishing that brings forth something that, manifest in and as itself, in its presencing brings into disclosure whatever is, with it, gathered forth. Like art, its prime instancing, it is always "the sapient setting of appearing," i.e., of Being, "into the work (*Ins-Werk-Setzen*)." It is the decisive setting of Being as unconcealment into some particular entity that has been knowingly wrought.[87]

Every work genuinely brought forth via *techne̅* partakes of uniqueness and hence of unexpectedness. Under the impress of the powerful arising that rules in the happening of the Twofold as *physis*, whatever genuinely appears—be it in what we would now call the realm of nature or in the realm of the works wrought by man—bursts into view.[88] Hitherto it was not, but now it is. Happened by Being, always in

---

[85]Cf. VA 17, QT 8.
[86]EM 122; IM 134.
[87]EM 133, 122; IM 146, 133f. Cf. EM 130, IM 142.
[88]"The Question concerning Technology," VA 19, QT 10f.

some particular on-going present, it is sudden and sui generis. In the case of a work that would be called a work of art, which would be brought forth simply to present itself powerfully in and of itself without reference to any utilitarian use purposed beyond it, this suddenness manifests itself directly in what may be a quite astonishing bringing to light.[89] In the case of another sort of work, i.e., of any equipment (*Zeug*) that properly plays its part as a tool, that work's own leaping to light and its immediate presenting of itself may go unmarked, as a looking beyond toward the purpose for which it is fitted eclipses what has been wrought and, offering it in an anticipatory context, robs it of power to astonish in and of itself.[90] Yet always the uniqueness of coming as something singular into a particular nexus of time and circumstance belongs to everything brought forth through genuine *technē*. In accomplishing the bringing-forth entrusted to it, the latter can never be a copying. It is always a fresh confronting of Being, always a fresh receiving of that which is being vouchsafed.

Citing Protagoras' assertion that "man is the measure of all things, of those that are that they are, and of those that are not that they are not," Heidegger finds in it testimony to the fact that, as genuinely Greek, man, affirming as the province of his surveying and assessing whatever is vouchsafed to him, obediently assents to being "limited to that which, at any particular time, is unconcealed." Greek man, he says, accepts the restrictive receptiveness "that," via the horizon of an ever particularized unconcealment, "always confines a self"—in its very transpiring as itself—"to this or that."[91] The doer of *technē* who shoulders the burden of letting unconcealing happen stands preeminently in this submissive,

---

[89]"The Origin of the Work of Art," UK 73, PLT 65. Cf. Ch. XVI, p. 630.
[90]UK 73f; PLT 65. On the very inclusive meaning that Heidegger early gives to the word *Zeug* cf. Ch. II, pp. 64-68.
[91]"The Age of the World Picture," HW 96f, QT 145. Behind Heidegger's use here of "horizon (*Horizont*)" for a limit provisive of disclosure, there lies, clearly, the meaning of the Greek verb *horizein*, which speaks of a bounding and hence of an ordaining or defining. See H. J. Liddell and R. Scott, *A Greek Lexicon*, Oxford, The Clarendon Press, 1968, pp. 1250-1251. On the importance of this complex of meaning for Heidegger, cf. this chapter, pp. 348 & 352. Cf. also, Ch. V, pp. 187f, and n. 69.

responsive relation to Being. He is skilled. He can foresee with exceptional clarity, from his vantage point of assured knowledge. The tradition of his people, at once borne up by its language and carrying that language forward as a guarantor of continuity of naming and knowing, lies ready to his thinking and perceiving. Whether he be temple builder or poet, the founder of a city or the shaper of a sacred vessel, he knows what is needed in the juncture of time and circumstance in which he stands. And he knows how to fashion it, how to bring it to appear as something now glimpsed only in intent and foresight but soon to stand ready as a fitting answer to the demand that he perspicaciously discerns and that he sets himself to carry out.[92] He is dexterous of mind and eye and hand, confidently able to bring forth the work required. And yet, for all this wealth of capability, he is not the final master of the event of bringing into appearance that he undertakes to achieve. His stance in knowing-seeing can never be wholly secure.[93] And more than this, that which is brought forth, through his skillful working, inescapably has about it an unpredictability. Only when it has been finally wrought out do its precise lineaments appear. It may answer closely to his skilled expectation. It may evade or it may surpass it. Always it discloses itself in an irreducible integrity that must simply be accepted, even by the one who bends all his powers to bring it forth. That which *techne̅* brings forth and brings to light is precisely that which "can look and turn out now one way and now another."[94] For whatever is wrought is governed first of all—as, indeed, is the working itself—by the happening of Being, the powerfully upwelling ruling that, as self-unconcealing self-concealing, happens into an endlessly varied wealth of nuance and into ever encroaching shifts of concealing that which it brings forth hither into presencing.[95]

Whoever accomplishes *techne̅*, then, hazards his skill, and, it may be, even himself, for the sake of the Being of what-is. When his intent is

---

[92]See "The Question Concerning Technology," VA 17, QT 8.
[93]Cf. Ch. III, p. 115.
[94]"The Question Concerning Technology," VA 21, QT 13.
[95]Cf. Ch. I, pp. 42f.

granted fulfillment, a pivotal disclosure is always set underway, for, moving powerfully out beyond the widely encroaching sphere dominated by appearance and misconstruing, he opens an arena of disposedness and meaningfulness within the on-surging manifold of what-is. In particular, he who brings forth a genuine work, in the full sense of the word, provides a locus of intense clarity, as he brings the accomplishing of *technē* to bear with peculiar singlemindedness and with devotion solely to the particular entity that his working is bringing forth into immediate presence.

As his intent yet receptive knowing cuts incisively into on-coming reality and sets forth in its midst the demanded foundational entity whose need he has foreseen, he brings forth something of intense specificity. In the definiteness of its uniqueness, in its appropriateness, in the immediacy of concreteness and the particularity of lineament that allow it to focus, illumine, and structure the on-going life of a people,[96] the determinative circumscribing that lets appear is crucially in play. Through it there is accomplished that fixing in place, that setting fast (*Fest-stellen*), not in the sense of an immuring into static immobility, but in that of the bounding, the setting forth in precise delineation, that is the necessary prerequisite for enduring in self-identity and hence for genuine disclosure. Even as the delimiting contours of a mountain permit it to stand forth in imposing presence, such bounding, intrinsic to the bringing forth of a "work," grasps, configures in specificity, and brings to rest in the immediacy of a particular on-happening present the work brought forth.[97]

Such rest (*Ruhe*) is no inert calm. It is, rather, the enforced tarrying in self-presenting of something that is intensely present in the radiance of self-disclosure that is the very happening of Being. Just in virtue of being released into this rest, the work throbs openly with the originative, unstemmable power of Being. The work's *rest* is brought to pass "in the pressing abundance of movedness (*Bewegtheit*)." It is this, Heidegger says, that the Greek word *ergon* (work) bespeaks; while the

---

[96]Cf. Ch. III, pp. 135 & 136f.
[97]"The Origin of the Work of Art," UK 96, PLT 83.

related noun *energeia* names the work's "'Being,'" the enabling Happening that rules mightily and discloses itself in such a work in self-presenting rest.   In that work, there surges "the initiatory movement (*Bewegung*) that rules in the lighting-clearing *and* concealing, more precisely in their union, namely the movement belonging to the lighting-clearing of self-concealing as such, out of which, again, all self-lighting originates."98

The movement here in question is nothing other than the sovereign on-thrust of Being toward its happening self-maintainingly as unconcealing.99 This it is that demands in some particular nexus of time and circumstance that a work be brought forth as a specific instancing of unconcealing.   It demands the bounding retaining setting-fast that takes place as bringing-forth-hither when the staunch receptiveness that is *working* as *technē,* the skilled knowing that shapes into specificity and sets forth to endure in unconcealment, is brought determinatively into play.100

Being, happening first of all as self-concealing, ever resists disclosure. So, likewise, does whatever is. That which—like the rock that is wrought into a Greek temple—is gathered into a work as its embodying element, is, as earth, already self-contained.101 In the tenacity of a stubborn inscrutability, it holds out against being drawn, through a "using," in the original sense of a freeing to self-manifestation,102 into the new interrelatedness that will let it genuinely appear. And Heidegger can portray the needed structuring, the intended "world" foreseen, as striving against incorporation into the recalcitrant materiality that can alone body it forth.103 Only through the waging-out of this contending,

---

98UK 96f; PLT 83f. On this clearing (*Lichtung*), cf. Ch. I, pp. 32f. On the provisive movement as which initiatory Happening as a way-opening be-waying (*Be-wegung*) carries itself into play in the unfolding of the Twofold, cf. Ch. IV, pp. 146-148, and Ch. III, pp. 123f, and n. 63.

99Cf. Ch. I, pp. 24-26.

100See "The Origin of The Work of Art," UK 95f, PLT 82f. For the discussion that underlies our inclusion of *technē* in the elucidation made here of a passage in which the word itself does not appear, see this chapter, pp. 347f.

101UK 47f; PLT 46f. On this contending, cf Ch. V, p. 179, n. 39.

102UK 72; PLT 64. Cf. Ch. VI, p. 229.

103UK 51f; PLT 49. In the passage here in question "world" has a meaning different

intense yet fruitful, can unconcealing be brought to a stand enduringly, so as to hold sway in formative power. He who brings forth a work must endure this struggle. In the sapient skill that is *technē* he pits himself against Being happening as self-concealing, in order that the happening of Being as unconcealing may prevail.

The unconcealing that can only thus be arduously wrested forth is, for Heidegger, what the Greeks in the original import of their language called truth, *a-lētheia*.[104] He who in a genuinely Greek manner confronts the onslaught of the happening Twofold and, bringing forth a work, opens in its dark arraying a place of lightedness and of meaningful relating where genuine discernment can take place and genuine enduring come to pass, through his able and compliant knowing lets truth happen in the work.[105] Via his working, truth, freeing-safeguarding unconcealing, sets itself into the latter.[106] Thus, he provides for the Being of what-is the only proper arena for its self-disclosure, namely an entity that belongs wholly to the complex on-going multitude of its fellows and that is exposed, as are they, to transiency and to the pervasive dominion of concealing, but that nevertheless stands forth for a while, presencing, imposing itself as altogether fitting and needful, lifting all the elements drawn into the constituting of itself into a clarity of manifestation conferred by its own uniqueness, and gathering everything touched by its self-disclosure to presence unifiedly in its light.[107]

---

from, but related to, that found for it in Heidegger's later thought. On these diverse meanings see Ch. V, p. 179, and n. 39.

[104]EM 77ff, IM 85ff; "*Alētheia*," VA 259, EGT 104. Late in his work ("The End of Philosophy and the Task of Thinking," SD 77f, TB 70f) Heidegger conceded, in response to criticism of his claim that for the Greeks *alētheia* originally meant unconcealment, that this meaning, while evident within the word *alētheia* itself, does not appear in the uses of it that are to be found from Homer onward. He admits that the meaning that appears there is correctness (*Richtigkeit*).

[105]Cf. "The Origin of the Work of Art," UK 65f, PLT 59f.

[106]UK 89; PLT 77. Cf. Ch. III, p. 135. On this character of truth, see Ch. I, p. 38, and n. 49, and p. 54, n. 109.

[107]Cf. UK 41ff, PLT 42ff. And cf. Ch. III, p. 135.

3

In the portrayal of Greek *technē* that we have just been pursuing we clearly find ourselves in the same realm of discourse as that wherein Heidegger sets forth *dwelling*, understood as nurturing building (*Bauen*), as the foundational comportment by way of which there is brought to pass the thinging of the thing as the unique centered happening within the manifold of what-is that, precisely in its immediate concreteness, in its presencing gathers to itself and illumines, through that very gathering, the accompanying particulars of that manifold that cluster to it.[108] Here once again it is man who, in heedful self-surrender to the happening of Being as unconcealing, accomplishes disclosure via some unique thing—now a thing specifically allowed to presence through his bringing to bear of a knowing, skillful doing, a keen-sighted bringing-forth—that, offering itself genuinely as itself, endures by way of some particular on-going present and therewith permits world, as the inclusive on-going structuring of his corporate life, to bring to pass its ordering sway.

Heidegger's discussions of dwelling and of the thing occur in writings subsequent by more than a decade to the work in which through a delineating of Greek *technē* he portrays what is for him the original, most genuine interrelating of the Being of what-is and man who permits Being's happening as unconcealing. In those discussions, we come upon a very significant movement on Heidegger's part to fresh insights that made more intricate and far ranging his understanding of that interrelating as carried out via an immediacy of human accordance with the happening of Being. Nevertheless, when Heidegger presents his thinking determinatively in terms of the dwelling of man, the thinging of things, and the happening of the Fourfold in the worlding of world,[109] he clearly remains, in a decisive sense, very close to his fundamental concern with that which is Greek, in the full sense of that word.[110] All that he says concerning dwelling, the Fourfold, and the thing elucidates, just as does

---

[108]Cf. "Building, Dwelling, Thinking," VA 153ff, PLT 154f. See Ch. V, pp. 187-195.
[109]Cf. Ch. V, pp. 172-187 & 187-195.
[110]Cf. this chapter, pp. 331f, and n. 6.

his discussion of Greek *technē*, a particular manner of relating subsisting between Being and man, that namely which allows man's providing of openness for the happening of the Being of what-is to come decisively to pass within and by way of the context of human corporate life. Although Heidegger does not himself make the statement, it would surely not be untrue to his thinking to say that among the Greeks, at the beginning of the self-sending-forth of Being that originated *history* in Heidegger's fullest sense of the word—Western, now become world, history, that is, as the bringing to particularized happening in changeful age after age of the unfolding happening of the Being of what-is[111]—genuine dwelling, accomplished as nurturing building (*Bauen*), was, through the human comportment named as *technē*, brought to pass; and, through the thinging of things that was therewith achieved, the unitary Fourfold of earth and heavens, divine ones and mortals was permitted so to happen determinatively as to bring to pass the worlding of world. In examining Greek *technē*, we may say, we are in fact examining *Greek* dwelling.

Striking parallels, in diversity, between Heidegger's discussions of *technē* and of man's dwelling in company with thinging things testify to this. In the discussion of *technē*, stress falls on the overpowering on-rushing surge of that which encounters man, offering itself to, indeed demanding, his disclosive intervention, and yet, thanks to the sovereignty of the power that rules within it and to the bent toward concealing in play therein, always eluding his total disposal and foredooming his disclosive work to forfeit its revelatory power. In the discussion of the thing, in contrast, stress falls on the illumining, ordering interrelating that governs in the thinging of the thing and on the fulfillment possible through man and to him as he carefully and intently participates in the crucial gathering that is underway and, through the things that he lets thing, allows genuine enduring to come to pass.[112] But here too, although the hues of the portrayal are brighter, the themes of precariousness and concealing again hold a decisive place. Men dwell *as mortals*. In order to allow the

---

[111]Cf. Ch. I, pp. 47-49.
[112]Cf. Ch. V, pp. 184-187 & 187f.

thinging of a thing, they must attain to a self-surrender to the rigor of their mortality that requires a submission to the ultimate happening of concealing, its happening as death; they must bear unflinchingly the inadequacies and distortions that continually beset their life; and they must willingly acknowledge that any *thing* that they nurture or establish in its thinging can endure only for its appointed while and then must lose its power to gather and disclose.[113]

Technē and dwelling clearly parallel one another. Dwelling is not directly identified, as is *technē*, as an able *knowing* that can as such bring forth that toward which it directs itself. Yet the need for a skilled comprehension that can discern, foresee, and assess is clearly implicit in Heidegger's characterization of dwelling as a saving-receiving-expecting-guiding with respect to the Fourfold.[114] In a relatively late passage, Heidegger speaks of nurturing building and thinking as constitutive modes of human comportment that together carry out dwelling. And in so doing he clearly suggests a skillful knowing such as *technē* immediately connotes, for he says that both nurturing-building and thinking can take place genuinely and can belong genuinely to dwelling only by pursuing their way in concert, and that means by knowing that they together come "out of the work-shop (*Werkstatt*) of a long experience and an unremitting practice."[115] Certainly dwelling, because it is defined in the fresh context in which the Fourfold is delineated and its arriving into disclosure by way of the thinging of things is portrayed, possesses aspects of connotation not found in *technē*. But it is striking that the cultivating of land and tending and taming of plants and animals, the bringing forth by some skilled fashioning of all manner of works from poems to cities, and the accomplishing of disclosive, founding thinking are all to be found within the human comportment of which *technē*, in its original inclusive sense, speaks. And such pursuits can readily be seen to be fundamentally the same as the primary modes of behavior that are later said to carry out dwelling, namely, nurturing building (*Bauen*) and thinking. Heidegger

---

[113]Cf. Ch. V, pp. 191-193.
[114]Cf. Ch. V, pp. 191f.
[115]"Building, Dwelling, Thinking," VA 162, PLT 161.

does not make clear just how widely the name "thing" should be applied
to concrete particulars that so hold sway as to gather the four unitarily
interrelated primary constituents of what-is together via their whiling and
therewith to permit, as determinative focal happenings within the manifold
of what-is, the disclosive worlding of world. Neither does he in his early
discussion of *technē* speak specifically of *things*.[116] But it seems clear
that the dwelling of mortals that can alone allow the thinging of things to
come to pass includes inherently within itself the prescient doing that is,
from within the context that is specifically Greek, identified as *technē* in
accordance with that context.

It is impossible not to conclude that Heidegger's eventual portrayal
of man's dwelling in disclosive relationship with things arose primarily
out of his understanding of the originative Greek comportment called
technē, which he saw as allowing meaningful disclosure to the unfolding
Twofold. The two aspects of his thinking, disparate though they be in
time of presentation and in detail of content, are intimately connected and
shed light on one another. Ultimately they tell us of the same happening,
although the one, dwelling, does so with an eye to what we may call
fundamental structurings belonging to the interrelating of man and the
Being of what-is, while the other, *technē*, concerns itself with a crucial
instance of that interrelating. In the on-goingness of the Being of what-is
as a Happening that brings itself to pass by way of time, Greek dwelling,
here glimpsed via Greek *technē*, passed away. Yet it is equally true to
say that in the self-sameness of the Happening ever in play as the Being of
what-is,[117] every coming to pass of genuine dwelling bespeaks in its turn
that which, at the beginning, bore the portentous name of Greek.

---

[116]But see "The Question Concerning Technology," VA 17, QT 8, where, in a
discussion in which *technē* is in question, Heidegger remarks, in explaining the
meaning of *telos*, that in the bringing forth of a chalice consecration and bestowal
"give bounds to the thing," and that "with the bounds the thing does not stop; rather
from out of them it begins to be what, after production it will be." It must be
remembered that "The Question Concerning Technology" was originally delivered as
the second lecture in a series that opened with "The Thing."
[117]Cf. Ch. I, pp. 29f. On the happening of Being, vis a vis on-going history, as the
"Same," see "The Word of Nietzsche, 'God is Dead,'" HW 196, QT 57, and note 6
there.

Seen fundamentally, Greek *technē* shows us a people *dwelling* at the beginning, at that *beginning* in which man fulfilled his appointed existing with peculiar immediacy, as he took his way poised in close, obedient juxtaposition to the ruling happening of Being and in direct meeting with what-is. Viewed thus, *technē* shows us too the precariousness that Heidegger sees as belonging to such dwelling—that precariousness that comes to the fore whenever, in awareness, man dares to meet and trust himself to and wrest to light the sovereign upwelling happening of the Being of what-is.

4

Greek man was imperiously summoned to accomplish Being's self-unconcealing. In the original immediacy of his relating to Being, he sustained resolutely the demanding confrontation with the happening Twofold. Time and time again, chosen individuals dared to fulfill that confrontation in clear-sighted venturings of power, brought Being to shine forth mightily in their varied works, and hence allowed the Greeks of that original age to live out with peculiar integrity and directness the openness-for-Being that was the genuine happening of their being human. The Greek language nourished and manifested that prevailing manner of relating to the Being of what-is. The poets disclosed it.[118] Thinkers— Anaximander, Heraclitus, Parmenides—spoke it forth in trenchant utterance.[119]

Yet always Greek man, intrepid though he was, found himself in

---

[118]EM 77, 112ff; IM 86, 123ff.

[119]Cf. "On the Being and Conception of *Physis* in Aristotle's *Physics*, B, 1," W 370, 349; Ph 268, 252. An important confirmation of Heidegger's view that one who undertook to accomplish *technē* might not succeed in achieving the full disclosure for which he sought is surely to be found in the fact that, crucially important for him as are the words of such early Greek thinkers, he can say that no Greek *thought* the happening of Being as unconcealment as presencing, ("The Word of Nietzsche, 'God Is Dead,'" HW 243, QT 108f) and he can allude, even with astonishment, to the fact that Heraclitus, in bringing to light Being's happening as *logos*, did not grasp the actual relation obtaining between that happening and the transpiring of human speaking as *logos*. Cf. "*Logos*," VA 228, EGT 77.

implacable encounter with the happening of Being itself; and Being happened as self-unconcealing self-*concealing*. The opening to light of Being that Greek man lived out and wrought into his founding works was not and could not be sustained by him. Even at its highest pitch his intrepidity was flawed. Again and again those who went forth to grapple violently with the impending Twofold so as to bring Being to self-disclosure recoiled, daunted in the face of the compelling need thus to press indomitably on, even while they were constrained resolutely to obey it.[120]

This fundamental drawing-back that befell the most genuine happening of Greek openness-for-Being played into the primal, ineluctable happening of Being as concealing. Through it, as Being maintained itself in self-disclosure via the openness that man was providing, a counter-possibility flashed forth in determinative power. This was the possibility that the openness-for-Being that man was living out via time might be accomplished not in affirmation but in denial.[121]

This is none other than that most extreme possibility belonging to man's existing as such, namely that it, *Being*-there-as-openness, happen rather as not-*Being*-there-as-openness[122] and therewith steadily direct itself toward self-obliteration. Hitherto we have caught sight of this, which is intrinsic to man's existing, primarily from an individual perspective, albeit always in a corporate context, as Being-toward-death.[123] Now we meet it overtly in the arena of history, where Being, in its self-sending-forth, is happening for itself a people that, standing in stalwart yet submissive juxtaposition to it, can let it happen as genuine unconcealing.

What if the shrinking-back that afflicts even the highest, most genuine happening of Greek openness-for-Being were to prevail, taking precedence over the unflinching grappling with the overpowering Twofold that is needed for unconcealing? Even this awesome possibility,

[120]EM 135; IM 148.
[121]EM 135; IM 148.
[122]EM 135; IM 148.
[123]Cf. Ch. II, pp. 98-101.

Heidegger says, came to utterance among the Greeks. Thus he translates Sophocles' words from *Oedipus at Colonus*: "Never to have entered into *Being*-there-as-openness prevails over the togetherness of what-is as a whole."[124]

This would be man's greatest victory vis a vis Being, pyrrhic indeed, but decisive. And precisely this victory begins in the faltering of Greek *Being*-there before the onslaught of Being. In that faltering, the possibility as which Being, happening as concealing, flashes determiningly into view strikes root in the Greeks' appointed manner of relating to Being. Now the mastery of the on-surging Twofold that is demanded of man in order that genuine appearing may take place begins to happen as an inverted negating semblance of itself. "Mastering" can no longer be an intent, keen-sighted confronting, in dedicated openness, of the irruptive, multitudinous happening of *physis* as such. It cannot be man's self-abnegating gathering of himself tenaciously toward the accomplishing of unconcealing. For now it is Being happening as concealing that is coming preponderantly into play. This transformed mastering of the Being of whatever is must instead rest upon a self-protective self-concern that manifests itself not in self-surrender but in withdrawn self-reliance. Ruled by the possibility now coming preponderantly into play, "the mastering of the overpowering will be securely and thoroughly battled out to a successful issue when concealment is absolutely preserved to Being, the uprising ruling that in itself enduringly holds sway as *logos*, the gatheredness of the opposed." Concealment "will be preserved," will be allowed enduring dominion, only if unconcealing is decisively thwarted and "in a sure manner every possibility of appearing gainsaid."[125]

---

[124]*Niemals ins Dasein getreten zu sein, obsiegt über die Gesammeltheit des Seienden im Ganzen.*—Heidegger's translation of Greek *me phynai ton hapanta ni/ kai logon.* EM 135; IM 148. Grene and Lattimore translate, "Not to be born surpasses thought and speech"; Watling translates, "Say what you will, the greatest boon is not to be"; (*Oedipus at Colonus*, 1224f).

[125]*In solchem Zurückschrecken und doch Bewältigenwollen muss für Augenblicke die Möglichkeit aufblitzen, dass die Bewältigung des Überwältigenden dann am sichersten und völlig erstritten wird, wenn dem Sein, dem aufgehenden Walten, das in sich als*

For this, the openness-for-Being that Greek man lives out by way of time must shut itself away from full encountering. Under the pull of that primal self-concealing that happens as self-maintaining keeping-to-self,[126] man stands out against that which meets him. His victory as the one who conquers Being by refusing to it its happening as appearing is underway. Self-confidently he does violence to Being, heedless of the cost, the vitiating of his own proper manner of existing. Retrenching thus in its self-launching-forth, Greek openness-for-Being—the responsive existing that is staunchly, immediately there to permit and sustain the latter's happening—evades the ultimate precariousness intrinsic to it. Firm in the power assured within itself, it goes forth to quell the power of the Being of what-is to happen overpoweringly, unpredictably, ungovernably, i.e., to happen self-maintainingly as presencing that lets appear.

From within itself Greek openness-for-Being alters. Yet man's existing, as the accomplishing of such openness, can neither bring about nor transform *itself*. It ever remains happened away, handed over by Being to happen of itself, and yet kept in hand by the Being of what-is.[127] It remains a carrying into play of the one Happening that, as the Being of whatever is, enables every taking place. At this crucial juncture, when the uncannily daring ones who constitute the vanguard of the happening of Greek *Being*-there-as-openness rise up audaciously in the supreme recognition that precisely in refusal of all openness to the happening of the power that brings to appear lies the path to standing-ground in power over against it and to mastery, Being is itself most compellingly holding sway.[128]

Logos, als die Gesammeltheit des Widerwendigen west, schlechthin die Verborgenheit gewahrt und so in gewisser Weise jede Möglichkeit des Erscheinens versagt wird. EM 135; IM 148.

[126]Cf. Ch. I, pp. 24 & 51f. Here the impulse toward self-assertion in isolative persisting that comes upon whatever is from out of Being in the urgent forcefulness of its ruling should certainly be brought to mind. Cf. Ch. I, p. 51, and n. 100.

[127]Cf. Ch. I, p. 53, and Ch. II, pp. 108f, and n. 190.

[128]We should here be reminded both of the impulse toward self-assertion that Heidegger later described as coming upon all particulars of the manifold of what-is from out of the happening of Being as presencing (cf. Ch. I, pp. 50-52), and also of

Man, presuming upon the strength of the manner of existing that he is urgently living out, assumes a ruling stance, in order that "the suitable site and abode (*die Stätte*) for appearing" proper to the imperious ruling that lets appear should "remain closed to the latter's all-ruling power."[129] But this not-*Being*-there, man's "highest victory," remains a victory "over Being." Man's existing, his *Being*-there as openness-for-Being, is nothing less and nothing other than a "constant need," a danger, a distress (*Not*), wrought out unremittingly at once via overwhelming defeat and via the unfailing reirruption of powerful, even violent deeds against Being for the sake of Being, and precisely at its behest.[130] Human existing pursues its course from within itself. Yet even when its adversarial comportment is most radical, it but corresponds to the "need (*Chrē, Brauch*)" for self-maintaining as which Being holds sway.[131] And always, in the crucial event of such corresponding, it must yield utterly to Being's dominion. Precisely in its happening in ever-resurgent power, its self-directing must give place to the ruling of Being, a ruling so imperious as to make of all of its self-determining nothing other than a channel for Being's own accomplishing of itself.

The all-ruling power of Being so rules, Heidegger says, as ultimately to destroy the integrity of man's *Being*-there-as-openness as a self-determining taking place. It violates it, breaks in upon (*vergewaltigt*) it, in such a way as to constitute it wholly as "the site and abode of its own appearing." Transformingly plundering man's existing

---

the self-depleting transpiring of human existing, in its transpiring in inauthenticity, that he portrayed in *Being and Time* as constituting an aspect of the taking place of that existing that evinces just such self-assertion. Cf. Ch. II, pp. 85f & 89-91. And we must recall Being's happening of what-is into disruptive deviance from itself, which can be seen to suggest the inception of all such self-assertive deviant happening. Cf. Ch. I, pp. 45f. For just these elements of Heidegger's thinking are seen in a crucially important historical instancing, in this portrayal of a falling away from the authentic carrying out of Greek *Dasein*. For another such instancing cf. Ch. VIII, pp. 301-305, and n. 42.

[129] . . . *seiner Allgewalt die Stätte des Erscheinens verschlossen bleibt.* EM 135; IM 148.

[130]EM 134f; IM 148.

[131]On this need, see Ch. I, p. 53, and n. 107.

into happening as that abode, it surrounds and permeates it with its ruling and retains it as so governed, holding it in Being.[132] The recognition of the possibility for a human mastery of Being that flashes upon Greek man in his *Being*-there-as-openness-for-Being at bottom does nothing other than manifest this ruling.

The altering of Greek man's manner of existing vis a vis Being springs out of Being itself. It happens as a "stepping-apart (*Auseinandertreten*) of *logos* and *physis*."[133] Man's discerning, provisive knowing that as *technē* intently, selflessly gathers and disposes after the manner of *logos*, and the powerfully upwelling Twofold that lets itself be unfolded into appearing through his doing begin to separate themselves from one another, drawing asunder the bond of contending-unitedness in direct meeting that had obtained between them.[134] And here it is, in this crucial sundering, that the import of Greek *technē* undergoes the change that sets it on its way via time to becoming the rapacious machine technology that now dominates our modern age.

Already the disparity between these two lies clearly before us. In contrast to *technē*,[135] in our technology no skillful knowing that foresees and wrests into play, now here, now there, the happening of Being as clear disclosure governs as a mode of "having-of-disposal." In it, the planning that is ubiquitously regnant does not freely transpire in accordant responsiveness toward that with which it has to do. The ordering that takes place is only the distortively wasting ordering of the standing-reserve, and every "mastering" lays hold only on the exploitively commandeered and on that which is provided and instituted apart from any concern for it as something that should have as its *raison d'être* a fresh provision of insight and meaning for a community on whose behalf it has been uniquely brought forth.[136] In our age, the fruitful contending

---

[132]EM 136; IM 149. . . . *die Allgewalt des Seins das Dasein zur Stätte seines Erscheinens ver-gewältigt . . . und als diese Stätte umwaltet und durchwaltet und damit im Sein einbehält.* Cf. EM 124f, IM 137.

[133]EM 136; IM 149.

[134]See this chapter, pp. 338f.

[135]See this chapter, p. 333.

[136]See Ch. VI, pp. 228-231, 241-249, & 259-264.

via which *techne* permitted clear determinative disclosure to the Being of what-is has shrunk away, under Being's happening in withdrawal, to a radically contentious challenging-forth of what-is[137] in which no disclosure of Being is vouchsafed. Whereas once unconcealing could be decisive, now only despoiling, deceptive hiddenness holds sway. Thus distant is modern technology from Greek *techne*, as Heidegger understands the two. Yet the former's origin lies in the latter. Out of the ruling Happening named as the Being of what-is, that disparity took its inception at the beginning. Out of that Happening, via specific human answerings to Being, it began to unfold.

---

[137] See Ch. VI, p. 249.

CHAPTER X

WITHDRAWAL

When we think with some attentiveness of the history of classical Greece, we are aware of swiftly changing phases—the time during the fifth century B. C., when, notably in Athens, extraordinary vitality and integrity were manifested in area after area of human life, in the political arena and in all the arts; the period of the Peloponnesian war late in the century, when traditional beliefs and wonted moral fiber were eroded, relativistic thinking and opportunistic action grew apace, and Socrates stood forth in a persistent effort to bring men to discover the bases on which their lives were lived; the fourth century, when, against a background of political disintegration, efforts to gain rational comprehension of the nature and structures of reality, groping and fragmentary hitherto, rose to remarkable heights in the successive work of Plato and Aristotle and laid the foundations for philosophical and theological thinking for centuries to come. We acknowledge the age of the Greeks to be preeminent in the wealth of its greatness, but we can also point to other epochs—the sixteenth and seventeenth centuries in Europe, for example—in which human intellectual and cultural accomplishments were of a remarkably high order and decisive changes in understanding and outlook took place with impressive rapidity.

For Heidegger, our ready assessment of Greek history, while catching sight of a few striking features, remains very far from a true understanding. For him every historical period, indeed every historical event, to be seen truly must be seen as unique. But more than this, as the time wherein the fundamental manner of relating[1] of Being and man was

---

[1]Here and elsewhere in this study the reader should be careful rigorously to separate "relate" from its usual meaning. Here "relate," in contexts where the Being of what-is is in any way under consideration, does not speak at all of a connecting of inherently discrete elements, one of which is to be subordinately referred to the other. The relating that is now in question is a *holding-toward-one-another* of participants that,

lived out in peculiar openness of encounter, "the Great Age of the Greeks"[2] that culminated in the early fifth century is unmatched in revelatory significance. Beyond it lay not a mere decline like any other but the very withdrawal of Being from that original relating, a withdrawal that left man adrift and striving to secure a footing amid a reality suddenly transformed. Socrates, Heidegger says, "did nothing other than place himself in the draft" of the "pulling away" of Being.[3] His ceaseless questioning manifests the endeavor of single-minded thinking resolutely to follow after Being in its self-withdrawal. Socrates' great successors, on the other hand, Plato and Aristotle, fled that demanding and dangerous place. They themselves withdrew from immediate relationship with Being and, through their work, grounded the withdrawal of the others of their age and of all who were to follow them in the West in the ages to come.

1

For the Greeks in their original compliant relating to Being, Being was the self-imposing enduring, arising powerfully of itself, that confronted them in whatever they encountered.[4] It was the self-validating self-presenting, the *appearing*, of that which showed itself to them. This appearing might be genuine disclosing, or it might be disclosing as mere semblance.[5] But always it was *self*-disclosing, sudden, peremptory, at bottom uncontrollable, leaping to light, fading or falling dark, in ways that ultimately evaded human disposal. Always it took its way by way of time, emerging, altering, vanishing, never untrammeled and unobscured by change and dissolution.

Here the Being of what-is is the latter's *appearing* in a dual sense: It is the intrinsic *élan* that throbs in that which appears, the impulse to self-disclosure, happening as gathering, that makes itself felt in the self-

---

while they are distinct from one another, are ultimately one. For we are using "relate" to speak of the happening of the single Twofold. See further, Ch. IV, p. 145, n. 10.
[2]"The Age of the World Picture," HW 84, QT 131.
[3]WHD 52; WCT 17.
[4]Cf. Ch. IX, p. 334.
[5]On this dual character of "appearing," cf. Ch. I, p. 44.

presenting of what-is; and it is equally the specific view of itself that whatever is offers for human apprehending, the appearance that it presents that both discloses it and marks it as itself.[6]

When, in the work of Plato, Greek thinking first decisively recoiled from an immediate relating to Being,[7] it was the apprehending of Being's happening as appearing that offered the fulcrum for the shifting away toward human self-securing vis a vis Being that was taking place. When Plato began to follow for himself the course taken by Socrates and to ask "What is the real?" an unprecedented answer came to him. He found the real in *appearing*, but in such a way as to take cognizance only of the second manner of happening of appearing as such. The Greek words *idea* and *eidos*, which Plato raised to crucial philosophical use, were, Heidegger avers, complementary. The one denoted the view (*Anblick*) of itself that something visible, in offering itself, directed toward seeing; the other the demeanor, the outward look (*Aussehen*) that was offered *in particularity* via time by that which thus came to encounter.[8] Plato audaciously used these words, which in ordinary speech referred to what was physically seen, to speak of nothing less than the Being of whatever was accessible to human apprehending.[9] He used *eidos* to speak of appearing, of the offering of itself via a particular outward look that pertained to particular entities at a particular time.[10] For him, *idea*, the specific outward look thus offered, took precedence over *eidos*. It was, Heidegger says, as though Plato had been overwhelmed (*überwältigt*) by the very manner of happening of appearing, the offering of a particular *eidos* as such, and had grasped it not as the ever-spontaneous manner of happening that it was but as something separable and identifiable, as "something that was presencing for itself," i.e., as *idea*, the sight, the offered look that was seen and

---

[6]EM 139; IM 153.
[7]Cf. Ch. IX, pp. 359-364.
[8]EM 138; IM 151.
[9]"The Question Concerning Technology," VA 27f, QT 20; EM 138, IM 152.
[10]N I (German) 246f; (English) 214f.

apprehended.[11]

Plato, continuing the Socratic quest for the selfsame reality that could be glimpsed in instances of itself but that was not to be identified with any of them, sought the "constant *(das Fortwährende)*" [12] that must constitute the true Being of the mutable and the imperfect.   Here permanence in the sense of such constancy must not be understood in terms of mere lasting. To be thought in a Greek manner, Heidegger says, it must be thought, rather, in terms of presencing. The permanent, the genuinely enduring, and that means that which "genuinely and uniquely is *(das eigentlich und einzig Seiende)*," offers itself, presences, solely from out of itself.[13]   On the other hand, that which is possessed of impermanence *(Bestandlosigkeit)*, that which lacks such enduring, i.e., that which *is* in a deficient way *(das Unseiende)*, is not adequate for its own self-maintaining. It takes its course, presenting or absenting itself, but always only on the basis of that which is of itself, "that which already lies before" and which, accordingly, can serve to undergird and sustain the innate insufficiency that pertains to it itself.[14] What, then, underlay the transient exemplars of a particular sort of entity?   How was it, specifically, that the craftsman could provide article after article of a particular genre, no two exactly identical, yet all nameable as this or that?[15] The likeness of such entities, whether tables or bedsteads or houses, lay in their look, their appearance *(idea)*. And to Plato that proffered look, discernibly persisting even in variance, showed itself to be the common, enduring determinant that constituted each entity as what it was.

This insight Plato extended to apply to every sort of encounterable reality, physical and non-physical.[16]   Therewith he succeeded at one

---

[11]"On The Being and Conception of *Physis* in Aristotle's *Physics*, B, 1," W 345, Ph 249.

[12]"The Question Concerning Technology," VA 38, QT 30.

[13]"On The Being and Conception of *Physis* in Aristotle's *Physics*, B, 1," W 338, Ph 244. On the meaning of *eigentlich* cf. Ch. I, p. 37, n. 46.

[14]W 339; Ph 245.

[15]Cf. N I, (German) 211f, (English) 182f.

[16]"The Question Concerning Technology," VA 27f, QT 20.

stroke in letting fall from view the inward depth of the happening that brought entities to appear as they did. As before among the Greeks, genuine knowing (*Wissen*) was a seeing (*eidonai*).[17] But now that seeing did not, as originally it had in the accomplishing of *technē*, confront that which met it in such a way as to grapple with it in the totality of its on-rushing, shifting, illusive complexity.[18]   Philosophical *seeing* now focused only upon the look that particular entities offered. And even here it was not the entities themselves that were of ultimate significance. Rather, it was the outward look that each instanced and that, imperfectly manifested in each, must itself lie beyond them. When accomplishing itself most purely, such knowing-seeing would lift itself clear of the realm of entities and occurrences and relationships that continually happened in deceptive changefulness and would direct itself toward the Ideas themselves, the formative aspects, that when embodied were seen only in semblance. This seeing could take cognizance only of the "what" of entities encountered and of the "how" of their self-presenting, whereas the "that," the stubborn particularity of their encountering each simply as itself, could be felt only as an impediment to knowing understood as such a seeing.[19]

This discernible, cognizable how and what of things was expressed by Plato in the word *ousia*, which was later to be translated into Latin by *essentia* and *substantia*.[20]   *Ousia*, Heidegger says, denotes both the self-imposing presencing of anything that presences and that presencing entity identified specifically in the "what" of the "outward look" that it offers.[21] *Ousia*, which meant in ordinary usage property, real estate, wealth,[22] manifests the altered relation in which human apprehending, characterized

---

[17]"Science and Reflection," VA 52, QT 163. Cf. Ch. IX, p. 333, and n. 7. Greek *eidonai* is akin to German *wissen*, with its underlying like connotation. Cf. Ch. VIII, p. 303, n. 36.

[18]Cf. Ch. IX, pp. 334f.

[19]EM 137-140; IM 151-154. Cf. "On the Being and Conception of *Physis* in Aristotle's *Physics*, B, 1," W 345, Ph 249.

[20]Cf. Ch. XI, pp. 407f.

[21]EM 138; IM 151.

[22]"On the Being and Conception of *Physis* in Aristotle's *Physics*, B, 1," W 330, Ph 238.

as seeing, now stands to the reality that meets it. The Being of what-is happens now as a presencing of what presences that is the offering of an outward appearance ruling in that which stands forth and endures as possessed of a particular outward look. This outward look, identified as a manner of appearing, permanent in itself—i.e., presenting itself with constancy directly from out of itself—and informing and making intelligible particular entities, is itself graspable by the apprehending that encounters it; and at the same time it renders graspable by that apprehending any entity that it brings to happen after its manner. Accordingly, *ousia*, the very how and what of whatever is, i.e., the very "Being" discovered for it, is, Heidegger says, laid hold of by apprehending as its possession, as that which it *has (dessen Habe)*.[23]

There is still discernible in Plato's thinking the Greek experiencing of the Being of what-is as *physis*, i.e., as the importunate arising that ruled in the unfolding Twofold and granted to anything that appeared in any manner—whether clearly or deceptively—both a disclosing and an enduring that impressed themselves upon him who encountered them as intrinsically constitutive of the encountering reality that yet remained ultimately beyond his disposal and spontaneously determined from within itself.[24] Plato himself understood *physis* as *idea*. In his so doing, ruling self-disclosure, specificity of appearance, and intrinsic stability remain from the original significance of *physis* to characterize the Ideas that he saw as the Being of whatever could be said in any sense to be.[25] When Plato came to feel the inadequacy of the Ideas, which were many, as realities possessed of this ultimacy of originative self-maintaining, he came to see beyond them the Idea of the Good—where "good," Heidegger says, has no moral connotation but means the "valiant" and refers to that which as such can and does accomplish whatever properly belongs to it, i.e., whatever is appropriate in its light. The Idea of the Good is the

---

[23]EM 138; IM 151. We should note the immediate parallel between the prephilosophical meaning of *ousia* and the present-day meaning of the German noun *Anwesen*. See Ch. I, p. 24, n. 8.
[24]Cf. Ch. IX, p. 334.
[25]EM 137f; IM 151f.

single supreme archetype (*Urbild*) that empowers the Ideas to stand forth as models (*Vorbilder*) in their turn. As the sun illumines all on which it shines, so this "Idea of Ideas" grants and maintains to the other Ideas, which it disposes, that capability of self-presenting as suitable prototypes that constitutes their Being.[26] But in all this these ways of holding sway that remain distantly suggestive of *physis* now belonged to that which was immutable, while, in contrast, *physis* itself as the restless, disclosive, disguising, changeful Being of what immediately presented itself in the on-going milieu of human life became for Plato, in contrast, the recalcitrant mode of happening of that which must be named *mē on*, "that which should not be and really is *not*."[27] Happening, as always, so as to particularize and embody itself via the happening of what-is,[28] this intractable Being of the immediate, the material, could but obscure and distort the Ideas themselves that yet exercised over and within that which was ceaselessly presenting itself their formative power.

The cleft that, with Plato, thus opened between the realm of the genuinely real, the Ideas, and the realm of familiar experience, with its imperfections in physicality, where no full reality could be found, now guarded philosophical knowing-seeing from the ultimate downfall that once inevitably awaited him who in skilled knowing dedicated himself to the accomplishing of the disclosing of Being.[29] Now Being was taken to happen only as disclosure. Concealment was relegated to that realm where nothing could be said genuinely to *be*. And philosophical seeing could, indeed must, eschew encounter with it. Moreover, the seeing, the apprehending that could discern and grasp the determinative outward look that constituted and let anything appear as what it was, therewith availed itself of a standard against which to measure other entities with respect to their Being.[30] The Idea becomes the pattern (*paradeigma*), self-disclosive

---

[26]EM 150; IM 164.
[27]EM 140; IM 154.
[28]On this manner of happening of the Being of what-is, which Heidegger sees as named with the word *physis*, cf. Ch. I, pp. 28-33.
[29]Cf. Ch. IX, pp. 342-345.
[30]EM 139f; IM 153.

but separate, for whatever in the material realm is shaped in conformity with it.[31] The embodied entity partakes of the form, the outward look (*idea*), that holds sway purely, but it does so distantly, shadowed by obscuring. The seeing-apprehending now in question confidently ranges itself with the Ideas, boldly accepting as definitive the distancing of the urgently happening realm of the seeming and the transient from the reassuringly stable realm of the delineate and the clear and from itself also as the devotee of that realm.

Thus does the self-withdrawing of Being begin to isolate Greek man in the comportment that forms the crux of his existing. Deceptively, self-concealingly, its potent ruling falls from view, while, yet undetected, it holds no less ultimate sway. Governed thus, Greek man's manner of existing is abruptly wrenched away from its wonted course and brought to happen in a way that, on its face, runs sharply counter to the new manner of Being's ruling. Even as the happening of Being as concealing hiddenly obtrudes itself, newly determining the manner of happening of all that is, man's apprehending assumes toward reality a stance in which disclosure alone is worthy to be heeded. And yet precisely at this juncture disclosure itself, wedded now to evaluative apprehending, is bereft of its genuineness. Once unconcealment, happening—by way of human working—precisely in and via some particular material entity belonging to the unfolding Twofold, had come to pass compellingly as truth (*Wahrheit*), i.e., as the freeing, safeguarding happening of Being in the midst of what-is.[32] Now, in contrast, what passes for genuine disclosing has been lifted clear of the Twofold happening in materiality. And the meaning of "truth," *alētheia*, as disclosure presided over by man, has changed; it bears now the import of that separation.

Heidegger finds in Plato's "allegory" of the cave in Book VII of the *Republic* testimony to a continuing awareness of truth as unconcealment. But for Plato, this is no longer decisive. Another

---

[31]EM 140; IM 154.
[32]Cf. Ch. IX, p. 354. On the meaning of "truth," see Ch. I, p. 38, and n. 49, and p. 54, n. 109.

meaning of truth has come to the fore.[33]   The seeing-apprehending that now looks toward the defining Ideas takes charge of the "form," the outward look, that in any case it discerns.   It sets it before itself and, by its light, judges of the reality, the Being, of particular entities with which it itself has to do.   Here "truth," which for Heidegger doubtless always carries in some sense its connotation of a safeguarding forth, happens as such judging.   It has become an "approximating," a "determining in accordance with," "a correctness (*Richtigkeit*)" of appraising seeing accomplished by apprehending taking place as conceiving, as setting-forth-before (*Vorstellen*).[34]

Thus does an apprehending that now finds itself the arbiter of disclosure and of Being, in undertaking the validation of what it sees and hence of itself as seeing, begin the carrying-out, unknowing, of the newly initiated dominion of Being as self-concealing.

2

Aristotle follows fundamentally in Plato's steps.   At the same time, moving primarily on Platonic ground, he thinks out and establishes in a further and decisive way the new manner of relating of Being and man that is now taking place.

Aristotle too thinks Being as *ousia*, i.e., as steady presencing (*beständige Anwesung*).[35]   Such presencing is again "appearing," but now in the sense of *eidos*, the self-offering outward look as seen in particular instantiation.   For unlike Plato Aristotle finds this *ousia* immediately inherent in the individual entities that present themselves on every hand. The entity that in some particular present is concretely encountered in

---

[33]"Plato's Doctrine of Truth," PL 33.   English translation by John Barlow in William Barrett and Henry D. Aiken, ed., *Philosophy in the Twentieth Century*, Vol III, New York, Random House, 1962, p. 261.   We should here recall Heidegger's admission late in his work ("The End of Philosophy and the Task of Thinking," SD 77f, TB 70f) that the Greeks' early use of *aletheia* does not in fact show the meaning "unconcealment" that he finds for it.   Cf. Ch. IX, p. 354, n. 104.

[34]EM 141; IM 154f.

[35]"On the Being and Conception of *Physis* in Aristotle's *Physics*, B, 1," W 336f, Ph 242f.

individuality is for him not to be overlooked as something simply deficient in Being (*unseiend*). The outward look that it offers, e.g., as this particular house here or that particular tree there, happens as the direct appearing of the entity itself.[36] Of itself, the entity presences as having the look of house or tree; this determinate presencing hither from out of itself (*von sich her Anwesen*) is its Being, *ousia*.[37] It *is* this or that.

Aristotle uses centrally the word *morphē*, earlier employed by Plato as synonymous with *idea*, for the definitive outward appearance here in question. The English noun "form," from Latin *forma*, is, like the latter, far too static in connotation adequately to carry the significance that Aristotle gives to *morphē*.[38] *Morphē* translates into the German word *Gestalt*. In it, the original Greek meaning of *thesis*, which Heidegger explicates through the use of *stellen* and its derivatives, lies close at hand. The *forming* with which we have to do in Aristotle's thinking must be understood from out of the happening of the Being of what-is as that setting-hither-and-forth into unconcealing which, ever bringing itself to pass via the bounding structuring (*Gestalt*) that could present in specificity, centered itself in ineluctable particularity in entity upon entity by way of the openness that Greek man, in his varied accomplishings, offered for its coming to pass.[39] For *morphē* speaks of an analogous setting-forth.

As employed by Aristotle, *morphē* denotes a structuring that is the offering of a specific outward look, a specific appearance (*Aussehen*). As such, it partakes of the original dual aspect belonging to appearing itself. It is both the "standing"—the steady self-maintaining in a particular appearance—and "the self-setting-into" that appearance—i.e., the prior self-presenting—that together determine anything that is as what and how it is and so allow it to present itself as itself. *Morphē* is the placing, the setting-forth, that gathers the entity in question toward its offering of a specific outward look and into that look (*die Gestellung in das*

---

[36]W 345; Ph 249f.
[37]W 331; Ph 239.
[38]W 344; Ph 248.
[39]Cf. Ch. IX, pp. 347f.

Aussehen).[40]

As this setting-forth-that-gathers-into-the-outward-look-offered, *morphē* is at once that via which any entity originates and that toward which it moves in coming to appearance as itself. In growing things that belong to what we would call "nature," *morphē* governs and accomplishes itself directly, moving of itself from itself toward itself, and so returning into itself, as in the repeated originating of a plant from a plant of the same species.[41] In things brought forth through human handiwork, the specific appearance, the form that forms—the form, that is, that lets something appear, setting it forth into itself as possessed of a particular outward look—governs as a pattern (*paradeigma*) for a bringing forth, a setting hither, that is accomplished by another precisely in such a way as to set forth something discernibly possessed of that look.[42]

Aristotle acknowledges no such cleft as did Plato between the realm of pure appearance, the domain of Being, and the realm partaking of materiality, where genuine Being never resides. But he does find there a twoness that is reflective of Plato's view. For him the counterpart of *morphē* is *hylē*. *Hylē* is not what we would understand as "matter," Latin *materia*. For us in a context such as this, the word "matter" characteristically connotes something that can be considered in isolation and that is as such, from beyond itself, equally susceptible of disposal in diverse ways.[43] It is something possessed of possibility, but of a possibility that can remain indefinite until some determining, some specific giving of form from without, "actualizes" it. In speaking of *hylē*, Aristotle uses the word *dynamis*, which is usually translated potentiality (*potentia*) and understood to refer to just this sort of indeterminate possibility. This meaning, however, is inadequate. *Dynamis*, Heidegger says, does not simply mean ability or capacity (*Vermögen*). He renders

---

[40]"On the Being and Conception of *Physis* in Aristotle's *Physics*, B, 1," W 346, Ph 250. On the force of the prefix ge-, see Ch. VI, pp. 258f.
[41]W 363; Ph 263. See also W 347, Ph 251.
[42]W 346; Ph 250.
[43]W 350; Ph 253.

it, rather, as *Eignung zu*, fitness or suitability for.[44] Aristotle makes no place for the intrinsically indeterminate; *hylē* is already, as such, determined. It *is* precisely as that which is governed by some specific outward look toward which it, as possessed of fitness, is already directed. For precisely that setting-into-the-outward-look-offered (*morphē*) that brings to fulfillment the fitness of that which is completed or shaped as this or that also precedes that fitness. Fitness takes place only as fitness for that which in some way already offers itself and which therewith makes possible and constitutes in advance just that directional relation that "fitness for" bespeaks.[45]

Aristotle uses for the constitutive completion as this or that that initiatorially and completingly governs in the self-maintaining self-presenting of a particular entity as itself the words *entelecheia* and *energeia*. Both *entelecheia*, having-itself-in-completion (*Sich im Ende Haben*),[46] and *energeia*, standing-in-the-work (*Im-Werk-Stehen*),[47] are words expressive of setting-into-the-outward-look-offered (*morphē*).[48] They speak of the way in which presencing (*Anwesung*) accomplishes itself in a particular entity.[49] But the entity in its particularity is brought to completion and hence to standing as a work—i.e., as something that "stands fully in completion"—via the shaping of that which, in order genuinely to constitute it, must move via some radical manner of transformation[50] from mere fitness-for to completion-as.

Fitness-for (*dynamis*) is itself also "a mode of presencing"[51]; but since it is so only in relation to the setting-forth-into-the-outward-look-offered that is antecedent to it and that can alone provide its proper realization, it is, as such, imbued with deficiency. It partakes of a not-

[44]W 356; Ph 257.
[45]W 357; Ph 258.
[46]W 354; Ph 256.
[47]W 354; Ph 256. On *energeia*, and on the meaning here of "work," see further Ch. IX, pp. 352f.
[48]W 357; Ph 258.
[49]W 345; Ph 249f.
[50]W 355; Ph 256f.
[51]W 357; Ph 258.

yet. Only as it is gathered and gathers itself up into the intended entity as completed does it come to disclosure as fitness for this particular entity presenting itself at this particular temporal juncture and offering this particular outward look. Fitness does indeed belong intrinsically to the self-maintaining self-presenting, i.e., the "having-itself-in-the-work and in-completion," that constitutes the *ousia*,[52] the Being, of the entity in question. But only as so gathered into specific appearing do both fitness for and the material, *hylē*, that that fitness characterizes genuinely assume their share in the concrete presencing of that entity.[53]

When we adopt Aristotle's perspective and keep individual entities firmly in view precisely in their particularity, we must still recognize that for him the Being of any entity resides in the *morphē* determining it.[54] The component that provides its concreteness in materiality cannot come into consideration apart from that *morphē* and has no standing, either in itself or as comprising the entity itself, except as determined by the *morphē* that sets it forth into the specific outward look that it bears.[55]

Aristotle understands the entire universe in terms of the interrelating of *morphē* and *hylē* in a mounting structuring of vastly varied, always individualized entities. Whenever that which possesses fitness for a particular outward look is transformed into that which stands forth as offering that look, the entity so set forth is not simply sufficient unto itself. It is, in its turn, possessed of fitness for some fresh configuring as another sort of entity offering another look. In the complexity of particularized appearing that thus takes place, gradations in excellence are to be found. Thus, for example, man, who alone possesses mind, stands higher than the animals, while the animals, having sensation

---

[52]W 357; Ph 258.

[53]W 350f; Ph 253f.

[54]W 363; Ph 263.

[55]Cf. W 363, Ph 263, where that which *is* after the manner of *physis* is specifically under discussion. In his essay, Heidegger is elucidating an Aristotelian argument concerning the manner of being of growing things as opposed to things brought forth through a human making. *Physis* here is not equivalent to *physis* as made manifest in the utterances of the first Greek thinkers. Cf. W 369, Ph 268.

and power of locomotion, must be reckoned above plants.[56]

But although the universe is taken to be governed by an interrelating structuring of originative "forms" that set forth into individualized appearing, that structuring does not simply take place directly.   *Morphē*, the gathering-setting-into-the-outward-look-offered, has a double character.[57]   It can indeed govern directly, bringing the entity under its governance to full completion and according presencing to it.   Or it can govern *in absentia*, i.e., it can make its presence felt in the entity by its very absence and thus can accomplish a presencing of the particular entity that involves the want of the specific look that has already determined and, as a referent bearing upon it, still determines the identity of that which has been brought to appear.   Aristotle uses for this, *morphē*'s absenting of itself, the word *sterēsis*.   *Sterēsis* is privation, but in the sense of denial, the denial of the specific setting-forth-into-a-proffered-look whose prior presence it bespeaks.[58]   The happening of *morphē* as *sterēsis* permits on-goingness in the succeeding of one individualized determination by another, when both are governed by the same setting-forth into an identifying appearing.   Thanks to it, for example, one plant follows another in a species through the succession of leaf and bud and blossom and seed.[59]   Seen more widely, that happening permits any entity either to stand forth into a particular outward look or to stand forth as denying that look, i.e., that specific determination of itself. Hence, as presencing via the absenting of that identifying determination, any entity can be either this or that, according to the determining form that governs it and sets it forth; or it can not be this or that, although its appearing thus deficiently still belongs within the original sphere of determination.[60]

This ruling of *morphē* both as itself and as the absence of itself

---

[56]Cf. Aristotle, *Nichomachean Ethics*, Bk. I, 1097b23-1098a19.
[57]"On the Being and Conception of *Physis* in Aristotle's *Physics*, B, 1," W 364, Ph 264.
[58]W 364ff; Ph 264ff.
[59]W 367; Ph 266.
[60]W 367f; Ph 266f.

finds its counterpart in *hylē*, that which is fitted for completion in particularity. That which is fitted for the self-accomplishing of a particular setting-forth into an outward look is fitted also for the accomplishing of the look constituted by the absence of that setting-forth.[61] Thus a range of possibility from full accomplishment to denial of accomplishment is constantly in play. In addition, Aristotle admits of chance and fortune as factors responsible for distortion and for the want of a proper setting-forth into the intended outward look.[62]

Morphē and *hylē* are not to be thought of as component parts of the entity that they characterize. They belong intrinsically together, and in concert they are "the way to be" (*Art zu sein*) that pertains to that entity.[63] But although Aristotle insists over against Plato on the unity and reality of the individual entity, for him this is by no means all that must be said. The same setting-forth into a proffered outward look that governs inherently within an individual entity is also in play apart from that entity. Its presence in an organ of sense permits sensation and its attendant phenomena to take place.[64] Its presence in the mind permits reasoning and judgment regarding entities perceived. For Aristotle, only like can know like. Hence in such occurrences *morphē* must answer to *morphē*.[65]

Again, *morphē* exceeds *hylē* in the fullness of its reality. It is the standing-in-completion of anything that it informs. *Hylē*, on the other hand, is possibility as suitability for coming to stand in completion, i.e., for being set forth into the offering of a particular look. *Morphē* happens always as itself, even in its self-absenting. *Hylē* may be brought to stand forth according to the fitness resident within it, but it may also fail to arrive at such fulfillment. This possibility of a failure of realization renders it inherently inferior.[66] For Aristotle, "the principle (*archē*) that

---

[61]W 364ff; Ph 264ff.

[62]W 366ff; Ph 266ff.

[63]W 351; Ph 254.

[64]Aristotle, *De Anima*, III, 2.

[65]*De Anima*, III, 4.

[66]"On the Being and Conception of *Physis* in Aristotle's *Physics*, B, 1," W 352ff, Ph 254ff.

originatingly and disposingly moves that which is possessed of fitness toward fulfillment" is always "superior" to what is so moved.[67] Thus *morphē*, the gathering-setting-forth-into-the-proffered-look that governs precisely in this way, is inherently superior to *hylē*. At the apex of Aristotle's universe, there stands pure mind (*nous*), unmixed with materiality and unconcerned with it as the inferior. It thinks only forms and is identical with what it thinks.[68] This mind, itself undifferentiated, enlightens the passive intellect of individual human beings, permitting among them identity of knowledge and understanding. And beyond this is the divine mind. As most perfect, it thinks nothing diverse, including forms, which are many. It thinks only its own thinking.[69] In all this, clearly, Plato's influence is close at hand.

Heidegger does not concern himself with the points in Aristotelian thinking just mentioned that could be seen to witness to the withdrawing of Being from its original all-including happening in the immediacy of the encountered as such in that they manifest a kind of dissecting of reality that finds gradations of degree in the beingness of particular entities. But that disruptive withdrawal is, although it remains unmentioned, itself centrally in question in his inquiry into Aristotle's thinking regarding the pivotally important meaning of *physis*. In considering in detail the treatment of *physis* given in a passage from Aristotle's *Physics*, he centers his attention on elements that portray both the unity and the inherent changefulness maintained for that which *is* as that which stands forth in particularity, but that also bespeak at its crucial point of introduction the disruption of unitary immediacy that Aristotle's analysis presents. Aristotle's portrayal exhibits *physis* as *physis-morphē* and in so doing depicts it as a mode of presencing (*Anwesen*) that is, as such, characterized intrinsically by self-withholding self-absenting. In so doing it shows the ruling character of *physis* as *morphē*. That is, it portrays *physis* as that which, happening powerfully as be-waying, *Be-wegung*,

---

[67]*De Anima* III, 5, 430a, 17-19. For Heidegger's rendering of *archē* as *ausgängliche Verfügung*, on which the above phrasing is based, see W 316, Ph 227.
[68]*De Anima* III, 4, 429, 30-32: " . . . the mind . . . is nothing until it has thought."
[69]Aristotle, *Metaphysics* XII, 9.

originatingly gathers and sets forth into a particular outward look the *hylē*
that, disposible as that which is fitted to move in answer to its
governance, is disposed into appearing as this or that.[70] For Heidegger
this elucidation serves to set forth aspects of Aristotle's thought that
mirror the original Greek apprehending of Being, but that do so in a
distortion born of distancing. Pointing to that distancing, he remarks at
the close of his discussion that "the *physis* that was brought by Aristotle
into the conceptualizing that disclosed its manner of holding sway can
itself be only a derivative of the original *physis*."[71]

In accordance with his concern—already evident in our discussion
of *morphē* and *hylē*—to bring to light for thinking the determinative
relationships and modes of happening that inform reality in all its
manifestations and that hence, when discovered, can render it surveyable
and comprehensible, Aristotle brings to philosophical statement, in a way
not found in Plato, a precisely delineated, inherent structuring taken to be
everywhere in play. Here the fourfoldness that Heidegger himself sees as
belonging primordially to the constitutive happening of Being as the Being
of what-is[72] comes, if hiddenly, into view.

Aristotle understands everything belonging to what we would call
the realm of nature or to the sphere of things brought forth through art or
craft to eventuate from the interplay of four determining elements. *Hylē*,
with its inherent fitness-for, is involved in whatever entity may be in
question; the *morphē* that sets that entity forth into a particular outward
look exerts its governance; some originative contributor provides a
specific impulse toward the entity's completion; and that completion
(*telos*) itself, circumscribing the latter in advance, decisively affects its
emergence as the entity it is to be.[73]

---

[70]See "On the Being and Conception of *Physis* in Aristotle's *Physics*, B, 1," W 351ff,
Ph 254ff. On Heidegger's use of *Be-wegung* to speak of the way-opening bewaying
that is the provisive movement as which initiatory Happening carries itself into play in
the unfolding of the Twofold, cf. Ch. IV, pp. 146-148, and Ch. III, pp. 123f. Cf. Ch.
IX, p. 353.
[71] . . . *die von Aristoteles in den Wesensbegriff gebrachte* physis *selbst nur ein
Abkömmling der anfänglichen* physis *sein kann.* W 370; Ph 268.
[72]Cf. Ch. V, pp. 173-175, 178f & 180-182.
[73]"The Question Concerning Technology," VA 16f, QT 7f.

Each of these four is an *aition*. That is, it is something that is responsible for the arrival of the specific entity into the presencing permitted to it. Yet, as *aition*, it must not be thought of as functioning, in its responsibility, detachedly as a discrete member of a series or group. It is, Heidegger insists, not what we mean by the word "cause." The responsibility of an *aition* for the appearing of a particular entity is inherently a co-responsibility. Always together, indeed in indissoluble concert and hence in responsible relation to one another, the four *aitia* play their roles in the bringing-forth into presencing of some particular entity that is, in its turn, in a unifiedness of dependent relationships, indebted immediately and simultaneously to the four.[74] Functioning as *aitia*, each in its own way, the four, through their carefully attuned interrelating, occasion (*ver-an-lassen*) that which they bring forth. They let it come toward presencing, freeing it to be itself.[75]

Here the gathering fourness as which Being fundamentally gathers to light what-is, which Heidegger discovers in the genuine thinging of a thing, is clearly to be discerned.[76] In Aristotle's consideration of the domain of *physis*, the realm beyond that of human doing—which remains identified as *technē*—the now veiled identities of the constitutive four components via which that ruling happens can be hard to discern. In the

---

[74]VA 18; QT 9. Heidegger speaks of this being responsible as a being guilty for (*Verschulden*). In *Being and Time* (SZ 282f; BT 328f), he sees the ultimate meaning of "being-guilty (*Schuldigseins*)" in a like way, identifying it—in his discussion of human *Dasein* as guilty precisely in being itself—as a primordial "*being the ground* for (*Grundsein* für)," "having responsibility for."

[75]On this occasioning, see "The Question Concerning Technology," VA 18f, QT 9f. It seems quite probable that the insight into that fourfoldness that Heidegger elaborates so detailedly (cf. Ch. V, pp. 173-175) may well have been indebted significantly to his thinking upon the fourfoldness that Aristotle presents. In the context in "The Question Concerning Technology" in which Aristotle's thinking regarding the four constitutive elements in question is under consideration—although not named—Heidegger asks, "Why are there just four causes?" (VA 15; QT 6.) May not that very question have served, in the course of his own thinking, to open new perspectives in his own thought concerning the happening of the Twofold of the Being of what-is? For a suggestion regarding another possible source for the opening of those perspectives, see Ch. V, p. 203.

[76]Cf. Ch. V, pp. 173f.

sphere presided over by *technē*, however, their lineaments are easy to trace. In both realms, earth, which provides and sustains, manifesting itself in plant and animal, water and rock, is readily discernible as *hylē*, the "stuff" imbued with appropriateness for whatever entities are intended. The heavens, the participant in the four that conjoins itself with earth, in fructifying and in giving the structure-providing movements of heavenly bodies, seasons, and day and night, meets us here as *morphē*, the gathering-placing into appearance that allows the inherent fitness of *hylē* to come to fruition and accords to the latter disclosive definition. In the realm of *physis*, notably in the spheres of plant and animal life, *morphē* is characteristically seen as fulfilling the remaining two roles also. It both delimits the particular entity to be brought forth, as the provider of the latter's completion, and, as resident in and setting into appearance its predecessor, provides the impetus for the said entity's arrival into a like appearing. When *technē* is in question, however, the divine ones, who proffer within the interrelating of the Fourfold the completeness (*Heil*) whose foreseen possibility defines with clarity the role of the gathering thing—as the presence of the patron saint of the bridge makes manifest the awesomeness of the latter's role seen in its fullness[77]—are readily discernible in the foreseen *telos* that presents to the mind of the workman the thing as completed, circumscribing it according to the realm wherein it belongs and the role it is to play, and thus showing it as what it is to be.[78] And mortals, who are men bent upon pursuing their precarious way as decisively determined by that primordial concealing that exerts its governance in ultimacy through death, are visible in the craftsman or artist who, in his bringing forth of the thing that requires his participation for its completion, takes thought, foresees, ponders, and, risking always some miscarrying of his endeavor, entrusts himself to the determinative gathering toward him of the other *aitia*, in giving himself to the achieving of his proper task.[79] More than this, the unitary, mutual interplay of the *aitia* as co-responsible antecedents that together occasion that which is

---

[77]Cf. Ch. V, p. 186.
[78]"The Question concerning Technology," VA 17, QT 8.
[79]VA 17; QT 8. On the uncertainty inherent in such working, cf. Ch. IX, pp. 350f.

brought forth in reliance upon them bespeaks clearly and strikingly the manner of ruling happening, as a unitary reciprocity of interplay that brings to pass disclosure, that belongs to Heidegger's primal Fourfold.[80]

In the four *aitia* that Aristotle sets forth as central to the understanding of what-is as concretely embodied, we once again catch sight of the withdrawal of Being. Heidegger does not elucidate this fact, but the fact is clear. *Aition* is, like *idea*, a "stamping" (*Prägung*) of Being.[81] The word, that is, names a particularized determinant something that is taken to be fundamentally constitutive and to disclose Being. But that something does not bear within itself the immediacy of Being's primal relating to man. A unity that simulates the unity ruling among the gathered, gathering four of earth and heavens, divine ones and mortals remains among the four *aitia*. But the new four do not now appear in a oneness wherein, at the same time, each can be named and known only as itself. Now the four—although inherently playing their roles in a unity, while yet distinctive in themselves—are being grasped definitively by means of one naming and one over-arching concept. They are "*aitia.*"

In the light of the new determination of whatever is that comes into play with Aristotle's presenting of *aition* as expressive of the manner in which anything is responsible for the coming to appearance of whatever

---

[80]In the four *aitia* of Aristotle we can thus glimpse, albeit as disguised, the Four of the primal Fourfold that, in their unity, gather and are gathered into the thinging of a thing. This connection would have been evident in Heidegger's original lecture series. There "Das Ding," with its portrayal of the Fourfold, was followed immediately by "Das Ge-Stell" (later published as "Die Frage nach der Technik"), with its elucidation in the introductory discussion of the four *aitia*. The connection was obscured, however. For the original sequence of the lectures was severed, when, in 1954, "Das Ding" and "Die Frage nach der Technik" were published together in *Vorträge und Aufsätze*. The latter stood first in the volume, the former seventh. "Die Frage nach der Technik" was further sundered from "Das Ding" when it was published a second time, together with "Die Kehre (The Turning)," in a volume entitled *Die Technik und die Kehre* (Pfullingen, Neske, 1962). In their English translations, the two essays have been radically separated, having appeared in two different volumes whose titles suggest a disparity of concern. "Das Ding" appeared in *Poetry, Language, Thought*, translated and with an introduction by Albert Hofstadter, Harper and Row, 1971; "Die Frage nach der Technik" in *The Question Concerning Technology and Other Essays*, translated and with an introduction by William Lovitt, Harper and Row, 1977.

[81]Cf. Ch. III, p. 131.

may be dependent upon it, the uniqueness that intrinsically marks what-is in its happening is again shut from view. That uniqueness does not bring itself to pass now here, now there, in the thinging of particular things.[82] Rather, what-is, taken as a whole, is seen—beyond all the diverseness of identity and role accorded to its components, definable and describable in many ways though they may be—as presenting itself for thought as a patterned if intricately varied interrelating of *aitia*, far-flung but graspable in a new kind of oneness. That oneness is not to be decisively seen via this thing or that, each uniquely manifesting itself by way of a juncture of time and circumstance peculiar to itself and gathering to light certain particulars that, in that juncture, in presencing cluster to it.[83] Rather, it makes itself known by way of a manner of relating, assumed to be fundamental, that is to be discerned as determinative throughout the whole complex of what-is. It shows itself in a sameness inhering in the evidently diverse, not in a meaningfulness that, as a gathering-together, is precisely an acknowledging of the uniqueness of this or that.

3

In our discussion of Aristotle's thinking to this point, we have had to do with thought that, while distinctively Aristotelian, is largely secondary—whether in acceptation or in rebuttal—to the thinking of Plato that, to Heidegger's mind, first brings to decisive statement the new manner of Being's ruling in self-disclosure in withdrawnness as the Being of what-is. But Aristotle's thinking too plays a crucial part in the determinative manifesting of Being's altered manner of holding sway. In elucidating this, Heidegger focuses his attention on the role that Aristotle finds for *logos*.[84]

Originally for the Greeks, the speaking accomplished as language that *logos* denotes was, as genuine, an accomplishing of disclosing. In such utterance, what is encountered is named and unconcealed, and the

---

[82]Cf. Ch. V, p. 183.
[83]Cf. Ch. V, pp. 185f.
[84]On *logos*, cf. Ch. III, pp. 124f.

Being of what-is happens as unconcealing precisely in the words uttered.[85]   As pertaining to man, "*logos* and *legein* mean," Heidegger says, "that relating (*Verhältnis*) on the basis of which what presences as something specific (*als ein solches*) first gathers itself around man and for him."[86]   Here, in what we have called the answering of *logos* to *logos*,[87] Being, happening as unconcealing, gathers and disposes what-is into meaningful disclosure through man's compliant work that is the word.[88]

Aristotle still stands in proximity to this original Greek experiencing of language.   For him the role of a statement remains "apophansis," disclosing, i.e., letting what and how anything is be seen from out of the thing itself.[89]   But a new relation between human speaking and that which is spoken of is now decisively in play.

Aristotle implicitly trusts the identifying word.   To him it is sufficient to point to a common manner of speaking in order to establish that things indeed are as habitual uses of language take them to be.[90]   An identifying word accords with the *morphē*, the setting-forth-into-the-specific-outward-look-offered, that constitutes particular entities as what they are, but in that according, the priority of identifying lies with the word itself.   Thus *morphē* must be understood, Aristotle says, as "the appearing that is in accordance with the spoken word," *he morphē kai to eidos to kata ton logon.*[91]

Here we come upon a fundamental statement in Aristotle's thinking that, as does the word *idea* for Plato, decisively manifests the newly constituted manner of relating obtaining between Being and man.[92] Human utterance does not now immediately reveal what-is in accordance

[85]Cf. Ch. III, pp. 124-126.
[86]"On the Being and Conception of *Physis* in Aristotle's *Physics*, B, 1," W 349, Ph 252.
[87]Cf. Ch. IX, pp. 338 & 342.
[88]For the discussions that lie behind this phrase, cf. Ch. IX, pp. 335-337, 347f, & 352f.
[89]"On the Being and Constitution of *Physis* in Aristotle's *Physics*, B, 1," W 349, Ph 252.
[90]W 346f; Ph 250f. See also Aristotle, *Physics*, Book III, 4, 195b, 31-35.
[91]W 345; Ph 249.
[92]W 371; Ph 269.

with the happening of Being. Rather, it has itself become the standard
whereby the setting-forth-into-specific-appearing, which itself decisively
determines the Being of an entity, is measured and judged as to its
adequacy. The specific appearing in question must accord with what the
speaker already knows it to be and hence with the meaning of his word.[93]

While Aristotle is confident that words will display that of which
they speak, his primary conception of language in fact holds it to be a
saying of something about something.[94] Always language had readily
slipped away from its happening as immediate disclosive utterance. What
was first truly said, in being handed on, said and said again, could, with
repetition, degenerate to a mere rehearsal by rote.[95] Aristotle, as it were,
laid hold of an early stage of this movement of speaking away from
immediacy of disclosure and made it fundamental for the relating of man
and whatever is and hence for the relating of man and Being.

For Aristotle, "assertion (*Aussage*)" became the configuring of
language that could suffice to disclose the what and how of things, which
was taken to be their Being. Asserting was itself a summoning. Indeed,
it accosted that which was addressed. It called it to disclose itself in
accordance with the prior knowledge of it that was resident in the words
spoken. It was, Heidegger says, in the earliest sense of the word, a
katēgoria. The word *katēgoria* originated out of the legal context in
which *kata agoreuein* meant to accuse a person of being the one
who . . . . Every statement partook of this arraigning character. It hailed
what was addressed forth as this or that. Beyond this, Aristotle gave to
the noun *katēgoria* a particular meaning as a designating term. The
categories—substance (standing-in-self [*ousia*]), quantity, quality, etc.—
underlie ordinary speaking. They constitute the prior wordless addressing
of that which is encountered. Each encountering takes place in terms of
fundamental determinations that render the entity met with
comprehensible and addressable in its particularity. Knowing "quality,"
for example, we can address something as red. Again, only inasmuch as

[93]W 347; Ph 251.
[94]EM 142; IM 156.
[95]EM 142; IM 155.

we have, in the apprehending that the word *ousia*, standing-in-self, bespeaks, antecedently confronted that which meets us and brought it before us as possessed of such standing, can we address it identifyingly as this or that, as tree or house or man.[96]

That which *is* in this Aristotelian sense, i.e., as that which is apprehended and identified as something confidently summoned and placed in view, lies before the speaker ready to be spoken about. The assertion, which, for Aristotle, in its stating manifests the how and what of that which is laid claim to and spoken about, grasps the latter precisely in this manner. Primary for the statement is that which already lies before and underlies the specific identification to be made, i.e., the "subject (*hypokeimenon, subiectum*)." All predication and attribution can but serve to let that precedent subject stand forth in its specificity. It is, then, in the appearing of anything as subject that its Being (*ousia*), i.e., the beingness belonging to it, is decisively to be seen. Stating assumes and declares the standing-in-self, the self-maintaining self-presenting, of its "subject"; while at the same time it itself fashions that self-presenting to happen as it does.[97]

Here we discern from another perspective the same manner of relating of Being and man that came to light with Plato's ascription of Being to *idea*. Now human apprehending accepts a type of utterance—assertion—as commensurate with itself and through it accomplishes a laying hold upon Being that secures the latter for knowing. Indeed for Aristotle language, as assertion, itself emerges as the arbiter of Being. No assertion admits of the validity of another that contradicts it. Two assertions with the same subject and contradictory predicates cannot both be true. Thus assertion, considered with respect to contradiction, serves to determine truth or falsity. In this extreme formulation of the role of assertion, we catch sight of the fundamental prerogative that has been

---

[96]See EM 142, IM 156, and "On the Being and Constitution of *Physis* in Aristotle's *Physics*, B, 1," W 322f, Ph 232. The parallel between this understanding of category and Heidegger's characterization of the mathematical should be noted. Cf. Ch. VII, p. 269.

[97]EM 142; IM 156. Cf. W 330, Ph 238.

arrogated to it. Asserting, properly accomplished, appears as the decisive arena of judgment as to whether and how that which is spoken about can be. The contradictory is precluded from being, while whatever is *not* contradictory is accorded at least the possibility that it may be.[98]

Heidegger points out that, like Plato,[99] Aristotle was aware of the original meaning of truth as unconcealment. In Book IX, Chapter 10, of the *Metaphysics*, for example, he maintains that whatever is the case for something that is known holds for it independently of true or false statements that may be made about it. Yet, as with Plato, the decisive thrust of his thinking lies elsewhere, as is evidenced in his statement that "the false and the true, namely, are not in the thing (itself) . . . but in the understanding.[100]" The assertion that the understanding propounds, as it were, carries into execution the judgment that has been made. Again truth happens now as correctness, here the correctness of assertion with respect to that which is in view. The latter is the point of reference for such correctness, but *not* as it genuinely *is* in and of itself, in virtue of its own self-unconcealedness, its Being as its truth. It *is*, rather, as that which already lies before the apprehending that confidently trusts it as its possession. As such, it is that to which true assertions rightly refer.[101]

On this basis, Aristotle works out the patterns of assertions whose types and interrelations can be specified in advance as constituting an ineluctable structure that displays the ways in which entities—both in themselves and in relation—can and cannot be. With this, *logos* in the sense of assertion becomes logic, the powerful standard of judgment upon which apprehending can rely.[102]

Here we must observe that for Aristotle the knowledge that is the proper basis of assertion as such cannot be mistaken. "The thinking of the

---

[98]EM 143; IM 157.
[99]See this chapter, pp. 374f.
[100]"Plato's Doctrine of Truth," PL 44, translated by John Barlow, in *Philosophy in the Twentieth Century*, edited by William Barrett and Henry D. Aiken, New York, Random House, 1962, Vol. III, p. 266.
[101]EM 143; IM 157.
[102]EM 143; IM 157.

definition in the sense of the constitutive essence is never in error."[103]
Such knowledge is always knowledge of the appearance, the gathering-
setting-into-a-specific-outward-look, that determines whatever is as what
it is and in the way that it is. Only the *morphe* character of particular
entities is knowable and thinkable. The *hyle* character, imbued as it is
with possibility that is only fitness-for and that may or may not be brought
to fruition, can in itself not be known and thought.[104] The "appearances"
that let appear, which are given to the individual human mind from
beyond itself, namely, from that mind that *is* only in its thinking of them,
can be known or not known. But when present they cannot be
misapprehended. It is this ontology, with its attendant confidence in
untrammeled knowledge, that underlies Aristotle's assurance that
assertion born of knowledge can be implicitly trusted to embody decisive
judgments concerning the Being of whatever is.

Aristotle, then, from his peculiar vantage ground, brings to light
and establishes the distancing now happening between Being and man, in
a way that grounds that distancing—taken from man's side—precisely in
the pivotal realm wherein man most centrally lives out, to the point of
concrete doing, his providing of openness for the happening of Being.
For he bases it in language, in the *word*, to whose utterance man as man
is centrally summoned in order that, through that word, he may allow to
what-is the self-disclosure vouchsafed it from out of Being.[105] In
Aristotle's arrogating to language—understood and exercised as logic—
and to the apprehending that regularly carries itself out by way of it the
power to judge concerning Being, the self-assertion of man vis a vis the
Being of what-is tellingly entrenched its position, and Being advanced
correspondingly its happening as self-concealing.

The paradox that we saw with respect to the work of Plato holds
here as well. Aristotle's salient concern was with the manifest. Aristotle
was permitted to some degree to discern concealing as playing a

---

[103]Cf. Aristotle, *De Anima*, BK III, 6, 430b, 30-32.
[104]*De Anima*, Book III, 4e, 4-8.
[105]"On the Being and Conception of *Physis* in Aristotle's *Physics* B, 1," W 348, 349;
Ph 251, 252f. On this role of man, cf. Ch. IV, pp. 157-159.

fundamental role in the Being of nearly all entities—excluding mind as fully realized and God—as he conceived them. He testifies to that role in his recognition that process, movement, and change are integral at nearly every level to the Being—as a coming into Being—of whatever is; in his insistence that Being must be seen to belong to the individual entity wherein the happening of *appearing* takes place as the bringing to fulfillment of the possible fitness for itself that characterizes the materiality that that appearing lets appear; and more specifically in his understanding of *physis*, taken as a particular manner of Being among others,[106] i.e., as a Being-on-the-way wherein self-maintaining happens as an appearing that is intrinsically a setting-aside and as a going-forth that is, as such, a going back into itself. But though all this be so, Aristotle is, in the last analysis, no more able than is Plato to conceive of reality understood in this way as that which, for Heidegger, it most truly behooves human apprehending to claim as its genuine counterpart. Unconcealing and concealing do not finally belong together as concomitant in the happening of Being.[107] It is only as standing-wholly-in-completion that the appearing that gatheringly sets forth a particular entity into the offering of a specific outward look is immediately present to pure apprehending (*nous*), which itself likewise stands wholly in completion. Individual human apprehending must needs concern itself with the various aspects and levels of reality that support and environ it. But its highest and most pleasurable comportment, its own manner of standing-in-completion, consists in sheer beholding of those appearances granted to it from the unembodied apprehending that *is* above and beyond it.[108] Once more apprehending ranges itself, finally, with the forming appearances that, on lesser levels of Being, let particular entities appear.

Again with Aristotle, Greek man, in the sudden self-assertiveness that now characterizes his living-out of the openness-for-Being incumbent upon him, finds himself assured of a direct apprehending in whose light

---

[106]W 369; Ph 268.
[107]Cf. SG 112f, and cf. Ch. I, pp. 24-26.
[108]Aristotle, *Metaphysics*, XII, 7, 1072b, e, 22-24. See "Science and Reflection," VA 52, QT 163f.

conceiving and speaking with respect to the complex components of the familiar world can be confidently carried out. As Being happens now so as to conceal its primal manner of holding sway, the clarity conferred by precise defining and delineation comes to the fore as a means of discovering the Being of what-is that is decisively within man's grasp. And language, once recognized as the ever shifting contexture of speaking wherein unconcealing and concealing vied with one another,[109] has become, in its definitive formulation as logic, a rigid structure supportive of the human self-securing now in play.

4

With the thinking of Plato and Aristotle, under the impress of the newly configured relating of man to the Being of what-is that arose out of Being's self-withdrawal, the bifurcating of reality into the forming outward appearance and the unformed carries itself out in a like bifurcating of human doing. No longer is the epitome of human knowing found in a skilled and resolute confronting of the on-surging manifold of what-is that could accomplish, precisely there, a unique happening of disclosure through the bringing forth of some work that, in its very concreteness, could let genuine appearing take place.[110] For Plato and Aristotle, the highest knowing—that of the philosopher—is a beholding of those forming outward appearances whose presence can alone provide defining identity to whatever is, whether the beholding be understood in terms of direct awareness (Plato) or of the direct grasping of disclosive propositions (Aristotle). At this juncture the word *technē*, to which Heidegger has pointed as the Greeks' name for skilled human knowing— and indeed of every sort—as confronting and withstanding the onslaught of the impending Twofold so as to accomplish Being's happening as disclosure through the bringing forth of whatever is needful for that disclosure in the context in question, undergoes a significant narrowing of application. It is as though this word, which, Heidegger tells us, speaks

---

[109]Cf. EM 141f, IM 155.
[110]Cf. Ch. IX, pp. 338f & 348f.

of the dedicated concentration, with the intent to bring forth something specific, of skilled human attention upon the Twofold as such in the full immediacy of its happening, can no longer suffice as a name for the newly defined knowing that is the prerogative of the philosopher who seeks ultimately to turn his yearning gaze away from the thronging, changing realm of particularity and uncertainty and toward the clarity and permanence of the Ideas. Now *technē* becomes the specific name for the skilled knowing practiced by craftsman or artist, as though only in the domains of craft and art can human knowing still be acknowledged to accomplish itself properly in the bringing-forth of particular works in the sphere of immediate presentness where concrete entities are encountered. The philosopher may indeed concern himself with activity in that sphere, out of responsibility to his community (Plato) or because, as a human being, he must needs function variously within it (Aristotle), but his knowing, for which such words as *epistēmē*, *sophia*, and *nous* must now be employed, has as its only proper goal not a concrete accomplishing but a self-sufficing beholding.

In both Platonic and Aristotelian thinking, *technē* as now delimited is understood in accordance with the ontological perspective that prevails. The craftsman and the artist are considered to exercise a knowing that directs itself toward and is fulfilled through skilled obedience to the determinative informing appearances that bring to disclosure whatever is. However, their glimpsing of those forming outward appearances lacks the clarity and thoroughness to which the philosopher may aspire.

Plato ranks the craftsman above the artist. The craftsman, through his peculiar skill, can succeed in bringing forth useful articles that participate with disclosive adequacy in the Ideas manifested to him, which govern their appearance.[111] The artist, on the other hand, can do no more than present Ideas through the copying of entities that are themselves but participant copies. The picture of a table is twice removed from the Idea table. It is a deficient, untrustworthy presentation that should not find a place, as does that provided by the craftsman, in the

---

[111]N I (German) 204ff; (English) 176ff.

well ordered life of the community.[112]

For Aristotle, on the other hand, all pursuits that might be considered crafts or arts belong together on a par as expressions of *technē*. All who follow these successfully are engaged in bringing forth works wherein the determining outward appearances appropriate to them are decisively brought into play. Here, in the context of the thinking that understands reality in terms of the interrelating of the four fundamental *aitia*, *technē*, the perspicacious knowing of one who brings forth a work, gathers to itself and gathers itself toward the responsible participants in that bringing-forth. The specific outward appearance that will inform the entity so as to set it forth as what it is, is present to his foreseeing knowing as the appearance of the entity in completion, as is the foreseen, now determinant, role that, as finished, the work must fill. Thus guided, that knowing, *technē*, in accomplishing itself, allows the *hylē* that is tending toward the outward appearance in question to receive fashioning as the properly completed thing, for it permits the craftsman or artist rightly to perform his initiatory, achieving work in the bringing-forth into presencing of some artifact or work of art. It is only this knowing, skillfully brought to bear by the prescient artist or craftsman, that can initiate and carry through the disposing that brings the four *aitia* unifiedly into play, that they may let the finished work appear.[113]

In both these contexts of thought the primary comportment requisite upon man is not a wholly dedicated, discerning and attentive openness to the encountering and bringing-forth into determinative disclosure of something inherently unique, thoroughly particularized in time and circumstance, whose prime significance would lie in its illumining and ordering of the life of some particular community.[114] It must be striking, even in the Aristotelian context—the richer of the two—how far, where the fundamental configuring of reality and the fundamental role envisioned for man are in question, we find ourselves from Greek *technē* as Heidegger portrays it, a comportment that has

---

[112]N I (German) p. 202; (English) p. 174.
[113]"The Question Concerning Technology," VA 17, QT 8.
[114]Cf. Ch. IX, pp. 339-341.

precisely this pivotal character and that embraces all manner of human pursuits,[115] let alone from *technē*'s subsequently adduced, more ample counterpart the "dwelling" that in its allowing of genuine thinging to things permits the worlding of world through the gathering forth of the unitary Fourfold of earth and heavens, divine ones and mortals.[116] *Technē* is markedly diminished now as the name for the knowing skill of craftsman or artist. The works that these bring forth might well still throb with disclosive power. The artist or craftsman, as one particularly fitted to provide for his community a work that could let genuine appearing shine forth, might well still endeavor to meet whatever met him in openness and, in so doing, might in his working allow the primal Four of earth and heavens, divine ones and mortals, their complex sway in the disclosive thinging presencing of this or that work. But the grounding thinking that now appraised his work could comprehend it only otherwise, obscuring and diminishing it in the seeing. For that thinking, a work might still have individuality, what we might call provisional uniqueness. It might serve, indeed in a lofty manner, as did the great tragedies, to illumine, purify, and order a community's life. But the highest endeavor of man is philosophical thinking. And just as this thinking considers the vast manifold of what-is with an eye to structure and patterns, to generally operative and applicable principles, to concepts that render comprehensible the multiplicity so fraught with particularity, and hence makes that manifold accessible on a plane where particularity has been transcended and subsumed, so it seeks to discover in such works the constituting principles common to them.[117] It can permit no work to stand forth in powerful revelatory singularity; for it views nothing solely as itself. Here it is not in the unrepeatable, time-bound appearing of some unique, concrete thing that the focal point of the disposing of human life, its insights and its comportment, is to be found.[118] It is rather in the

---

[115]Cf. Ch. IX, pp. 347-349 & 335f.
[116]Cf. Ch. V, pp. 187-195. On the correspondence between *technē* and dwelling, cf. Ch. IX, pp. 355-359.
[117]See Aristotle's *Poetics*.
[118]Cf. Ch. V, pp. 183 & 185f.

forming and informing appearances that set forth the multitude of particulars and that can be apprehended, seen in relationship to one another, and brought to bear on that multitude that the focal point lies. It is, in the last analysis, only from the cognitive grasping of the determinative, disclosive "forms" that the ordering of the many aspects of human life must properly begin.

With the sundering of *technē* and knowing thinking that takes place in the work of Plato and Aristotle, what would pass for dwelling is grounded not in self-surrender to that which immediately encounters and demands to be knowingly husbanded forth but in a secure apprehending of the permanent and stable that lies at once within and beyond the particular and offers security over against the elusive, ever-shifting manifold to those who grasp it via the disciplined exercise of thought. At the same time, when that work is considered in the light of Heidegger's portrayal of the thing, it can be seen, in distinction from it, to cast a veil over the happening of what-is. Particular things are not left to thing in their uncapturable immediacy in order that the primal participants in the unitary Fourfold—earth and heavens, divine ones and mortals—may be directly, caringly brought forth into their own, so as to play in concert their peculiar roles in letting world world genuinely for those who dwell attentively in company with things.[119]   Now things are analyzed into components. Unity is found in terms of some overall structuring where repetitive likeness of relationship is taken to be primary. The stable manifesting appearances, *idea, morphē*—together with the explanatory doctrines that they engender—so recommend themselves to thinking and find embodiment within it as fundamentally to interpose between the thing and man the distancing, unnoticed barrier of structured thought.

In this changed relating of man to the Being of what-is, the primal Fourfold, decisive in the happening of what-is as centered in things,[120] masks itself as the fourfold, ubiquitously happening *aitia*. Through the intervention of the thinking that seeks security, man is sundered from the

---

[119]Cf. Ch. V, pp. 187-191.
[120]Cf. Ch. V, pp. 173-187.

thing; the wholly singular and unitary thing is sundered into elements; and man, correspondingly, sunders from one another the functions belonging to himself, considering them to be higher or lower, according as they have to do with the permanent or the transient in things. Here, with the disrupting of the immediate belonging-together of Being and man that originally obtained, unities intrinsic to the on-going happening of the Twofold fall from view. Nevertheless, hidden, they remain. Even now, in the unity that we have found Heidegger to posit among modern technology, modern science, and modern metaphysics,[121] the original inclusiveness belonging to *techne* is in play. And in the ravaging intercourse that is the mode of meeting of man and what-is in this technological age,[122] the original unitariness of the power-fraught meeting of the two that could permit the direct self-disclosing of Being still transpires, if in unrecognized guise.

In the context of the "stepping-apart" of Being and man[123] that is brought to overt accomplishment in the work of Plato and Aristotle, division begins everywhere to obtrude. Under its dominance a self-assertiveness of what-is as such in relation to Being, which is akin to the self-assertiveness that has now begun to characterize man in his fundamental relating, sets in as well.[124] In his stepping-away from the encountering of the unfolding Twofold in the fullness of its on-coming immediacy, Greek man in his decisive comportment in thinking and speaking loses touch with the happening of Being as self-unconcealing

---

[121]Cf. Ch. VII, pp. 280f, and Ch. VIII, pp. 323-325.

[122]Cf. Ch. VI, pp. 241-252.

[123]Cf. Ch. IX, p. 364.

[124]Here the belonging of man within the manifold of what-is as a whole, albeit as one possessed of a manner of Being—his ek-sisting—that is unparalleled throughout that manifold (cf. Ch. II, pp. 58-61, n. 5) is in question. Cf. Ch. I, pp. 28f. The impulse toward isolative self-assertion, which Heidegger depicts as arising mightily within the very happening of Being as presencing, (Ch. I, pp. 50-52, and n. 100) is ever, ultimately, the same impulse, however and wherever it may manifest itself. It ever happens correspondingly in man and in what-is as it meets him as the happening Twofold. Cf. Ch. II, pp. 84-86, and n. 101, & pp. 91-93, and Ch. VIII, pp. 304f.

self-concealing that was originally experienced via the apprehending that, in compliance with the governance of Being, variously brought the manifold of what-is to light. Now Being has decisively withdrawn behind what-is. And what-is presents itself as though it were Being.[125] The new philosophical thinking, even at the highest reaches of its inquiry, concerns itself only with particular entities, even if those entities be seen to be self-manifesting "Ideas," governing "forms" that stand wholly in completion, or a god who in his perfect self-maintaining thinks only his own thinking. Being is ascribed, its role as the governing determinant of whatever is, is recognized, but Being is misapprehended as a clearly discernible property of that which is. It is at the disposal of apprehending thinking. It is no longer to be radically encountered and acknowledged in the directness of its ruling, precisely in every manner of meeting with what-is. The latter alone has become man's fundamental concern.

More than this, within the manifold of what-is that has now thrust itself forward and usurped the place of Being as the primary concern of thinking, individual particulars have in like self-assertiveness come to stand out inordinately from one another.[126] The concern of the thinking that is taken to be man's highest doing[127] is now not with the thinging of things and the presencing together of particular entities in meaningful configuration. It has now to do most fundamentally with "Ideas" or "forms," each sufficient unto itself, standing forth in isolation one from another. If individual entities are seriously considered, they are viewed first of all as this or that. Each is identified according to the specific outward look that it offers. None is known most truly by way of the self-presenting that discloses it, through some limited present, precisely in its compliant gatheredness among its fellows.

Here the self-withdrawal of Being from its happening as *logos*, as the disclosive, selective gathering that lets compliant presencing-together take place,[128] has not only left man unquestioningly to lift himself up in

[125]See "The Anaximander Fragment," HW 310f, EGT 26. Cf. Ch. I, pp. 44f.
[126]Cf. Ch. I, pp. 50-52.
[127]"Science and Reflection," VA 52, QT 164.
[128]Cf. Ch. III, pp. 124f, and Ch. IX, p. 338.

superiority in his exercising of *logos* as assigned to him, but it has left what-is as a whole and in all its particulars to an individualized self-assertion, a claiming of priority in isolation,[129] that once again profoundly masks the genuine manner of Being's ruling. Here too seeming clarity of disclosure serves but to conceal the newly encroaching dominion of Being's happening as self-concealing.

With Plato and Aristotle, Greek philosophy did, as we customarily suppose, enter upon a beginning. But, to Heidegger's way of thinking, that which was thus a beginning was in fact philosophy's end.[130] With the inception of Platonic thought, Greek thinking had begun that long fulfillment[131] that would culminate, in time, in the far other thinking of our very different age.

In its original unaltered beginning, philosophy had made manifest via the often aphoristic utterances of attentive, receptive thinking immediate glimpses of the manner of Being's happening as the Being of what-is and of man's genuine manner of existing in relation to it. But Being, ever happening most primally as self-maintaining self-concealing, changed in its self-sending-forth. Therewith it masked itself by way of a new beginning, a beginning that must needs see the demise of genuine openness and the first movement toward the ascendancy of self-withdrawal and self-assertion, as man began to live out a newly constituted relating between himself and what is in Being. Being was beginning, on the vast scale of that which we call Western history, to fulfill hiddenly, ineluctably, the selfsame manner of happening that Greek man had originally daringly descried. The forgottenness (*Vergessenheit*) of the primordially constitutive Difference obtaining between Being and what-is is, Heidegger can say, as the Bringing-to-pass provisive of metaphysics—itself a carrying into play of the history of Being (*Seinsgeschichte*)—"the richest, most far-ranging enowning Bringing-to-pass (*Ereignis*)," in which, indeed, "Western world-history" attains its

[129]Cf. Ch. I, pp. 50f.
[130]EM 137; IM 150.
[131]*Vollziehung.* For this meaning of "end," see "The End of Philosophy and the Task of Thinking," SD 63, TB 57. Cf. Ch. VIII, pp. 324f.

carrying-out.[132]   Nevertheless that oblivion, which now begins to hold sway, is, precisely as such, likewise a playing out of an impoverishment of the on-happening Twofold.   For at its very inception, concealing—invincible, all-pervading concealing—had begun to encroach upon unconcealing, wasting and distorting it with its own darkening, disguising power.   The "end" had begun to take place that would bring itself to completion only with the modern age.   And with Plato and Aristotle, Greek philosophy, lofty in its unwitting obedience to the changed manner of Being's governance, brought to utterance the new manner of relating now underway and thus grounded, by way of thinking and its determinative words, that altered happening of man's existing vis a vis Being in the midst of the unfolding Twofold that had now decisively begun.

---

[132]*Die Vergessenheit des Unterschiedes, mit der das Geschick des Seins beginnt, um in ihm sich zu vollenden, ist gleichwohl kein Mangel, sondern das reichste und weiteste Ereignis, in dem die abendländische Weltgeschichte zum Austrag kommt.* "The Anaximander Fragment," HW 336, EGT 51.   In the passage in question Heidegger twice uses *Ereignis* in a way that might easily be supposed to suggest a meaning close to that usual for the word, namely "event."   The import of his statements should, however, clearly not be so understood.   In those statements, which allude to the on-going disclosing of Being via metaphysics through reference to the manner of ruling of the originative Happening that Being names, the peculiar meaning that he elsewhere accords to *Ereignis* as a Bringing-to-pass that so discloses as to bring something into its own (see Ch. IV, pp. 143-151) should again be heard.   See also "Overcoming Metaphysics," VA 71, EP 84f, a roughly contemporary passage concerned similarly with metaphysics as the history of Being, where *Er-eignis* is hyphenated so as to make clear the meaning intended.

CHAPTER XI

IN DEVIANCE

Between the time of the ancient Greeks and our own lies the long sweep of events that we call Western history. That phrase "Western history" can instantly call to our minds salient aspects of its course: Alexander the Great and the Hellenistic Age, the Roman Empire, the appearance and spread of Christianity, barbarian incursions and the "Dark Ages," the flowering of culture in the high Middle Ages, the Renaissance, the Reformation, the rise of modern science, the industrial revolution. Nearer at hand, events in political, economic, intellectual, and artistic life in our own and other Western countries present themselves in greater detail, down to the moment in which we live in the extended aftermath of the second World War. Here fear of atomic destruction inescapably haunts us; environmental threats confront us world-wide; financial insecurity threatens us on a global scale; our disparate intellectual outlooks and modes of behavior and expression jostle and supersede one another with bewildering speed; and "high-tech" products and processes crowd in upon us in such number and variety that we can grasp neither their extent nor their significance and can indeed scarcely afford them attention, so that they throng upon us almost unaware.

We recognize that perils beset us. Nevertheless, when we look back over Western history, we see in it progress, betterment in realm after realm of human life. We can readily discern throughout that history specific instances of excellence in intellectual, moral, and religious insight and in practical or artistic achievement. But when we judge of past periods as a whole, systems of thought and belief or political organization and social structure habitually seem to us outmoded and inferior. In particular, we pride ourselves on our humaneness, on the gains in human rights that have been achieved in the past few centuries and on the alleviation of human suffering that has been made possible through the aid of modern science, modern medicine, and modern technology. And we consider the science and technology that are the hallmark of our age to have opened to us vistas of possibility unknown to any other and to have brought an enriching of every aspect of life that gives to our time, despite its distresses, the undeniable stamp of superiority.

To Heidegger this our customary view could appear as nothing other than a self-aggrandizing delusion that affects us from out of the actual manner of happening of our history, precisely in order to blind us to that history's deceptively spoliative course. "Greek" is, after all, for him the adjective that names the delicately poised relating of Being and man wherein man alertly attends upon Being, receptively and responsively allowing it free play in its ruling as the Being of whatever is.[1] And just those aspects of our modern epoch that suggest to us improvement and superiority manifest to him the deviance (*Irre*) that we should in fact perceive to be the proper measure of our distance from Western beginnings in what is Greek. There is and there has been among us a moving toward completion, taking place by way of time, but, to his mind, it is a movement rife with concealment and distortion.[2] To him, in the history of the West that we so confidently survey, there is no ascending "progress" at all.

1

With the growing ascendancy of Rome in the late pre-Christian era, a marked shifting away from Greek origins took place. Of necessity the Latin-speaking Romans undertook to find equivalence in their

---

[1]Cf, Ch. IX, p. 332. Here and elsewhere in this study the reader should be careful rigorously to separate "relate" from its usual meaning. Here "relate," in contexts where the Being of what-is is in any way under consideration, does not speak at all of a connecting of inherently discrete elements, one of which is to be subordinately referred to the other. The relating that is now in question is a *holding-toward-one-another* of participants that, while they are distinct from one another, are ultimately one. For we are using "relate" to speak of the happening of the single Twofold. See further, Ch. IV, p. 145, n. 10.

[2]"The Anaximander Fragment," HW 310f, EGT 26f. Cf. Ch. IX, p. 332. On the happening of what-is into deviant divergence (*Irre*) from Being by Being itself, which is here in question, and on the manner of happening of what-is as historical that arises therefrom, cf. Ch. I, pp. 45-47 & 48f.

language for the words that had become fundamental for Greek thinking, and therewith the transformation in the Greek manner of relating to reality that was both manifested and established in the work of Plato and Aristotle was transformed in its turn, and that in a decisive way.

The Greeks' manner of existing, the openness-for-Being peculiar to them, was molded by their language.[3] The changed understandings of what-is and of man wrought out by Plato and Aristotle rested still, Heidegger says, on the experience fundamental to the Greeks, and the original import of the words provided to them, although it might be veiled, remained significantly in play.[4] Roman experience was other than Greek, for it was shaped by a language of a different stamp. Hence, even when the Romans took care to translate Greek words into literal Latin equivalents, they carried the words away into a new and different sphere of discourse. "Roman thought takes over the Greek terms (*Wörter*)," Heidegger says, but without a like experience of their import, "without the Greek word (*Wort*)."[5]

Inexorably Latin, a fresh yet kindred language employed by a distinctive people whose peculiar manner of existing it had shaped, superseded Greek as the language of the thinking that, heir to the work of Plato and Aristotle, sought to establish itself and ground its world by means of its comprehending of the what and the how that characterized the reality with which it had to do. Over the centuries the use and dominion of Latin spread. Diverse currents—notably those from the Judeo-Christian tradition—which were born out of alien sources of experience flowed into its domain, only to be taken up and molded according to the pattern that it had fashioned from the Greek original. The Latin language and the way of naming and of summoning reality into view through speaking that was intrinsic to it determined and served thinking in the accomplishing of its fundamental task.

As for Plato and Aristotle, so for the Roman, the reality of

---

[3]Cf. EM 120, IM 131f. On the central role of language as enabling human self-carrying-out see Ch. XIV, pp. 573-582.
[4]Cf. "The Origin of the Work of Art," UK 15, PLT 23.
[5]UK 15; PLT 23.

anything was thought in terms of disclosiveness. Reality did not, as reality, partake of diminution or distortion, i.e., of concealing; hence such negative conditions could only imply a falling away from realness or "being." In this basic context a sharp shift within men's apprehending of what-is came into play. The Greeks' encountering of what-is always in some way took place and was apprehended in terms of presencing. Accordingly, it was characterized by a sense of unitariness pervading all diversity.[6] The awareness of the Latin-speaking Romans, on the other hand, was an experiencing of a self-contained standing-forth that lacked the unplumbable vitality and innate interrelatedness that Heidegger takes "presencing" to connote. The Romans' primary encountering of what-is was an encountering of discrete components united via a network of agency. All their translating of Greek words and understanding of Greek thinking was carried out on that basis. In instance after instance this new perspective makes itself felt.

Physis, whose primal meaning as imperious, protean emerging and standing-in-self had already fallen from view even in Aristotle, now becomes *natura*. For Aristotle *physis* remains a ruling going-forth that governs directly in whatever, partaking of *hylē*, appears of itself without human intervention.[7] The noun *natura* is formed from the verb *nasci*, which means "to be born" or "to originate." Hence, in it a new emphasis on the dualness evident in the relationship of something to something else from which it derives, rather than on the character of eruptive power provisive of enduring,[8] stands as a starting point, although that emphasis is scarcely evident in *natura*'s philosophical use. There the word speaks of a continuity, a self-identity, conferred through the governance of an inherent configuring power. *Natura* is, Heidegger says, "what allows [something] to derive from itself."[9] Wherever *natura* is in play, it is

---

[6]Cf. *"Logos,"* VA 218ff, EGT 68ff. See Ch. I, p. 55.
[7]See Ch. X, p. 382.
[8]EM 11; IM 11f.
[9]"On the Being and Conception of *Physis* in Aristotle's *Physics*, B, 1," W 309, Ph 221. This is an understanding of nature that is close to that of Aristotle. Cf. Ch. X, pp. 376f & 382f.

assumed to rule disclosively. It specifically identifies and defines that in which it inheres. Accordingly, as is illustrated in Heidegger's portrayal of the Christian view of human nature, distortion in its self-manifestation renders "nature" suspect and less than itself, so that it becomes "that which ought not to be."[10] Here even the vestige of Being's happening as concealing that can be glimpsed in Aristotle's attributing of *sterēsis* to *physis* in its happening as *morphē* has fallen from view.[11]

　　Hypokeimenon, which for the Greeks speaks of lying before, i.e., of "that which presences of itself, and hence of presencing (*Anwesung*),[12] and *symbebēkos*, which speaks of an aspect of such presencing of a particular entity that is incidental to it, become respectively *substantia*, that which stands under, i.e., the core of an individual thing, and *accidens*, "that which always already stands along with" and "comes in along with" that core,[13] literally, that which falls to it. As *substantia*, *hypokeimenon* becomes the bearer of properties.[14]　　More directly translated—although again with a noun rather than an equivalent participle with its active verbal force—*hypokeimenon* becomes *subiectum*, that which is thrown or laid under, in a rendering that substitutes for its connotation of self-presenting an implication of positioning by another. In this form it persists, notably in grammatical usage, as the subject of predicates.[15] *Hypostasis*, standing-in-self, which Heidegger identifies as the complementary equivalent of *hypokeimenon* (lying before), is, like it, although now with very immediate linguistic correspondence, rendered with *substantia*.[16] The word *ousia*, which Plato and Aristotle drew into philosophical discourse and used to denote "beingness" (Being)—and which Heidegger takes to mean a manner or mode of presencing

---

[10]W 309; Ph 221.

[11]Cf. Ch X, p. 380.

[12]"On the Being and Conception of *Physis* in Aristotle's *Physics*, B, 1," W 331, Ph 239.

[13]FD 26; WT 34.

[14]FD 26; WT 34.

[15]"On the Being and Conception of *Physis* in Aristotle's Physics B, 1," W 330; Ph 238.

[16]W 331; Ph 239.

(Anwesung)—is equated with *substantia* and also with *essentia*,[17] that which, in a manner of thinking that accords ultimacy to unchangingness, constitutes the being as the "what" of anything.[18]

Morphē, which for Aristotle is a gathering-placing-into-appearance that is a mode of Being, i.e., of presencing *(ousia)*, and which is as such the fulfilling "movedness" that brings anything to appear as itself, now becomes *forma*. Aristotle's *hylē*, which, as possessed of fitness-for, is in each particular instance disposable by some disclosively governing *morphē* into a particular appearing, becomes *materia*. To Aristotle the relating of *morphē* and *hylē* is not that of constituents conceived as brought together in a way that achieves a conjoining of factors that, although radically disparate in character, complement one another as standing, if viewed purely structurally, in a relationship of parity. As presencing, Being, happening and disclosing itself for thinking as *morphē*, sets forth and governs the fittingly responsive *hylē*, letting it stand forth and present itself as whatever it is.[19] The indissoluble singleness happening in relation that characterizes the primal Twofold as Heidegger portrays it,[20] here, despite the inception of Being's self-withdrawal, remains fundamentally in play. In contrast to this, *forma* and *materia* denote commensurate components both of which, Heidegger says, belong within the sphere of what-is. *Materia* names a matter or stuff that, apart from any consideration of the particular entities constituted of it, can, as such, be disposed and brought to appearance as whatever is to be.[21] And *forma* (form) has become a "property *that is* (seiende *Eigenschaft*)." Inhering in matter, defining and disclosing anything material as whatever it is, this form, itself immaterial, is as much on-hand *(vorhanden)* as is matter itself.[22]

---

[17]W 331, Ph 239; EM 138, IM 151. Cf. Ch. X, pp. 371f & 375f.
[18]"The Question Concerning Technology," VA 37, QT 29.
[19]For Aristotle's understanding of *morphē* and *hylē*, see Ch. X, pp. 376-378.
[20]Cf. Ch. I, pp. 27-33.
[21]"On the Being and Conception of *Physis* in Aristotle's *Physics*, B, 1," W 334, Ph 248f.
[22]W 346; Ph 250. On on-handness, cf. Ch. II, pp. 65f & 88.

The transposing of the word *aition* into Latin evinces especially tellingly the new perspective in which what-is is being grasped. In it both the replacing of an interrelating grounded in unitariness by an interrelating of the inherently discrete and the defining of that new relating in terms of an agency brought to bear by one discrete participant upon another are saliently apparent. *Aition*, which Aristotle brings to prominence as that which in an integral fourfoldness of co-responsibility permits something to come forth into presencing, now becomes *causa*. *Causa*, which like *accidens*, derives from *cadere*, to fall, and which signifies that which falls out in a particular way, that which befalls, denoted originally something of concern.[23] In philosophical usage it came to mean that which, considered discretely, i.e., in itself and in a singleness of relation to that for which it is *causa*, so plays its part as to give rise to a result directly dependent upon it. *Causa* implies agency. It relates forward toward that which follows from it. Whereas *aition* bespeaks a gathering-together, in mutuality, to enable, *causa* bespeaks a bringing about achieved via a linear sequence of the inherently disjoined.[24]

The alteration here evident is paralleled, decisively, in the translating of *ergon*, work. For the Greeks, working as such is a happening of the gathering placing as which Being as presencing brings itself to pass, and the human working of a work is a readying, a completing (*ver-fertigen*) that permits whatever is gathering itself toward completion to arrive into presencing as itself.[25] Thus Aristotle, in whose thinking in this regard the original Greek experience lies close at hand, can use a noun formed from *ergon*, *energeia*, which is paralleled by *entelecheia*, to mean "self-holding-in-consummation (i.e., consummation of presencing)," therewith allowing *energeia* to speak of a self-maintaining in completeness of presencing and letting it name the presencing of anything that, self-maintainingly offering itself as itself,

---

[23]"The Question Concerning Technology," VA 16, QT 7. "The Thing," VA 173, PLT 175.
[24]On *aition*, cf. Ch. X, pp. 383f. On *causa*, cf. "Science and Reflection," VA 50, QT 161.
[25]Cf. Ch. IX, pp. 347 & 352f, and see "The Question concerning Technology," VA 21, QT 13.

fully is.[26] The Romans, however, "translate, i.e., think, *ergon* in terms
of *operatio* as *actio*, and they say, instead of *energeia, actus.*"[27] *Actio*
and *actus* derive from the verb *agere*, to drive, to do. "Work" and
"working" belong now within a decisively new realm of meaning, that of
a *doing* that is an operating upon.[28]

This transformation in the meaning of working and work was
momentous. Originally signifying as it does the *doing* that is the
fundamental mode of laying-forth-before as which presencing happens in
whatever stands forth enduringly of itself,[29] "working," however it may
be understood, shows us the crux of the interpretation of reality that is at
any juncture holding sway. Heidegger emphasizes this in pointing out that
the German word for reality is *Wirklichkeit*, from *wirken* (to work) and
the real is *das Wirkliche*.[30] In the newly constituted realm of Latinized
thinking, working becomes an effecting (*efficere*).[31]   Now, whatever
presents itself self-maintainingly as itself presents itself and is understood
as a result, a consequence (*Erfolg*).[32] The working that is now in view is
not an immediate bringing of something hither to stand forth self-
maintainingly from out of itself in disclosive presencing. Instead, it is an
acting, an operating, that lets anything stand forth only as its
consequence. And the manner of presencing belonging to a work so
worked, its manner of self-maintainingly standing forth as itself, has the
character now of an *act*. That is, that which is worked appears as the
result of an antecedent operation, and at the same time it appears as that
which must itself relate to whatever depends in turn upon it precisely as
something in respect to which consequences follow. Here the primal
ruling of a disposing wrought out through the gathering happening

---

[26]"Science and Reflection," VA 50, QT 160: *das Sich-in-der-Vollenduung (nämlich
des Anwesens)* -halten. For meanings of *energeia* and *entelecheia* earlier given by
Heidegger, see Ch. X, p. 378, and Ch. IX, pp. 352f.
[27]VA 50; QT 160.
[28]VA 51; QT 161.
[29]On working, see Ch. IX, pp. 347f & 352f.
[30]"Science and Reflection," VA 49, QT 160f.
[31]German *Bewirken*. VA 49; QT 160.
[32]VA 50f; QT 161f.

together of the intrinsically related, which is still visible in the *aitia* that Aristotle saw as everywhere providing a structuring in the happening of what-is, has wholly vanished. The relating that is now seen as determinative is sequential and takes place as the effecting of something by something else.

The manner of happening that now comes to reign supreme is that causality with which we ourselves are so familiar. In this altered context, Aristotle's four *aitia*, identified as *causa materialis, causa formalis, causa efficiens*, and *causa finalis*, were accorded a central place and seen as distinctive in meaning, but the notion of agency, overt in the *causa efficiens*, colored the understanding of all the others. Although *materia* as such was conceived as pure potentiality, the *causa materialis* could easily be seen as having an initiatory role.[33] The *causa formalis* was clearly effective in function, for the *forma* that constituted it was seen, in its relation to materia, as *act*. The *causa finalis* was understood as goal or purpose, but here too the thought of agency readily intervened. Thus, in the first of his proofs for the existence of God, Aquinas, arguing within a framework that assumes a final cause, illustrates the action of such a cause by alluding to the hand's wielding of a staff,[34] and Duns Scotus finds it necessary to defend the reality of the distinction between the final cause and the efficient cause.[35]

With the rise and spread of Christianity, new, very significant elements entered to augment those provided to philosophical thinking from its Greek heritage. Indeed, philosophy could be said to be, in the familiar phrase, "the handmaiden of theology." Heidegger, however, would state the relationship in the opposite way, for to him it was philosophy that was carrying itself out via the medium of Christian theology.[36]

---

[33]See, e.g., Boethius, *On the Consolation of Philosophy*, Book V, in John F. Wippel and Allan B. Wolter, O.F.M., ed., *Medieval Philosophy from St. Augustine to Nicholas of Cusa*, New York, MacMillan, 1969, pp. 84f.

[34]St. Thomas Aquinas, *The Summa Theologica*, in Anton J. Pegis, ed., *Basic Writings of Saint Thomas Aquinas*, New York, Random House, 1945, p. 22.

[35]John Duns Scotus, *The Oxford Commentary on the Four Books of the Sentences*, in Arthur Hyman and James J. Walsh, ed., *Philosophy in the Middle Ages*, Indianapolis, Hackett, 1983, p. 642.

[36]Cf. "The Onto-Theo-Logical Constitution of Metaphysics, ID 122f, 55f.

For Aristotle, God, thinking only his own thinking and utterly removed from all the reality beneath him, maintained himself as perfect completion.[37] By his very perfection, while himself wholly unconcerned, he drew all toward himself. In this god, *aition* happening as completion found its highest exemplar.[38] The God of the Judeo-Christian tradition, on the other hand, is an initiating creator. This character accorded well with the manner of thinking that was now underway. And God became, "for theology but not for faith," Heidegger says, "the highest cause."[39] Thus causality, the structuring manner of happening that had come to be seen by the thinking of the Latinized West as the interrelating that characterized what-is throughout all its diverse elements, was taken to have its inception and, as it were, its sanction in the divine creativity itself.

In accordance with this thinking, the particular entities of what-is, which were ranked in a hierarchical structure according as they were nearer to God or farther from him, i.e., according as they partook less or more in an admixture of materiality, existed in a tiered configuration of Being that, as created, was by definition a causal structure. Now each entity owed its reality, the standing that it had in and of itself, not to the uniqueness of the Happening that set it forth and clearly or deceptively manifested itself via it, but to its place in the hierarchy of created beings.

Latin provided to philosophical thinking a specific word *res* for the designating of an entity as a thing. In *res*, the meaning that Heidegger finds for thing (*Ding*), understood as a gathering center of constitutive presencing,[40] is not in view. Like *causa*, *res* meant originally something that was of concern. Roman thinking weakened its relational connotation,

---

[37]See Ch. X, p. 382.

[38]We should note that in Aristotle's God we may surely glimpse, if in the diminished guise of the purely distant, the divine as such, as Heidegger understands it, i.e., as that which, exercising a beneficent power, embodies itself in particular "divine ones" that beckon toward it as, possessed of wholeness, it holds dominion as that toward which mortals make their way. Cf. Ch. V, pp. 176 & 186. For Heidegger's portrayal of the divine as at once distant and near—in this case as instanced in "the god," discloser *par excellence* of the divine—see Ch. V, pp. 206-208.

[39]"Science and Reflection," VA 50, QT 161.

[40]Cf. Ch. V, pp. 173-187.

characteristically taking it to mean that which stands here (*im Sinne des Herstandes*).[41] *Res*, so understood, became a term to designate any entity. Eventually, its German counterpart *Ding* (*Dinc*) even became applicable to God himself.[42] In this context, the "thingness" of a thing became for thinking the "substance" (*substantia*), i.e., the beingness (*ousia*), that stands under, that is, that to which accidental qualities and characteristics fall.[43] Such properties are added to it. Indeed, the thing's existence could itself be thought of as such an addition. Here agency, and not a spontaneous, intrinsic self-showing belonging to the thing itself, is again decisive. And there is no hint of such powerful, disclosive happening as Heidegger takes to be the thinging of a thing.

Thus did the withdrawal of Being that had begun, even among the Greeks, to cloak its original manner of happening with a multitudinous panorama of the definable and the graspable, illumined by likeness and shot through with self-sufficiency,[44] intensify itself in the succeeding epoch. Now, in the arena where human apprehending and what-is encounter one another in a meeting grounded and made manifest through philosophical thinking,[45] in a transforming upsurge of the isolative self-assertiveness that had already begun so significantly to hold sway,[46] vestiges of primal unities are severed, diverseness of relating is eroded toward sameness, a cooperative enabling is superseded by a doing that, in the last analysis, compels the appearing of that which follows from it. Causality is now the decisive guise in which Being rules. Happening by way of the Latin language and governing the experience and thinking of the age thereby, the Being of what-is conceals yet more deeply the primal manner of its holding sway. The primordial divided singleness of the unfolding Twofold[47] and the primordial keeping-in-hand of man by the

---

[41] "The Thing," VA 174, PLT 176.
[42] VA 175; PLT 176.
[43] Cf. "The Origin of the Work of Art," UK 15f, PLT 23f, and see this chapter, pp. 407f.
[44] Cf. Ch. X, p. 398.
[45] Cf. Ch. VIII, pp. 293-295.
[46] Cf. Ch. IX, pp. 360-362, and n. 128, and Ch. X, pp. 392 & 398-400, and n. 124.
[47] Cf. Ch. I, pp. 27-29.

Being that hands him over to pursue his existing in opening for Being an arena for its happening as unconcealing[48] remain in play, but as newly masked. What-is, every sort of particular within it, even to God himself, and man appear as possessed of an innate autonomy and causally determinative power. And Being, the immediately initiatory Happening that all clear duality and simple agency can only belie, offers itself to view fundamentally after the manner of the causality that permits only these defining traits to come to light. Happening yet more urgently as concealing, Being unconceals whatever is in increased sharpness of detail and simpleness of relation, thereby veiling beneath an undiscerned opacity of disclosure the impelling, fragmenting disposing as which it itself now holds unsuspected sway.

2

The emergence of causality as the decisive mode of happening of what-is as apprehended and made manifest through Latin-governed thinking brings to a newly spare and narrowed manifestation the tendency first determinative in the thinking of Plato and Aristotle, the tendency, namely, to ask when encountering what-is: How are these phenomena to be understood, what determines them to be what and as they are? In pursuance of that tendency, Plato sought to discover the genuinely real, and Aristotle, presupposing that concern, inquired beyond it regarding the interrelationships and processes that structured reality and bound its components to one another. Both undertook to discover and disclose the bases that underlay the phenomena, themselves so frequently equivocal, that confronted them and their fellows, and in so doing they brought to bear a thinking that could, they were sure, be implicitly relied upon to present explanations of that which was itself intrinsically explicable in such terms.[49] When in a time centuries removed from theirs causality came to rule as the determinative mode of such explaining, the inquiry concerning the grounding that could render immediately encountered

---

[48]Cf. Ch. II, pp. 108f, and n. 190.
[49]Cf. Ch. X, pp. 370-372 & 389-391.

phenomena comprehensible was constrained by a simplified and highly focused outlook that effectively blocked from view the Greek origins that, in their contrasting amplitude and complexity, underlay it. Nevertheless, the inquiry was in fundamental character the same.

The impulse to discover grounds that came to the fore at the momentous juncture when the relating of Being and man was transformed through the "stepping-apart" that brought into play a divisiveness in the fundamental relation of the one to the other[50] manifested that sundering at its deepest level and remained primary as, inexorably, the latter carried itself out. In the disclosive executing of that impulse, the original happening of Being as *logos*[51] both wrought itself out determinatively in the interplay of Being and man and effectively concealed itself from view.

In the original Greek relating of the Being of what-is and man, Being's happening as *logos* is precisely its happening as presencing (*Anwesen*) in the sense of *physis*, letting-arise enduringly from out of itself. This is the selective letting-lie-forth-before (*Vorliegen Lassen*) that *logos* most immediately names.[52] But precisely in this character, *logos*

---

[50]Cf. Ch. IX, p. 364.
[51]Cf. Ch. IX, p. 338.
[52]SG 180. Cf. Ch. III, p. 124. There is to date no complete English translation of *Der Satz vom Grund* (*The Principle of Ground*). Soon after the publication of *Being and Time* Heidegger inquired into the meaning of ground and grounding, in *Vom Wesen des Grundes*, literally, *The Essence of Ground*. There, while setting his entire discussion in a context defined by his portrayal of the "Ontological Difference" between Being and what-is, he elucidated the question of grounding primarily from the point of view of its transpiring via human existing (*Dasein*) as transcending Being-in-the-world that happens as freedom. (See especially WG 26 & 100-130; ER 27 & 101-131.) In *Der Satz vom Grund*, written years later, he approached the same question by taking as his point of departure the happening of the Being of what-is, understood as the presencing of what presences, which *logos* names; and the human role, which is extensively considered, appears without reference to man's existing as openness-for-Being (*Dasein*). This difference in perspective is very similar to that obtaining between *Being and Time* and "Time and Being," works that, again, date from the beginning and near the end of Heidegger's work. Both pairs of writings illustrate especially clearly the change in approach that from out of a thinking undergirded by a singleness of perspective, carried Heidegger away from concentration on what we might call the inward manner of happening of human existing and toward inquiry centered directly upon Being as the Being of what-is, the Twofold for which man, in corresponding to its ruling, provides openness. (Cf. Authors' Prologue, pp. 9f.)

bespeaks the happening of Being as that which, thus uniquely initiating, brings itself to pass antecedently to anything that, although happening from out of itself, yet depends on that initiating. Hence, Heidegger says, *logos*, in naming Being as presencing, simultaneously manifests it "as that which lies before *(das Vorliegende)*," i.e., "as the matter to be considered *(die Vorlage)*" in respect to whatever lies in dependent relation to it. Seen in this second regard, *logos* names Being as "the ground *(den Grund)*." Thus "*logos* is most particularly presencing and ground."[53]

Concomitantly, in Being's disclosive sundering of itself so as to happen as the disposing that informs man's apprehending in its answering to the original ruling of Being, *logos* names also that disposing through which the presencing of what presences, the unfolding of the single Twofold as *physis*, is brought to a stand in the particularity of enduring disclosure in some particular on-happening present. *Logos* is the name for human speaking, the pivotal avenue for the bringing to appearance of what-is in its Being;[54] *logos* is the governing mode of happening in *technē* as the many-faceted skillful knowing that rightly discerns requirements, possibilities, and means of proceeding and, pitting itself against confusion and concealing, permits presencing to come disclosingly, foundingly, into play; *logos* answers compliantly to *logos* when Greek man, in his peculiar openness-for-Being, is permitted to respond in unique immediacy to Being's pervasively directing sway.[55]

When, with the self-withdrawal of Being, Greek man stepped forward in unwonted self-assertion vis a vis Being's ruling happening, the delicately poised complexity of relation carried in the word *logos* was disrupted and recast. In the utterance of Heraclitus the happening of Being as the spontaneously initiating arising that grounds and, by way of man's accordant living-out of openness, disposes whatever is had come

---

[53] SG 179. In this characterization of Greek *logos*, the reciprocal relating that Heidegger sees as definitive of the happening of the Being of what-is is clearly evident. Cf. Ch. I, pp. 30-32.
[54] Cf. Ch. III, pp. 124f, and Ch. X, pp. 387f.
[55] Cf. Ch. IX, pp. 338-340, 342f, & 348.

specifically if fleetingly into view.[56]   Now it was concealed. Being had withdrawn from a direct bringing of itself to bear on the apprehending that as philosophical thinking was in a determinative way to disclose its governance.   That thinking, in however lofty a manner, could concern itself only with what-is. *Grounding* obtruded itself as the province of the permanent.[57] But it could be discerned only as pertaining to the realm of what-is.   Whether discovered in Platonic Ideas, in Aristotelian inhering, determinative "forms," in *aitia*, or in the Unmoved Mover, the grounds of immediately perceived phenomena were found always within that realm of the particular.[58]   More than this, human thinking, presupposing its adequacy to discern the grounds that it sought, unwittingly had begun to put itself forward as the arbiter of Being, until in Aristotle's formulating of *logos* as language into logic, it took directly to itself the right of definitively stating, and hence of grounding in its utterance, the structures of reality that it discerned.[59]

The Romans translated *logos* with *ratio*.   With this step, the connotations of a letting-lie-before that was intrinsically a selecting-gathering-forth into manifestation, which were never lost in *logos*, were no longer evident. In *ratio*, what we may perhaps call the mutual linear interrelating of something with something else, which was implicit in the second significance of *logos*—i.e., as ground—in its naming of Being and which had been drawn forward with the inception of metaphysical thinking, comes decisively to the fore.[60]

Ratio is derived from the verb *reor*, to deem or suppose. Deeming involves an imputing (*Unterstellen*), a taking of something for something. Such imputing brings that toward which it is directed into a subordinate relation to itself and to whatever it imputes, for "that to which something is imputed is," Heidegger says, "prepared and shaped (*zugerichtet*) with

---

[56]SG 180.
[57]On the meaning of permanent, see Ch. X, p. 370.
[58]Cf. Ch. X, pp. 339-401.
[59]Cf. Ch. X, pp. 389-391.
[60]Cf. SG 179.

respect to that which is imputed to it."[61]   Such regulative arranging
(*Richten*) of something in accordance with or pursuant to (*nach*)
something else is *reckoning (Rechnen)*.   Reckoning is, with respect to
anything that may be in view, at once a "reckoning with something," a
keeping of it in sight that involves a "self-regulating (*Sich Richten*)" in
accordance with it, and a reckoning on that something, a taking account
of it, that expectantly awaits it and, so awaiting, serves rightly to prepare
and order it as something of a sort whereon something is to be built.[62]

    Ratio, carrying the dual implication resident in the adjudging that,
as   a   reckoning,   definingly   and   determiningly   imputes,   bears
correspondingly a double meaning.   On the one hand, *ratio* signifies that
which, as what is imputed to something, is, antecedently, itself precisely
that which underlies the manner of self-presentation found to characterize
the entity in question.   The fact that "it stands with something as it does
stand" depends (*liegt . . . an*) on that defining imputation.   What is thus
reckoned out (*das Errechnete*) in advance and imputed is, as "that on
which it depends [that . . .]," precisely "that which lies before (*das
Vorliegende*)."[63]   As the already reckoned (*das Gerechnete*), i.e., the
determiningly   imputed,   that   belongs   intrinsically   to   reckoning
(*Rechnung*), as the latter accomplishes itself, *ratio* is "foundation, base,
i.e., ground."[64]   On the other hand, *ratio* signifies as well the reckoning
that, relying on its own accounting and attributing, judges according to
the underlying assumption, regarding whatever it has in view, that it has
reckoned out for itself and imputed as fundamental to that with which it is
concerned.   In its imputing (*Unterstellen*), that reckoning sets that with
which it has to do forth (*vor-stellt*) before itself.   This bringing before
itself at some particular juncture in time "of something as something"
takes place as an encountering that is decisively shaped and accorded
detail by the reckoning itself.   In its presupposing imputing, that
reckoning "takes hold in advance of something that lies before it (*ein

---

[61]SG 167.
[62]SG 167f. On reckoning, see also "Science and Reflection," VA 58f, QT 170.
[63]SG 174.
[64] *. . . die Basis, der Boden, d.h., der Grund.* SG 174.

Vorliegendes vor-nimmt)." It takes hold, that is, of something that it has, in its imputing, itself grounded as that which it takes it to be.[65] Here, only in the context of such predetermining *taking-hold* does constitutive, human apprehending (*Vernehmen*) happen; for apprehending is here circumscribed by the reckoning that is underway. That reckoning, as the setting-forth-before that brings determiningly before itself whatever it has in view, "apprehends (*vernimmt*) how it is ordered and disposed (*bestellt*) with that which is reckoned with and reckoned upon." As the reckoning that thus apprehends, *ratio* is the apprehending reason (*Vernunft*) that discovers and answers to the determining grounding that reckoning as such carries out. Thus "*ratio* is, as reckoning, both ground and apprehending reason."[66]

In thus bringing variously to embodiment in itself an adjudging, *ratio*, in translating *logos* for Roman thinking, carries over into decisive dominance an aspect of meaning that is only derivative in *logos* itself. The verb *legein*, in speaking primarily of a selective gathering-together that is a laying of something toward something else (*Legen zu*), can likewise speak, derivatively, of such laying of one thing toward another, not as a genuine selecting that highlights the uniqueness of each through the disclosing of all the gathered in their mutual relation, but as a laying-together of elements wherein "one is dependent upon and conforms to (*sich richtet . . . nach*) the other." *Logos* can, thus, derivatively, denote reckoning, i.e., "to rectify something with respect to something else (*richten zu*)," to bring something, that is, into dependent conformity toward another. It can mean relation (*Beziehung*) in the sense of Latin *relatio*, literally, a carrying back in the sense of a reconciliatory

---

[65]Within this structure of *ratio* we can discern the basic structure that Heidegger finds for man's carrying out of his existing by way of a seeing of that which comes to meet, that takes place always on the basis of an orientation, a seeing, already given. Cf. Ch. II, pp. 73f, and Ch. III, pp. 113f.

[66]SG 174. On the structural similarity between such representing, reckoning thinking and the manner of transpiring of human existing as such, which is itself a manifesting of the structure intrinsic to the happening of the Being of what-is, cf. Ch. VIII, p. 300, n. 24.

juxtaposing.[67]   This is that relating, the arranging or setting right
(*Richten*) that takes something as congruent with something already
assessed, which is the *reckoning* that is conceptualized and made
determinative in *ratio*.[68]

In *ratio* the predominance of human thinking that began in Plato
and reached early even to the confident discovering and ascribing of
Being is brought to a new, more intense pitch. For it comes to focused
statement in a word that, with all its implications, however undiscerned
they might be, assumes a central role in Western thinking, at once
unobtrusively manifesting the latter's character and shaping that character
in accordance with itself. In *ratio* the original interplay of Being and man
that *logos* in its most intrinsic meaning bespeaks has been distorted into an
interplay between apprehending reason and the determining ground that it
reckons out for itself and unhesitatingly ascribes as fundamentally
explanatory to that which it thus brings into its field of view. The self-
subordinating, self-accomplishing answering of *logos* as man's
apprehending disposing to *logos* as Being's sovereign disposing that, as
presencing, primordially grounds whatever is and brings it to self-
manifestation by way of the attentive apprehending that man brings to
bear has now been transformed into a meager inversion of itself. Now it
is man's apprehending, as *ratio*, to which all disposing implicitly falls.
That apprehending reckons out and ascribes the ground that, as *ratio*,
underlies and determines a correspondence of one thing with another that
is, in the last analysis, but a subtly contrived accommodation wrought out
by human apprehending among the particulars of what-is with which
alone the latter has overtly to do. In the putting forward of *ratio* as the
definitive successor to *logos*, Being as immediately ruling presencing has
been directly obscured, and, concomitantly, Being as imperiously
initiating ground has been wholly shut from view.

We should here note that, with the change in the manner of

---

[67]Our frequent use of the verb relate intends a meaning other than this. On this point,
see, *in extenso*, Ch. IV, pp. 145, n. 10, and the briefer comment in n. 1, above, which
has been repeated as needed, chapter by chapter.
[68]SG 178f.

happening of the Being of what-is that is here in question, the spontaneous setting hither and forth of which Greek *thesis* spoke via a particular emphasis—as a word that gave utterance to Being's happening as *logos*[69]—has now been superseded. Greek *thesis* happened as a placing that, bringing into delineateness, in every instance of its ruling provided something that was possessed in its peculiar context of time and circumstance of an irreducible uniqueness of presencing. Now, in the manner of manifesting that *ratio* bespeaks, the counterpart of *thesis* has become an evaluative setting-in-place for the purpose of adjudging, which, as a kind of taking of command, banishes both originative spontaneity and the self-imposition in uniqueness that is ultimately unforeseeable. Heidegger's use of verbs compounded of *stellen*—*unterstellen* (to impute), *vorstellen* (to set forth before), *bestellen* (to set in order or arrange)—in his explication of *ratio*[70] points, if without comment, to this very significant change in the character of the setting-forth-into-place that has become newly decisive and should surely remind us of the historically impending rule of Enframing (*Ge-stell*), in accordance with which a use-oriented setting-in-place is so ubiquitously in play.[71]

The realm of meaning opened up for Latin thinking in and through *ratio* as the appointed bearer of the meaning present in *logos* is of decisive ultimacy in the epoch wherein that thinking carries itself out. The character and primacy of the causality that comes to dominance as the central tenet of the metaphysical thinking of the age spring directly from it.[72] The effecting that, as a manner of working, brings forth what is effected as what is consequent, by way of a doing understood as a self-sufficient providing, displays in a particular guise the prejudging subordinating that *ratio* bespeaks. For in the particular kind of relating

---

[69]Cf. Ch. IX, pp. 347f.

[70]For a similar repeated use of verbs formed on *stellen*, see "Science and Reflection," VA 56, 61f, & 67; QT 167, 173f, & 178. Cf. Ch. VII, pp. 283f.

[71]Cf. Ch. VI, pp. 258f & 247-249. For the use of *stellen* in a Greek context, see Ch. IX, pp. 347f.

[72]Cf. SG 167.

that obtains between cause and effect, the determining basis is precisely the unquestioned presupposing of commensurability between something and something else that is viewed in relation to it, which is fundamental for the reckoning that *ratio* names. The apprehending that reckons with and reckons on that with which it has to do tends everywhere to discern effecting causality in play, thereby catching sight, unknowingly, of the changed manner of happening now governing it itself, according to which it has become the overt arbiter of disclosure as an arena wherein an immediate impingement in intrinsic unity has yielded to manipulative ordering in extrinsic interrelation.

In disguising itself in its happening as causality, Being, in its deepening self-withdrawal, as it were focuses its changed manner of happening in a particular manifestation that, by its very centrality and prominence for thinking, obscures the sharp shift in the interplay of Being and man that has taken place. The vestiges of the original interplay between them that clung always within the Greek words determining Greek thinking have been withdrawn. Now man and what-is meet in an encountering that man's reshaping disposing structures and carries out. All that is—the creator God and all that he creates—presents itself and comes to be seen in the light of causality. Being is conceived as a qualification to be predicated of particulars viewed according to the structuring that that causality fundamentally defines. And, more fundamentally still, in the translating and replacing of *logos* with *ratio*, both grounding via commensurability and the reckoning account-giving apprehending that carries it out are named in one and put unobtrusively underway. Under the dominion of a self-sending-forth of Being that is a self-withholding as which Being reduces the gathered particulars of what-is to the conceivable and reckonable that man by his power of presupposing and imputing definingly sets forth and orders for himself, man pursues his thinking confident that he is glimpsing basic relationships and ways of happening pertaining to that with which he has to do. Heedless, he thinks and propounds, in self-confident oblivion toward Being, while Being at once governs his manner of thinking and the appearing of what-is that it vouchsafes *and* leaves him alone in encounter with what-is, as though Being itself in its imperious bringing to disclosure were nowhere in play.[73]

3

Throughout the centuries of Western history that culminated in the high Middle Ages, the philosophical thinking that Heidegger sees as determinative and as implicitly indicative of the manner of relating obtaining between Being and man was for those who pursued it secondary to the Christian revelation based in authoritative scripture and tradition. It was the latter that was looked to as determinative. The tenets, concepts, and terminology of philosophy were utilized with the intention of explicating the faith and were considered by the thinkers who employed them to function in its service. With the beginning of the modern age, when the authority of church and scripture so long assumed and relied upon was radically questioned and superseded by a new reliance on the authority and security vouchsafed in man's representing thinking itself, the metaphysical traits long resident in Christian thinking in the West sprang into the foreground.[74] At this point the changed relating of Being and man that underlay the Latin-Christian epochs changed yet again, intensifying and bringing to pervasive, admitted dominance the manner of man's relating to reality that had been set underway with the translating of Greek into Latin so many centuries before.

With Descartes' pivotal enunciation of his *ego cogito ergo sum*, the configuring inherently resident in the word *ratio*, with respect to the interrelating of thinking and that with which it has to do, comes to light in a direct statement. Human subjectivity, in thinking, simultaneously presents to itself itself as aware and whatever it marshals within its awareness. The structure of the thinking that as *ratio*, i.e., as reckoning, brings before itself that which it encounters from out of an assumed ability to adjudge comes here clearly into view.[75] In this context of self-

---

[73]Cf. Ch. I, pp. 44f.
[74]Cf. Ch. VIII, pp. 296-298.
[75]Cf. Ch. VIII, pp. 297-301.

sufficient self-dependence, Descartes presupposes, as his philosophical forebears had presupposed before him, the capacity of human thinking as *ratio* to discover through the tracing of logical connections the way in which reality is to be rightly understood.[76]   Now no reliance on the authority of revelation interposes to make subordinate the role of *ratio* as the reason that discerns and exhibits to itself connections that may confidently be taken to characterize reality precisely because they accord with the structures of thought that reason has acknowledged and claimed as normative for itself.

The reliance on logic as definitive of that which can or cannot be, which first came to direct statement in the work of Aristotle,[77] here comes unquestioningly into play in the strikingly divergent context in which man as thinking subject has become the only recognized foundation or locus of the acknowledging, evaluating, and establishing of anything as something that is.   For Descartes, logic serves as an instrument for ordering into clear and distinct relational sequences the conceptualized contents of a conscious awareness.   For Aristotle, the Being (*ousia*) of anything is found in the *morphē* via which and as which it presences, and whenever human conscious awareness is in question the selfsame *morphē* informs the encountered entity belonging to the realm beyond man, man's organs of sense, and his intellect at every level of its functioning.[78] Therefore logic, for all its formal embodying of the new human temerity that dared decide concerning Being, is taken, in its working with concepts, to be anchored immediately in the external reality in which man participates and to which he belongs.   Descartes, on the other hand, discovers his Being and the Being of whatever he conceives via a representing thinking that relies first of all only on itself.[79] He endeavors to move through argument to the incontrovertible establishment of God's existence and hence to the guaranteeing of his own existence and that of the world, but the fulcrum of his certainty remains in his own original act

[76]See Descartes' *Meditation III*.
[77]Cf. Ch. X, pp. 389-391.
[78]Cf. Ch. X, pp. 381.
[79]Cf. Ch. VIII, pp. 298f.

of thinking. Through the exercising of logic, he establishes, always only from within the sphere of that thinking, the indubitable dependability of the foundation in thought that he has already with assurance made primary to his entire enterprise.

With Descartes, *ratio* as reckoning thinking tacitly assumes control over everything with which and on which it reckons. Subsequently, in the work of Leibniz, the manner of happening intrinsic to all such thinking and implicit in the word *ratio* itself comes at last to clear manifestation, in his *principium rationis*.[80] Here the *ratio* in question is in the first instance *ratio* as ground. Leibniz's principle of ground states: "Nothing is without ground (*Nihil est sine ratione*)."[81] Positively put: "every being has a ground (*omne ens habet rationem*)."[82] Leibniz also phrases his principle, "Nothing happens without cause (*Nihil fit sine causa*),"[83] where "cause" has the broad meaning, reaching even to purely formal logical relations, of an underlying condition that decisively determines that which depends upon it. Everything, according to the principle, is underlain by some such cause (or causes) that constitutes the ground, the basis, of its being what and as it is.[84]

Here the ground alluded to is assumed to be sufficient to account for the existence of whatever depends upon it. More than this, it is assumed to be accessible to the reason that, as *ratio*, discovers grounds. Accordingly, Leibniz, in a more rigorous delineation of his principle, designates it as the principle concerned with "the rendering of the sufficient ground (*principia reddendae rationis sufficientis*)," and, so considered, the principle states "that for every truth" (where "truth" means a true statement [*Satz*]) "the ground can be rendered (*quod omnia veritatis reddi ratio potest*)".[85]

---

[80]SG 31f.
[81]SG 16.
[82]SG 16.
[83]SG 52.
[84]SG 54f.
[85]SG 44. In considering as the locus of grounding the freedom as which human existing happens (see this chapter, p. 415, n. 52), Heidegger finds the validity of Leibniz's principle of sufficient reason to lie in the character of transcendence as that

The immediate coincidence that obtains for Leibniz between the account of grounding connections rendered by thinking and the connections as they pertain to the entities thought about is shown by his "common" formulation of his principle, "Nothing happens without cause," or, as Heidegger paraphrases it: "Nothing becomes something that *is* without cause."[86] For that principle is intended, in its proclaiming of a dependence that is causal in character, to state succinctly both the way in which logically governed reason must view anything whatever in relation to that on which the latter depends *and* the dependence itself that is in question.

At this juncture, causality, so long fundamentally determinative of the manner of happening of what-is, becomes decisive in a new context and in a new way. Leibniz's thought is of efficient causality. There is, he says, in the very nature of things, a "ground . . . because of which something exists rather than nothing." This ground, which manifests itself in an inclination toward existence, must reside either in the real entity in question or in that which, as antecedent to it, adequately determines it, its cause. In order that the causal configuration here envisaged may possess the solidity belonging to it by definition, a being that is a first cause must necessarily exist. That ultimate, i.e., highest "ground of things" is ordinarily called God.[87] Within the confines of the structure that is thus delineated, Leibniz undertakes, through the rigorous application of logical thinking, confidently to detail and account for the character of all reality in what, by the light of his logic, he clearly sees to be "the best of all possible worlds."[88]

Here the correspondence between what is logically discovered and the reality to which the conceptualizing of logic refers, which was presupposed by Descartes, is by Leibniz directly discerned and accorded unquestioned validity. Here Hegel's subsequent dictum, "the rational is

---

freedom. WG 122; ER 123.
[86] . . . *nichts wird zu etwas Seiendem ohne Ursache.* SG 52.
[87]SG 53.
[88]Cf. Leibniz's "The Twenty-Four Statements," in N II (German) 454ff, EP 49ff.

the real and the real is the rational,"[89] has been anticipated in a succinct epitomization in Leibniz's principle of ground and in the thinking that he founds upon it so as to disclose definitively the character and structure of whatever is.

This causal structure with God at its apex, which defines reality, at once accords with and itself substantiates the principle of ground. But beyond this seeming circularity, it is in fact the principle itself that is decisive. That principle requires a rendering of a sufficient ground in order that anything may be seen to be, and that rendering transpires in a context in which only causality, rigorously established in terms of cause and consequence, can adequately support it and allow it to take place. Hence the causality ascribed to the interrelating of real entities is itself fundamental for the carrying out of the thinking that concerns itself with detecting it and following it out.[90]

Leibniz's principle, rigorously stated, requires a rendering, a providing, of the ground for anything whatever that is to be acknowledged as something that is. This "rendering" is of a peculiar sort. It is no simple providing. The verb that Leibniz uses, Heidegger points out, is *reddere*, to give back, to restore. And the rendering in question is, he says, a delivering (*Zustellen*) of the *ground* back to the thinking subject that thinks in quest of it. Leibniz has in view thinking as *representatio*, i.e., as the thinking subject's re-presenting of what is thought to itself in a reflexive inner movement whereby that which is encountered is grasped and set before it as object, as that which stands fixedly over against (*Gegenstand*).[91] It is precisely such representing thinking (*Vorstellen*) that is called upon to provide to itself sufficient grounds for whatever can be said to be. The principle of ground, when so presented as to require the rendering of the sufficient ground, asserts that for anything that is,

---

[89]Georg Wilhelm Friedrich Hegel, *Philosophy of Right*, translated with notes by T. M. Knox, Oxford, Clarendon Press, 1942, p. 10.
[90]SG 55f.
[91]SG 45. In his discussion of this aspect of Leibniz's meaning, Heidegger uses the word *Gegenstand* (object), that which stands over against, although the word *Gegenstand* was first coined by Kant.

there is a ground adequate to account for it and holds that every such ground can be discovered by human thinking and made evident by the latter to itself.[92]

Leibniz's concern in this presentation of his principle is with the application of the principle with respect to assertions whose truth can be established, and can be established only, through the following out of the grounding sequences of statements that underlie them. For him, his principle of the sufficient ground that can and must be given ranks with the principle of contradiction in its power to identify and demarcate statements that are expressive of what is. The primary role of his newly propounded principle lies in the spheres of philosophic and scientific knowing, where the proving toward which the principle rigorously directs the inquiring mind establishes definitive knowledge concerning the reality that is under consideration.[93]

In providing to itself the grounds of what it knows, via demonstrably true statements, such knowledge grounds itself. It establishes itself as proven knowledge. But the establishing that takes place, when the requirement is met that the grounds that underlie the statements constituting such knowledge be delivered to the knowing, does not, Heidegger avers, pertain only to the knowing in question. It pertains also to that with respect to which knowledge is being gained. The proven knowledge with which Leibniz is immediately concerned is a manifestation of representing thinking (*Vorstellen*). And the delivering (*Zustellen*) of grounds to itself that it undertakes involves the setting forth before itself of whatever is known, *as* object. For representing thinking, it is only as object—i.e., as something standing stably over against,

---

[92]In the rendering of the ground via representing thinking, which Leibniz's principle requires, we see again the fundamental structure, that of a going-forth that is simultaneously a gathering-toward, that for Heidegger characterizes human existing itself and, within and beyond that instancing, characterizes intrinsically the provisive Happening that is named as Being. With respect to the structure inherent in the thinking that is here presupposed, cf. Ch. VIII, p. 300, n. 24. On the ultimate structure in question, cf. Ch. I, p. 33, n. 36.

[93]SG 45. The very close parallel between the discussion that immediately ensues here and Heidegger's portrayal of the manner in which modern science carries itself out should be noted. Cf. Ch. VII, pp. 270-275.

presented to thinking by itself—that anything *is*. Representing thinking that sets the object forth before and the object set forth before constitute an indissoluble unity in polarity. Hence the establishing of the one establishes equally and concomitantly the other. Leibniz's principle that requires the delivering of grounds for knowing involves equally the grounding of knowing and of what is known.[94]

The principle of ground, which in this strict delineation might seem to be concerned only with the establishment of knowledge, has to do, then, instead at once both with knowing and with whatever is as what is known. Here the identity of this characterization of the principle of ground with the principle's "common" formulation, "Nothing happens without cause," can be seen. The latter formulation alludes to the objects of knowledge, whose grounds are to be delivered to knowing thinking. The allusion to causality, Heidegger says, itself manifests the interrelation of subject and object that is fundamental for the thinking to which the principle is addressed. The principle requires that knowledge be thoroughly grounded through the delivery of the grounds for whatever is known, i.e., for the objects that representing thinking sets before itself. The most thorough-going grounding possible is that characterized in terms of efficient causality. This it is that can establish the object of knowledge, bringing it to a firm and fully grounded stand. Therefore it is this also that can thoroughly establish the knowledge that depends on such grounded stability. Only when considered in the light of such causality can both objects known and the knowing of them be made secure.[95]

Leibniz's principle of the sufficient ground that must be rendered to thinking voices a demand for demonstrable knowledge. In it, beyond the securing of the object and of the knowing of the object, the securing of the knowing subject is at stake also. In validating the assertion which a configuration of grounds establishes, the thinking subject offers to itself that which, through such validation, establishes the subject in the rightness of its understanding. The latter makes itself secure as the

---

[94]SG 46f.
[95]SG 47.

representing subject by confirming the adequacy and effectiveness of its approach and the competence of the perspective that it has presupposed for the discovering and definitive disclosing of that which can now confidently be said to be.[96]

Leibniz's assertion that for every truth, i.e., every true statement, the ground can be rendered, can now be seen to have ramifications that extend its domain to everything whatever that is. The thinking-knowing subject that presents to itself its knowledge in grounded, demonstrable statements establishes itself in establishing its knowledge. More than this, its assertions, when true, indubitably establish that about which they are made—which is conceived and represented as object—as something that *is*. Thus statement, governed by the principle of ground, possesses the capacity to confer or withhold Being, and Heidegger can write that here "something 'is' only, i.e., is identified as what is, only when it is asserted in a statement that is adequate to the principle of ground as the principle of establishing and accounting for (*Begründung*)".[97] Even as, centuries before, Aristotle in his propounding of the principle of contradiction had manifested and given sanction to the dominion of human thinking carried out as logic, with respect to the acknowledgement of the happening of Being, so Leibniz, in the changed modern context, in enunciating the principle of ground arrogates to and claims for logic as the vehicle of thought the thorough-going capacity to accord and establish Being.

Leibniz's principle of ground, which overtly extends its sway via representing thinking to everything that can claim to be, sets forth the reckoning thinking originally implied in *ratio* as solely and unequivocally adequate to the determination of Being. That which is, is that which as stable object stands over against thinking. It is there made fast through being grounded via the efficient causality that can alone securely establish the thinking that pursues and displays the grounding causal structure in unquestioned reliance on its adequacy as a perspective and means through which to discover and establish validity. Representing thinking reckons

---

[96]Cf. Ch. VIII, pp. 301f.
[97]SG 47.

with whatever it encounters, by reckoning on it in accordance with the manner of viewing which it, that thinking, has already assumed and which it unhesitatingly employs for the adjudging and disposing of the encountered something with which it has to do. In the sphere of representing, reckoning thinking made manifest in Leibniz's principle, the presupposed ground (*ratio*) already reckoned out and held as definitive, namely verifiability via causality, is imputed to whatever is encountered. Relying upon that presupposed ground, the reckoning thinking (*ratio*) that has accepted it as fundamental views everything in accordance with it. Whatever is met with, i.e., whatever representing thinking re-presents to itself as object, is deemed to be verifiable as causally grounded. It is ordered in accordance with that grounding presupposition; it is considered and adjudged only in its light. And its Being, precisely as the verified, is attested in statements that give concrete expression to verifiability and that, insofar as they accord with the principle of ground, can provide, for the reckoning thinking that confidently accounts for, an impregnably secure structure of knowledge and of what is known that secures and emboldens it as, in self-assured self reliance, it carries itself out.

When viewed in accordance with Heidegger's understanding, Leibniz's principle concerning the rendering back to thinking of the sufficient ground discloses with impressive clarity the way in which the reckoning, representing thinking characteristic of the modern age takes its course. The principle, which was stated by Leibniz in full assurance of its loftiness and commanding power,[98] manifested the new self-confident perspective of modern thinking with inclusive and compelling force. In it, Leibniz brought to trenchant expression the fundamental assurance that was everywhere operative in the modern age. In his principle (*Grundsatz*) he spoke a grounding word. For in that word, Being itself, as the ultimate and now unguessed ground, disclosed in the power of a decisive utterance the changed manner of its dominion.[99]

In free obedience to the governance of Being,[100] Leibniz

---

[98]SG 47.
[99]SG 47.
[100]SG 47.

propounded a proposition that as a "principle" possessed priority in relation to all propositions that might be founded upon and determined by it. It brought to light and centered in itself dominant tendencies that had not hitherto been brought together into a clarity of statement that could bring to determinative appearance—if not to direct recognition by those who responded to that statement—long prevalent aspects in Being's manner of ruling which had now shifted away into a yet deeper hiddenness and a yet sterner intensity in this new age.[101]

With the new withdrawal of Being and the new self-assertion of human subjectivity as that which could guarantee for itself its own Being and could assume unrivaled disposal over whatever might be said to be, the manner of human comportment named long since in *ratio* comes to unquestioned dominance.[102] Similarly, efficient causality, long decisive in the context where reliance on the authority of revelation circumscribed the sphere and intent of human thinking, now emerges as of pivotal significance for the human self-securing that has come self-confidently to the fore; for it serves crucially to secure in the stability of the demonstrable the objects with which the fundamental knowledge of the self-securing subject has to do.

Within this gathering-forth of tendencies already variously prevailing into clear statement and rigorous dominion in Leibniz's principle of ground, the fundamentally changed manner of Being's happening in withdrawal, which marks the inception of the modern age, manifests itself, though it is not yet named. The Being of whatever is holds decisive sway now as objectness (*Gegenständigkeit*), i.e., as the manner of presencing of that which stands fixedly over against.[103] What-is does not now merely so offer itself to human apprehending as to eclipse the happening of Being.[104] Rather, what-is presents itself as wholly dependent on the thinking that it encounters. It carries in itself no claim

---

[101]For Heidegger's discussion of the meaning of "principle," i.e., a generally accepted concept (Greek *axiōma*), see SG 32-34.
[102]Cf. Ch. VIII, pp. 302-304.
[103]Cf. Ch. VII, pp. 282, and Ch. VIII, pp. 304f.
[104]Cf. Ch. X, pp. 399f.

to self-presenting, to an appearing from out of itself. Being has so fallen away into self-concealing as to hide all traces of its ruling as the sui generis initiating Happening that lets what-is happen of itself. Ruling thus, withdrawn in disguisedly manifesting itself as objectness, Being concomitantly happens man away in his juxtaposition to what-is in a new manner, that of the self-assertive I-subject.[105] Being has ever ruled as subjectness (*Subjektität*). But whereas once, for the Greeks, ruling thus, Being manifested itself as the powerful lying-before (*hypokeisthai*) of whatever presenced as what lay-before (*hypokeimenon*),[106] now, in order that Being may rule as objectness (itself a mode of subjectness, as a self-presenting), man too has been brought to present himself in a new sort of subjectness, i.e., in the self-preoccupied, self-assured subjectivity (*Subjektivität*) that boldly carries itself out in commandeering what-is, leaving to it no self-presenting and making it subservient to the I-subject's securing of itself. Such subjectness discerns and acknowledges no referent beyond itself. For here too Being, in its happening as that which initiatingly underlies, has, in its withdrawnness, completely hidden itself from view. Through the representing thinking that is his peculiar mode of apprehending, modern man as subject provides the arena via which Being as objectness can carry itself out.[107] He encounters that which encounters him, grasps it, and sets it forth before himself to stand fixedly over against him. Alone and wholly oblivious of Being as Being, he accomplishes Being's dictate that only as something brought self-concealedly to a stand as object can anything whatever now be.

4

Leibniz's principle of ground exemplifies with particular clarity Heidegger's pronouncement that "metaphysics grounds an age."[108] In the philosophical thinking that followed upon Leibniz's work, the validity

---

[105]"The Age of the World Picture," HW 81ff, QT 128ff. Cf. Ch. VIII, pp. 302-304.
[106]Cf. Ch. VIII, p. 301. See "The Word of Nietzsche: 'God Is Dead,'" HW 207, QT 68, and n. 9.
[107] Cf. Ch. VIII, pp. 304f.
[108]"The Age of the World Picture," HW 69, QT 115. Cf. Ch. VIII, pp. 293-295.

of the asseveration that nothing is without ground is presupposed. Whether the formulation be that of Kant or Fichte, Schelling or Hegel, the fundamental assumption is made that whatever thinking can consider as something that in anyway *is* belongs within a structure decisively characterized by grounding and groundedness. Leibniz's granting of a pivotal place to efficient causality does not again appear. But the relation between knowing subject and known object that his pointing to causality as central implicitly bespeaks becomes, with Kant's *naming* of the object, a primary metaphysical concern.

Always the reality with which thinking has to do in the light of this polarity is possessed of a configuring marked by grounding and groundedness. That configuring accords with thinking. Indeed, it is seen as provided to whatever is, either by the individual thinking subject (Kant) or by the transcendent subject in which the individual thinking subject participates (Fichte, Schelling, Hegel). Here the manner of happening characteristic of reckoning, representing thinking remains definitive. *Ratio* as ground, presupposed and imputed by that thinking to whatever it sets before itself as something that is, undergirds and bears forward *ratio* as the thinking that, in its initiatory and grounding role, unquestioningly reckons with and reckons upon the ground. Even when, with Schopenhauer, in the "will to live," the self-assertiveness constitutive of modern subjectness rises to primacy, a role is retained for a pervasive rational structuring, although here that structuring is dependent on the will that stands outside it and provides it. Only in Nietzsche's thinking is such rationality no longer assumed. There reason has become an instrument of the will to power, and the comprehensibility of that with which reason concerns itself, seen in terms of grounding and groundedness, is no longer presupposed. Nevertheless, in Nietzsche's portrayal of the will to power as arrogating to itself and ordering in its service whatever it requires, in order to mount to an ever higher level of power,[109] the fundamental structure of *ratio* is clearly to be seen. For *ratio* is a reckoning that reckons with and reckons upon precisely in such a way as to adjudge

---

[109]Cf. Ch. VIII, pp. 311-313.

according to its own presuppositions, just in order that it may in its own terms account for and take account of whatever it encounters and so may successfully carry itself out.

Beyond the specific domain of modern metaphysical thinking, the definitive grounding power of the dictim enunciated in the principle of ground is strikingly displayed, not only in the areas where objectness is the decisive mode of happening of what-is,[110] but also in areas where what-is presents itself definitively only in terms of the standing-reserve in which the standing of anything as an object is no longer of any account.[111] In the first case the power of the principle shows itself in the comportment of modern science. Science establishes its ground plan (*Grundriss*) so as to pursue its on-going activity in the light of that plan and on its basis. It demarcates its object spheres, viewing that with which it has to do in accordance with the presuppositions and procedures already selected to facilitate the viewing and always assuming the existence of a coherent network of interrelations—a nexus of grounding and groundedness—that it can properly strive to discover and on which, as the structuring intrinsic to reality, it can confidently rely.[112]

Again, in the machine technology that rests upon modern science and provides, if hiddenly, the basis for the prior arising of that science,[113] the far-ranging power resident in Leibniz's principle makes itself evident. Heidegger can characterize technology, understood in a properly technological manner, as "the highest form of rational consciousness."[114] Ceaselessly it carries itself out by way of the grounding of one thing upon another. Taking its stand upon its axiomatically held conviction that the proper way of dealing with whatever comes to hand is to marshall it so that it may be put most expeditiously to use, technology assesses and orders everything it encounters in accordance with that unquestioned goal. Forever it amasses

---

[110]Cf. Ch. VII, pp. 270f & 282.
[111]Cf. Ch. VI, pp. 244f, and Ch. VII, p. 280.
[112]Cf. Ch. VII, pp. 270-272.
[113]Cf. Ch. VII, p. 281.
[114]"Overcoming Metaphysics," VA 87, EP 99.

its standing-reserve in order that whatever has been gathered into that reserve may serve as the basis of some further amassing and the achieving of some further end.[115]   Technology too accomplishes itself as a reckoning, and that in an open and audacious way.   It unabashedly takes account of and counts on anything only in its, technology's, own terms, accounting for it and counting it significant only as something fit to serve the need that it, technology, discerns and sets itself to meet.[116]

In Leibniz's principle that demands the delivering of the ground, there sounds a claim that lies uncompromisingly upon the modern age.[117] It spreads as widely and pervasively as the rule of modern technology itself. Being, which, happening among the Greeks as *logos*, happened primally at once as presencing and ground,[118] happens in the modern age, in its withdrawnness, as quite another manner of presencing and ground. It rules now as *ratio reddenda*, i.e., as the initiatory ground that lets-lie-before through happening as a presencing that comes to pass solely as the delivering, the rendering back to thinking, of the ground, which alone lets anything presence as something that is. The insistent claim that goes forth in the principle of ground is the claim of Being in this its modern manner of happening as the ground that, in accordance with its original character as *logos*, initiatingly gathers and brings forth into place whatever is. As such, it is a summons that is everywhere in play as the manner of happening that belongs to modern technology in all its guises. Indeed, it is precisely that enframing summons, the manner of holding sway of technology, that is named with the name *Ge-stell*.[119]

Antecedent to this *Ge-stell*, is that other happening of Being that is so named in Heidegger's consideration of Greek origins. That original *Ge-stell* was the gathering-setting-forth intrinsic to bringing-forth-hither understood as "letting arrive hither into configuring," into configuring, that is, that "as limiting outline (*Riss*)," rends open the undifferentiated

---

[115]Cf. Ch. VI, pp. 229 & 244-249.
[116]Cf. Ch. VI, pp. 241-243.
[117]SG 13f.
[118]Cf. this chapter, pp. 415f.
[119]Cf. Ch. VI, pp. 258f.

and so unconceals.[120]   Heidegger's use of *Ge-stell* for the peculiar
gathering disposing now holding sway in modern technology is thought
from out of that early meaning. The line of descent linking one Being-
manifesting word with the other is direct, yet tangential. Succinctly and
centrally it bespeaks the on-going epochal withdrawal of the Being of
what-is that we have been tracing out.[121]   *Ge-stell* as the gathering
disposing that is the peculiar mode of enduring of modern technology
"arrives (*kommt . . . her*) from letting-lie-forth-before, *logos*, experienced
in a Greek manner, from the Greek *poiēsis* [bringing-forth-hither] and
thesis [placing]."[122] But how marked is the discrepancy that is in play.
In the *Ge-stell*, the enframing summons that now rules, the gathering into
place that lets lie forth before does not, as once among the Greeks,
accomplish itself by way of the immediate encountering of human
apprehending with what-is in the imperious arising that is its Being.[123]
Rather, it does so via the self-securing presenting-to-itself in which
modern man in his I-subjectness encounters what-is precisely as that
which he has set forth securely before himself and hence as that which, so
it seems, stands firmly in his control. This gathering-setting-in-place as
ground of something to be grounded, this taking account of everything
and accounting for it just on account of something beyond it, is precisely
the demand resident in the principle that requires the delivering of the
ground, and it is likewise the manner of happening of Enframing. "In the
setting-in-place (*Stellen*) characteristic of Enframing (*das Ge-stell*), i.e.,
now, in the challenging forth of everything into securing, the summoning
claim belonging to the *ratio reddenda*, i.e., belonging to the *logon*

---

[120]See "The Origin of the Work of Art," UK 97, PLT 84. Cf. Ch. IX, pp. 347f. In
this context, where disparate meanings for *Ge-stell* are in question, we may note that
the disclosive, configuring delineating outline, *Riss*, important in the portraying of the
Greek *Ge-stell*, has a distant counterpart in the *Grundriss* (ground plan) of modern
science, which so delimits, delineates and configures an object area as to open it up to
scientific bringing-into-view as governed by the modern *Ge-stell*. Cf. Ch. VII, pp.
270f.
[121]On "epochal," cf. Ch. I, p. 48.
[122]*Das Ge-Stell als Wesen der modernen Technik kommt vom griechisch erfahrenen
Vorliegenlassen*, logos, *her, von der griechischen* poiēsis *und* thesis. UK 98; PLT 84.
[123]Cf. Ch. IX, pp. 337-339.

didonai, speaks so thoroughly that now this claim, in Enframing, assumes
sovereignty over that which is absolute and unquestioned, and representing thinking (*Vorstellen*), as setting-forth-before, from out of
Greek apprehending gathers itself in the direction of the setting-in-place
that secures and makes firm."[124]   From out of its distant origin, the
presently ruling *Ge-stell*, the enframing summons that gathers everything
into place as supply, as a mode of the happening of Being in self-
withdrawal happens as the ground that delivers the ground via the
seemingly self-sufficient human apprehending that it has hiddenly claimed
to accord with itself.

Here the import of Heidegger's assertion that we are not only at a
chronological distance from the Greeks but are, if our present
circumstancing is to be genuinely understood, in deviance in relation to
them,[125] comes to striking embodiment.   In the original, *Greek*, relating
of Being and man, the happening of Being as *logos* called forth an
answering response of *logos* as vouchsafed to man through it, in order
that, through the skillful knowing that was called *technē*, man might
accomplish disclosure, letting Being happen in immediate acknowledged
dominion as the Being of what-is.[126]   Now, in the modern age, in
accordance with an opening-out of time that can be rightly understood
only in terms of Being's self-withdrawal, the ruling happening of Being,
which, in its self-masking instancing of itself as self-aggrandizing will via
the grounding thinking of metaphysics, summons forth as man's
answering response the comportment of thought and deed that is
everywhere technological, governs—and governs no less directly for
governing in concealment—precisely as the manner of enduringly holding
sway belonging to technology.[127]   Yet it reigns undetected and

---

[124]*Im Stellen des Ge-Stells, d.h. jetzt: im Herausfordern in die Sicherstellung von
allem, spricht der Anspruch der* ratio reddenda, *d.h. des* logon didonai, *so freilich,
dass jetzt dieser Anspruch im Gestell die Herrschaft des Unbedingten übernimmt und
das Vor-stellen aus dem griechischen Vernehmen zum sicher- und fest-Stellen sich
versammelt.* "The Origin of the Work of Art," UK 98, PLT 84.
[125]"The Anaximander Fragment," HW 310f, EGT 26f. Cf. Ch. IX, p. 332.
[126]Cf. Ch. IX, pp. 337f.
[127]Cf. Ch. VIII, pp. 315-323.

unacknowledged. For the "stepping-apart" that disrupted the original unitary answering of *logos* to *logos*[128] has now become, through a deepening sundering wrought out via time, a radically self-assertive, adversarial standing-over-against that inherently drives on toward total separation, i.e., toward becoming an annihilating of the other, in the unperceived void as which Being rules in self-withdrawal. Modern man, making his way in undiscerning encounter with what-is as that which refuses disclosure of itself as itself, is astray in self-securing self-assertiveness, having been led thither by the Being of what-is precisely in its self-withdrawal into hiddenness via its self-dissembling self-imposition as all-consuming will. The ruling transpiring of that will, in its heedless, obliterative on-striving, is for Heidegger solely a deviant erring into nullity, away from the genuine self-bringing to pass of Being.[129] Under its sway, what-is now merely presses forward unconditionally "toward being used up in consumption."[130] And modern man, locked into his indirect meeting with what-is through the medium of his representing thinking, in his endless planning and calculating pursues a course that must be seen as nothing but an erring, a deviant straying vis a vis the withdrawnness of Being, which takes place amid an "unworld of errancy (*Unwelt der Irrnis*)" where incessant, all-consuming putting to use of the undifferentiated, the Being-bereft, is rampant upon a desolated earth that can itself be called—as viewed via the history of Being wherein Being rules in withdrawal—"the star astray (*der Irrstern*)."[131]

In all this, "nihilism,"—here to be thought of solely from out of Heidegger's understanding of Being's ever more disguised self-disclosure via metaphysics as the "history of Being"—holds sway disposingly, in the

---

[128]Cf. Ch. IX, pp. 364.

[129]"Overcoming Metaphysics," VA 72f, EP 86. Cf. Ch. VIII, pp. 316f.

[130]VA 92; EP 104.

[131]VA 97; EP 108f. Elsewhere ("The Word of Nietzsche 'God is dead,'" HW 198f, QT 59f) Heidegger includes centrally in his discussion a long quotation from Nietzsche that contains a significantly similar portrayal. On the meaning that Heidegger himself finds for "world" understood as the self-initiatory happening of the Twofold as gathered meaningfully to disclosure, cf. Ch. II, pp. 61-63; Ch. III, pp. 136f; and especially Ch. V, pp. 174-179.

culmination of a centuries-long drift, after the manner of the oblivion as
which Being rules in self-negating withdrawal from genuine disclosure.
Ruling thus, "Being looses itself into deception" and in that self-loosing,
"takes man into an unconditioned service" that is thus radically pervaded
by erring, i.e., by a commitment to the accomplishing of non-disclosure
which, from out of non-disclosure, is rendered blind to its own character.
Heidegger emphasizes that the governing happening from out of which
this "erring" in deceptiveness and despoilment takes place is in no sense
whatever to be thought of as something negative in any ordinary meaning
of the word. It is not a "decline or decay (*Verfall*)," for it is nothing less
than a bringing into its own of a mode of Being's ruling.[132]

Thus in the modern age has the ultimate tendency of Being to
happen as self-concealing brought itself to powerful dominance. In the
rule of Enframing it thrusts to the fore, and its dominion grows. Once,
among the Greeks, the press of the on-coming Twofold met man with an
urgency of all but undecypherable presence that could overwhelm but that
could also be arrested and brought into the clarity of orderedness by some
responsive, perspicacious deed that could achieve disclosure by setting
forth in illumining uniqueness precisely that which was needed to gather
to itself the particulars of the Twofold in manifesting-interrelation.[133]
Through an immediacy of meeting and response unconcealing triumphed,
if always provisionally, over concealing.[134] Now immediacy is gone.
Modern man, in his confronting of reality, has no fundamental awareness
of the inherent play within it of a demanding self-presenting; for that self-
presenting, the ruling happening of Being, is deeply disguised in
withdrawal. Governed thus, man can do no more now than carry into

---

[132]*Das Seinsgeschichtliche Wesen des Nihilismus ist die Seinsverlassenheit, sofern in
ihr sich ereignet, dass das Sein sich in die Machenschaft loslässt. Die Loslassung
nimmt den Menschen in eine unbedingte Dienstschaft. Sie ist keineswegs ein Verfall
und ein "Negativum" in irgend einem Sinne.* "Overcoming Metaphysics," VA 91, EP
103. On the happening of Being as Nothing as in no sense negative, see "The Age of
the World Picture," HW 104, QT 154. On "erring," see "Overcoming Metaphysics,"
VA 72f, EP 86. Cf. Ch. VIII, pp. 316f. Cf. also Ch. I, pp. 44-46, and Ch. II, pp.
91-93.
[133]EM 113ff; IM 125ff. Cf. Ch. IX, pp. 339f & 348f.
[134]Cf. Ch. IX, pp. 342f.

play Being's puissant absence by, seemingly of himself, letting whatever is *be*. Possessed of an assured self-reliance that assumes that his thinking can and must of itself provide the defining perspectives in accordance with which whatever is can properly be recognized and dealt with, he lays down the conditions for disclosure and in their light confidently takes charge of everything that is. His logic is determinative. His representing thinking takes cognizance of what-is only by way of the skeletalizing re-instancings of the latter, as defined in terms of generalized attributes, that it gathers and offers to itself in the conceptualizing that is its mode of carrying itself out. His science, likewise, in sphere after sphere of investigation, observes that with which it has to do only through a meeting bereft of direct encounter, setting it before itself in an explanatory framework of hypothesis and law. And his technology, although seeming to concern itself directly with the things with which it has to do, in fact invariably encounters them entirely by way of its own mediation, always presenting them to itself only as depicted and evaluated in the light of its own calculative purposes. Whatever *is* in our age presences solely in this way. It presents itself only as once removed from itself, via some mode of re-presenting thinking. Indeed, so all-inclusive is the range of such thinking and so fundamental is its obtruding of itself as the mode wherein all reality is properly to be grasped, that Heidegger can point definingly to our age as "the age of the world picture (*Weltbildes*)." It is a time when men so confront the "world"—understood by them, with no thought of unreality, as everything whatever that is—as to set it before themselves, from out of an unquestioned readiness and overriding intention to set it thus forth, precisely as a surveyable totality that, albeit as vast and complex, can be brought to a stand before the mind like a picture on view and that in fact *is* only as thus set in place and systematically seen in their conceiving which, in effect, in advance takes charge of what-is.[135]

And, in all this, the happening of Being as self-concealing presencing, from out of which whatever is in this manner self-

---

[135]"The Age of the World Picture," HW 82f, QT 129f.

concealingly presences of itself, rules withdrawn and forgotten, the unconcealing via which it is surmounted and maintained serving only constantly to shut it away from bringing itself to bear discernibly on that arena of comportment that man takes to be exclusively his own.

In this modern context, man and what-is confront one another in adversarial opposition. The original happening of Being as *Hervorbringen, poiēsis,* i.e., as the bringing-forth-hither that, as an immediate doing, places into the open, via the responsive bringing-forth that, as *technē,* is the working of man, whatever is appropriate in some particular on-going present,[136] has been succeeded by the happening of Being as the despoiling enframing summons that orders man arrogantly to set upon what-is and to set it forth only as that which serves his passion to calculate and make secure. Thinking, as *ratio,* adjudges; via a setting fast it disposes into reassuring interconnectedness what-is as encountered in objectness. Science strives after what-is, refining and entrapping it in accordance with its own preconceived perspectives.[137] Technology challenges forth what-is, demanding it out hither in order that it may be relentlessly amassed for ends that lie always beyond it and that take no cognizance of it in itself.[138] What-is, for its part, happened by Being into its encountering with man, resists his predations. Its presenting of itself is a self-exhibiting, a setting of itself out hither (*Sich Herausstellen*)[139] that is a self-withholding wherein it offers only a semblance of itself. Whatever is provides itself to thinking only by way of the secondariness belonging to the object securely delivered via representing. "Nature," the modern successor and distant manifestation of *physis,* that primal imperious arising that as the Twofold originally encountered Greek man, now shows itself to physics as "natural science" only as a "space-time determination of the motion of points of mass,"[140] or, in the peculiar case

---

[136]Cf. Ch. IX, pp. 347-349.
[137]Cf. Ch. VII, pp. 268 & 283.
[138]Cf. Ch. VI, pp. 247-249.
[139]"Science and Reflection," VA 56, QT 167f.
[140]FD 72; WT 93. Cf. Ch. VII, p. 271.

of contemporary physics, as a network of relations.[141] History (*Geschichte*), originally the complex, unitary taking place into which Greek man was sent forth via the founding opening-up that his responsive able working accomplished for Being's ruling of a people,[142] now displays itself to historical science only as a readily surveyable pattern of categorizable events causally connected and devoid of the uniqueness that defies reduction to generality.[143] In every science a like self-disguising of what-is in its self-presenting is underway.[144] And the manifold of what-is offers itself to modern machine technology only as a host of entities, whether natural or man-made, that lie immediately at hand as commodities to be exploited and incorporated as secure, unindividualized components into the ever-shifting, ever-growing, ever-expendable standing-reserve.[145]

This transformed manner of meeting that now obtains between man and what-is manifests on the vast scale of an historical era the innate tendency, imparted from out of Being itself, that inclines all particulars that constitute the manifold of what-is to disrupt the compliant unity in gatheredness that characterizes genuine presencing and to assert themselves in their traversing of the particular present opened to them by endeavoring to make themselves each more present through a contentious self-maintaining that is as such intrinsically a falling-away from presencing into a mere persisting in isolatedness, which exchanges genuine enduring for mere lasting-on.[146] Now the radical self-withdrawal of Being is writ large in the self-isolating self-withholding into the inertness of availability that everywhere characterizes that which is, including man himself. The only disclosive manner of happening that is in play is a mere shadow of the supple, spontaneous, immediate, yet elusive unconcealing that ruled originally among the Greeks. Attenuated

---

[141]"Science and Reflection," VA 56, QT 168.
[142]Cf. Ch. I, pp. 47f; Ch. III, pp. 136f; and Ch. IX, p. 331.
[143]"Science and Reflection," VA 63f, QT 175.
[144]VA 64ff; QT 175ff.
[145]Cf. Ch. VI, pp. 263f & 241-249.
[146]Cf. Ch. I, pp. 50-52, and Ch. VI, p. 264.

and sucked back into rigidity, it isolates and represses through its ruling. The peculiar assertiveness of the subjectness of the I confronts the stubborn self-refusal of the object in its objectness. The high-handed commandeering that is characteristic of the technological approach to whatever is confronts the obdurate self-withholding that permits what-is to offer itself only as standing-reserve.

The magnitude of the restrictive depletion that has here taken place appears with telling clarity when the fundamental mode of manifesting named in working and work is in view. Working remains, as it has so long been, an effecting. But now, as the primal placing that lets the real lie forth as what-is, working has the character of the gathering-setting-in-place-as-supply made manifest in the name *Ge-stell*. The defining arena for the meaning of work is the sphere of science and technology. Thus, in the basic portrayal of nature in terms of measurable masses in motion within a nexus of loci and time-points whose relations can be determined, the motions in question are defined in terms of forces that work, i.e., that effect (*wirkende Kräfte*).[147] The German word for work that is regularly used in mechanics is *Arbeit*: work, labor, toil. The fundamental law of the conservation of energy (*Kraft*) is expressed in terms of work (*Arbeit*) considered as expenditure or consumption, a formulation that, Heidegger says, is reminiscent of economics with its stress on the calculation of results, i.e., of the successful effect (*Erfolg*).[148] And, strikingly, when Heidegger considers the manner of proceeding that characterizes modern science as such, he uses, in speaking of its manipulation of the real (*das Wirkliche*) by means of its entrapping theory, a series of verbs based on *arbeiten*, to execute, to work, to fashion.[149]

Carrying itself out in the modern context, where causality reigns as determinative and is everywhere presupposed, science achieves a working that is far removed from that original working which, as *technē*, i.e., as

[147]FD 72; WT 93.
[148]FD 73; WT 94.
[149]*bearbeiten* (to work over or refine), *zuarbeiten* (to work toward), *umarbeiten* (to work around or recast), and *herausarbeiten* (to work out). "Science and Reflection," VA 55ff, QT 167ff, and n. 19.

an attentive, obedient bringing-forth, accomplished the placing into appearance and uniqueness of that which was itself uniquely coming into appearance and directly offering itself to stand forth as a work.[150] In carrying out its refining, i.e., its working-over (*Bearbeitung*), of the real,[151] so as to set the latter forth in accordance with its own structure of preconceptions, science, answering to the manner of happening in play in that with which it has to do, accomplishes a working that is, rather, a coercive taking-in-hand. In its domain, "that which has been brought about, in the sense of the consequent, shows itself as a circumstance that has been set forth in a doing—i.e., now, in a performing and executing (*Leisten und Arbeiten*)."[152] Modern machine technology functions overtly in this same milieu. Technology, as the means by which modern man carries into effect his purposes with respect to the reality that environs him and therewith continuously strives to render ever more secure the mastery that he claims, can be spoken of by Heidegger precisely as that by which man "establishes himself in the world in that he recasts it, works it over (*sie . . . bearbeitet*) in the manifold modes of making and shaping."[153] And for technology, "working" is very evidently the achieving of results.

So ubiquitous, indeed, is such heedless goal-serving working in the domain of technology that Heidegger can see there an identity of function for man and machine and can aver that in the sphere where technology rules, the appointed role of the *worker* (*Arbeiter*), no less than that of his machine, is to perform a specific task or process that will provide, i.e., result in, a stipulated product. And in testimony to this identification he can say that the hand of the worker is but an extension of the machine.[154] Thus far, as the minion of technology, is the modern worker removed,

---

[150]Cf. Ch. IX, pp. 347f & 352f.
[151]VA 56; QT 167, and note 19.
[152]VA 51; QT 162f.
[153] . . . *in der Welt einrichtet, indem er sie nach den mannigfaltigen Weisen des Machens und Bildens bearbeitet.* "The End of Philosophy and the Task of Thinking," SD 64, TB 57.
[154]WHD 54f; WCT 24f. See also William Lovitt, "*Technē* and Technology," in *Philosophy Today*, Spring, 1980, Vol 24, No. 1/4, p. 69.

not only in time but in the fundamental manner in which he plays his needed part, from that spontaneous, intense, dedication of every capacity for knowing and doing to the accomplishing of a work as a thing of inner integrity and determinative self-presenting in power that the word *technē* originally bespoke.

The manner of working that must now properly be named with *Arbeit* has, indeed, nowhere what we might call the independent status of a fulfilled achieving through which something of intrinsic significance, something possessed of the stability of a self-presenting in self-contained disclosure, can be brought forth. Rather, this working is a peculiar mode of useful labor that can take place only in such a way as to subserve that "securing of ordering (*Ordnen*)" toward which the performing that is technological in character incessantly strives.[155] Accepting as presently determinative the metaphysical defining of man as *animal rationale*, Heidegger identifies man of the modern, the technological, age as "the laboring animal (*das arbeitende Tier*)."[156] Not only the ordinary laborer, who is to be understood in his numbers as participant in a "subhumanity" wherein "animality," with its confinement within the sphere of the animal's own self-concern, is clearly evident, and who should, accordingly, be disposed of and even bred as "human material" to serve intended goals, is pointed to in this designation. But the "leaders," in whom *ratio* is powerfully instanced, and who are called upon calculatingly to take full disposal over everything, including human beings, in the service of the all-inclusive planning that the implementing of the technological ultimately demands, although they can be called "superhumanity," also instance animality through their own peculiar laboring. For the "intellect" via which they accomplish their far-ranging ordering is in fact "instinct" and, as such, is once again restricted to a self-preoccupied meeting of needs. The work of those leaders belongs, if at a superior level of activity, to precisely the same realm as that of the merest worker. They are, in the sphere of expert planning, "laborers for

---

[155]"Overcoming Metaphysics," VA 97, EP 108.
[156]VA 72; EP 86. See also VA 72, EP 85, where Heidegger's phrase is *das arbeitende Lebewesen*.

preparedness (*Rüstungsarbeiter*)," i.e., laborers dedicated to the provision of every armament in the human campaign for dominion over everything that is.[157] Just as dual aspects of a single being, man, are named in the ancient designation "rational animal," so, Heidegger insists, superhumanity and subhumanity, while differentiate, belong together as the same.[158] Modern man, "metaphysical humanity," protagonist as the executer of technology of unconditioned setting securely in order for use, is, whatever his specific individual role, simply the "laboring animal" who, bereft of any discernment of Being's happening, is being given over solely to ever more frantic, self-destructive preoccupation with the "giddy whirl" of products of his own making.[159] Here, for all, "work" in its former sense has been succeeded by a self-engrossed, ultimately fruitless working that is only a laboring carried out at the behest of a setting in order for the sake of consumption that is nothing more than a consuming of everything so that some intended order—itself immediately requiring to be superseded—may be pursued and pursued ever again.

This shift in the significance of "working" as the undertaking wherein man wittingly or unwittingly carries out the ruling of the working-placing that is the happening of Being has its equally crucial counterpart in the changes in the manner of happening of what-is as such, as that which is placed hither into unconcealment. Once Aristotle's use of *energeia*, standing, in pressing vitality, into-the-work (*ins Werk Stehen*),[160] with its parallel meaning of *entelecheia*, having-itself-in-completion (*sich selbst in die Ende haben*),[161] manifested with clarity the character of what-is as the real, i.e., as "the working, the worked (*das Wirkende, Gewirkte*)."[162] *Energeia* bespeaks the manner in which that which is, whether standing forth directly of itself or through the

---

[157]VA 94f; EP 105f. On the meaning of *Rüstung* see VA 91, EP 103.
[158]VA 91; EP 103.
[159]VA 72; EP 86.
[160]Cf. "On the Being and Conception of *Physis* in Aristotle's *Physics* B, 1," W 354, Ph 256. Cf. Ch. X, p. 378. On the full meaning for Heidegger of *energeia*, see Ch. IX, pp. 352f.
[161]W 354; Ph 256. Cf. Ch. X, p. 378.
[162]"Science and Reflection," VA 49ff, QT 160ff.

intermediary achieving of man, gathers itself into an appearing, a self-presenting, in which movement, stilled into its highest intensity as rest, is poised as the working that lets the entity in question impose itself as itself, from out of the working that has set it forth, i.e., from out of the presencing whereon its own self-presenting depends.[163] Now *energeia*, having, in the Latin recasting of Greek thinking, already been translated by *actus*, with its connotation of agency, reappears as energy, *(Energie)* which is understood definitively in physics—and hence in science generally and in technology—as the power to do work, i.e., to produce an effect.[164]   The   self-containedness-in-power-of-self-presentation   of *energeia* has given place to the driving-forward-toward-something-consequent that characterizes "energy."[165]   Under the rule of the *Gestell*, the enframing summons that gathers everything into place as supply, modern science, with physics in its vanguard, assumes precisely this energy to be the most fundamental reality, whose manner of happening in its far-flung, ever-changing manifestations in countless embodiments and interrelations is to be discovered, measured, made comprehensible in verifiable laws, and hence made accessible to purposeful application. No less centrally does machine technology occupy itself with energy. Antecedent to all its other concerns stands its concern with nature as a vast repository of a standing-reserve of energy *(Energiebestand)* whose availability is ceaselessly presupposed and whose resources are avidly exploited, conveniently laid by, distributed, and utilized in accordance with technology's ever proliferating ends.[166]   Heidegger, indeed, describes vividly the way in which the recent discovery of atomic energy can be expected to eventuate in now undreamed-of advances in

---

[163]Cf. "The Origin of the Work of Art," UK 96f, PLT 83. Cf. also Ch. IX, pp. 352f.
[164]*The Compact Edition of the Oxford English Dictionary*, Oxford University Press, 1979, p. 864.
[165]A similar change can be seen in the transmuting of Aristotle's *entelecheia* into the entelechy of modern philosophy, "a vital force urging an organism toward self-fulfillment." (*The American Heritage Dictionary of the English Language*, Boston, Houghton-Mifflin Co., 1971., p. 436.)
[166]"The Question Concerning Technology," VA 29, QT 21. Cf. Ch. VI, pp. 245-247.

technological achievements.[167]

Via the fundamental self-presenting of what-is as energy and the decisive human grasping of what-is in this guise, the adversarial relationship subsisting between man and what-is under the rule of Being's happening as Enframing has come in our own day to fateful manifestation. Through his science and technology, modern man has searched out and "harnessed" the foundational energy of the atom. Now, via the very self-assertiveness of human comportment that has disclosed it, that energy has been brought into menacing encounter with man, continually threatening him by the very overtness of its presence in his hands. Heidegger points to the fact, which ought to strike us as significant and ironic, that we have named this present time, in which man is constantly presupposing and seeking to complete his mastery over that which is, "the atomic age (*die Atomzeitalter*)," therewith identifying it primarily not with some preeminent human accomplishment but with the decisive embodiment of reality as which what-is now shows itself to us.[168]

This naming is no accident. It in fact displays, Heidegger says, the claim of the principle of the rendering of the ground that in the ruling of Enframing everywhere unfolds in an increasingly unbridled manner by way of the determinative mode of existing of modern man. Modern science, seeking ever for a wholly secure explanation and hence pressing always back behind what has been brought to light, in order to discern a yet more ultimate ground, has arrived at the atom. Eagerly it has striven to discover the elementary particles that constitute it. Even as those particles proliferate, science resolutely seeks for an underlying unity that will securely gather and ground them together and that will, at the same time, securely and ultimately ground all phenomena whatever, to which

---

[167]"Memorial Address," G 20ff, DT 50f. Presumably because he understood the development of "atomic technology" to have come to pass thanks to the provisive happening of Being as Enframing (cf. G 25f, DT 55), Heidegger could assert confidently, in the mid-1950's, that the needed "taming of atomic energy" would in time be successful. (G 21; DT 51.)

[168]SG 57f. Our idiomatic English translation cannot match the directness belonging to Heidegger's phrase: (literally) "the atom-age." Cf. also SG 58f.

they, by definition, give rise.[169] In present-day physics this search for a unified theory stands as a preeminent concern.[170] The name "atomic age" testifies to the puzzling and dismaying fact that the investigation into the atom, which has succeeded in placing in jeopardy everything that is, has been and remains an investigation intended to provide to knowing and doing an ultimate secureness through the providing of the ultimate, calculable, and disposable ground of whatever is.

Seen even more profoundly, the name "atomic age" bespeaks the fact that in his intent, even heedless, scientific technological search for the most fundamental ground of what-is and the most fundamental source of effective energy by means of which to assure the carrying out of his plans for an ever more successful putting-to-use, modern man, in subjecting what-is to himself by his manipulations, has concomitantly accomplished his own subjection to the prevailing dominion of the mind-set and procedures of science and technology. He has undercut the possibility that he should maintain himself in any meaningful way except as the one who pursues science and technology with single-minded, confident perseverance. In so doing, he is bringing to puissant expression and living out, if unaware, what Heidegger calls "the counterplay (*Wiederspiel*) between the summons to delivery of the ground and the withdrawal of the undergirding basis (*Boden*)."[171] The delivery of a ground discovered within a causally secured structuring of what-is can only evince and put into play the happening of Being in self-withdrawal as the enframing summons. Being, *the* ground, which undergirds and yet happens uniquely beyond all grounding sequences within the manifold of what-is, has withdrawn from immediate, wholly enabling governance of man and what-is; and the demand for grounding and groundedness that characterizes its ruling happening of man and what-is is but an evidence of that distancing and of the distorting attenuation of presencing, i.e., of spontaneous self-bestowing in compliant unifiedness, that is now relentlessly holding sway. Accordingly, man's proper mode of existing

---

[169]SG 56 & 58-60.
[170]Cf. Ch. VII, pp. 280.
[171]SG 60.

can find no genuine basis. And what-is yields itself to man only as that which has been all but stripped of its capacity to happen of itself as itself, and which can therefore only so encounter him as in fact to despoil him in the same radical way.

Behind and within this implacable confronting of man and what-is, which is fashioned from man's self-assertive pursuing of his self-securing and from the insistent self-withholding that allows what-is at once to persist in face of man's importunate depredations and to offer itself precisely as that which is ready—even explosively ready—to accord with his insatiable demand, there rules the "stamping" of Being—the manifest mode of its happening that masks the manner of happening primary to it— that characterizes metaphysical thinking as, in accordance with the extreme withdrawal of Being now in play, in unguessed impoverishment that thinking accomplishes its completion.[172] As, under the intensifying rule of modern technology, what-is comes to stand forth decisively vis a vis man, not as object (*Gegenstand*) but as standing-reserve (*Bestand*), the manner of Being's happening that governs this changed relationship shifts from objectness to the sheer lastingness that obliterates the individuality still remaining to the object and leaves that which *is*, in accordance with this changed happening, possessed only of the tenuous identity of that which is taken into cognizance merely as something useful toward the achieving of an end that lies beyond it and has nothing to do with it taken in itself.[173] Metaphysics reaches its "end," its fulfilling completion, in the work of Nietzsche, and there the attribution of Being accords with just this technological appraisal of everything that is. In Nietzsche's metaphysics of the will to power, with its depiction of the on-thrusting will that mounts always to greater and greater power on the basis of that which it has appropriated to itself and holds fast as its secure possession, Being presents itself as value. Here that which is, that which receives acknowledgement as something possessed of significance and realness, is

---

[172]On "stampings" of Being, cf. Ch. III, pp. 131f. On the completion of metaphysics, cf. Ch VIII, pp. 324f.
[173]"The Question Concerning Technology," VA 24, QT 17. Cf. Ch. VI, pp. 244f.

accorded this standing not in itself, but merely as something that has value with respect to the will to power's pursuing of its self-designated course. That which *is* is that which the will values for its constancy and availability. It is that which the will can count on for the serving of its own self-chosen ends.[174] And value in this sense expresses precisely the manner of happening of the standing-reserve that modern technology everywhere single-mindedly commandeers for its use.

In this conceiving of Being as value, the long history of Being wherein it has offered itself to metaphysical thinking comes to its fulfillment. With the sundering of the immediate, genuine relating of Being and man that originally happened among the Greeks, Being, concealing itself as itself—i.e., as the powerful arising that ruled primordially in every sort of appearing of what is—began, via the thinking of Plato, to offer itself to thinking, rather, as a mere semblance of itself that in guise after guise would, in the overt relating of the two, permit to man seeming mastery over Being. Variously Being allowed itself to be named as something that was not Being itself but some component of the manifold of what-is. For Plato, Being is identified as *idea*. As such, it becomes not the wholly spontaneous and uncategorizable, but the prototypical.[175] It then lies at the disposal of *logos* as carried out in man's thinking, a relationship that Aristotle's categories and logic bring to precise formulation. Logic at its inception begins to judge of appropriateness.[176] Epoch by epoch, as Being ever more disguisingly presents itself, the ascendancy of logic over Being likewise grows. *Ratio* as such proceeds on the basis of causality, but it, as it were, encompasses and subordinates the latter. The presupposing and prejudging that define *ratio* as reckoning thinking are antecedent to the structures of perceived linkage by way of which it carries itself out. In the modern age man's thinking self-consciously puts itself forward as determinative. Objectness has its place as supportive of the self-securing of the thinking I-subject, which sees it to be, as ground-providing, indispensably useful for the

---

[174]Cf. Ch. VIII, pp. 312f.
[175]Cf. Ch. X, pp. 370-374.
[176]Cf. Ch. X, pp. 389-391.

validating of the thinking itself. With Kant's setting forth of the categorical imperative as a necessary manifestation of reason, the often tacit assumption that whatever conforms to and accords with human thinking as *logos-ratio* not only is but also ought to be is carried into the sphere of the moral. This perspective Nietzsche rejects. For him, what is as what ought to be, i.e., as that which corresponds to and makes viable the decisive comportment of man, has been diminished to value in the sense of that which is provided to the will to power by itself for the will's own sake and precisely as that which it deems to be appropriate for itself.[177]

Thus does Being come, seemingly, to yield itself to man's control by way of what we must still here see as the belonging-together of Being and thinking.[178] Via the history of that belonging-together, the original, sui generis happening of the Being of what-is, which it was man's appointed role to encounter in intrepidity and in accordant acceptance to bring to unconcealment, is superseded and superseded yet again in age after age by self-masking modes of Being's happening, which ever more evidently have as counterpart a human comportment wherein an evaluative adjudging takes place as central to the accomplishing of whatever manifesting of itself Being is permitting to happen via the disclosing, through self-disclosure, of whatever is. With its fulfillment in Nietzsche's thinking, marked as that thinking is by a thorough-going if veiled consummating of the mode of relating of man and what-is in Being that was so long regnant in *ratio*, Plato's Idea of the Good, i.e., of the

---

[177]For this discussion, cf. EM 149ff, IM 164ff. Heidegger's portrayal of every sort of thinking that assumes a standard according to which a judging is in any way carried out as manifesting the withdrawing of Being from full ruling disclosure of itself must remind us of the portrayal in *Being and Time* in which, with reference to what he takes to be the proper understanding of being guilty, he considers all judging to be a manifestation of an inauthentic manner of human existing. Since judging, by definition, has recourse to an adducing of standards and makes appeals to what ought to be, it fails to accept as aspects of existence that are to be acknowledged and affirmed the intrinsic limitation and specificity, the intrinsic nullity, and hence the intrinsic "guilt," belonging to thrown *Being*-there-as-openness-for-Being as Being-in-the-world. Cf. SZ 281-288, BT 326-335, and cf. Ch. II, pp. 95f.
[178]Cf. Ch. III, pp. 129f.

"valiant," that accomplishes appropriateness[179]—a notion obviously close to the happening of Being as *physis*—is distantly bespoken in the concept of *value*.[180] And the human apprehending that, with Plato, had just begun to assert its assumed capacity to lay hold of Being as its possession and to concern itself with the Idea as a norm for adjudging the real[181] has given place, in a like distancing, to the technologically determined will to power that accords Being by assigning value solely in keeping with its own enhancement of itself.[182]

Conceived by Nietzsche as value whose disposal would lie wholly within the province of the will to power as the latter might find manifestation in man, Being came, subsequently to Nietzsche's work, to be all but forgotten by what passes for thinking in our day. Heidegger never tires of alluding to the fact that "Being" is now regularly dismissed as a mere "empty concept" which it is surely absurd to take into consideration at all.[183] So confident is man, then, in the juncture of our present, of his mastery over Being as to dismiss it from view. Thus has Being's disclosive masking of itself proceeded yet one step further, until it scarcely presents itself at all with definitive power within the arena of discerning thought.[184]

Once, for Plato and Aristotle the highest endeavor of thinking was considered to be the pure beholding of that which, whether as *idea* or *morphē*, constituted the Being of anything that could be said to *be*.[185]

---

[179]Cf. Ch. X, p. 372.

[180]Heidegger sees this distance, once again one of deviance, to be very great. For he insists that the notion of valuing, i.e., the notion of considering something to have worth, was unknown to the Greeks. They had no such concept. For them, he says, to esteem meant "to bring something to appearance in the look [the self-presenting look] in which it stands and to preserve it therein." (*Würdigen heisst: etwas in dem Ansehen, darin es steht, zum Vorschein bringen und darin bewahren.*) SG 34.

[181]Cf. Ch. X, pp. 371f.

[182]Cf. EM 149f, IM 164f.

[183]SZ 2, BT 2; EM 27 & 58, IM 30 & 64; "Time and Being," SD 6, TB 6.

[184]Heidegger gives no importance to the uses of Being as a significant concept in the writings of the Neothomists, or, more importantly, of the existentialists Sartre and Marcel. He would see these as derivative and as constituting no fresh *Prägung* of Being. Cf. Ch. VIII, p. 325.

[185]Cf. Ch. X, pp. 371 & 393.

Now, with the out-working of the adjudging seeing named as *ratio* to the extremity of its ruling, no catching sight of whatever passes for the Being of whatever is, as the latter is confronted by thinking, is in any way sought. Only a thinking-seeing whose preoccupation with itself has risen now to a veritable failure to notice whatever is met and looked at *in itself*, i.e., in its Being, remains. Thus, contemporary philosophy turns its attention to various aspects and components of reality, but always its seeing is determined first of all not by any open intention to see and comprehend them in and as themselves but by an underlying predisposition to view and assess them *as* they relate to man, to his understanding and maintaining of himself.[186] Scientific observing "strives after" that with which it has to do, to the end that it may, via it, confirm its own antecedently projected ground plan, verify the hypotheses and laws that it has adduced, and carry itself forward to ever fresh "entrappings" and secure explications of the reality with which it is concerned. The purposeful discernment characteristic of technology calculatingly grasps everything that comes within range of its seeing and sets it fast as something seen only as available to serve whatever end is in view. And the seeing as which thinking transpires in these kindred realms has, accordingly, no inquiring eye for the Being of whatever is.

The goal of man's original self-assertion vis a vis Being has, in our time, been achieved.[187] The fundamentally ungovernable, ultimately shattering dominion of Being has been thoroughly gainsaid. Man need not even concern himself with the Being, the inviolable self-presenting, belonging to that with which he has to do. His concern should be with his own purposes and with the calculable, elucidative conceptualizations and orderings of the latter that render it disposable by and serviceable to himself. Man stands alone—able, determined, summoned by his thinking and his habitual comportment to shape reality to his needs. Yet precisely here, Heidegger says, the relentlessly imperious happening of Being still rules. In the great sweep of history, as ever in the small, it moves, by its

---

[186]Cf. Ch. VIII, p. 325.
[187]Cf. Ch. IX, pp. 360f.

own inexorable onward happening, to overwhelm with the opacity of the undifferentiated and the meaningless the configurings of meaningful disclosure that man has brought forth and that, in his modern self-assurance and arrogance, he self-satisfiedly calls his own. And it moves likewise to shatter man in his overmastery, poising him in his self-exaltation above an abyss of groundlessness that can deny to him the very possibility of being himself.

As the ultimate impulse for the initial stepping-apart of Being and man arose from out of Being itself, so the consummation of that estrangement remains determined ultimately from out of Being. The opposition in self-asserting self-refusal that now characterizes the relating of man and what-is bespeaks the primal opposition in self-assertive self-withholding that is now the manner of relating vouchsafed to man in his relation to Being, from out of Being itself. Happening via the tangential pull of self-withdrawal, in its primal happening as self-concealing, Being has happened man away into a withdrawnness from itself, while yet keeping him in hand in an ever more hidden, ever more stark and importunate happening of its own all-determining rule.[188]

5

Modern machine technology, as seen from the perspective of Heidegger's thinking, shows itself to us as a vastly ramifying present-day phenomenon that both arises out of and gathers powerfully forth as determinative of itself complex modes of happening whose origins lie temporally and qualitatively far distant from itself. We have been considering that technology as Heidegger considers it, namely, *historically*, i.e., as a phenomenon belonging within the on-going, changeful manifold of what-is, as, wide-ranging by way of time, that manifold is provided via Being's changeful self-sending-forth to the governance of people after people. In speaking of modern technology thus, we must be careful to recognize that, although with almost no direct

[188]On this manner of Being's ruling, Cf. Ch. I, p. 53, and Ch. II, pp. 108f, and n. 190.

acknowledgement of the fact on Heidegger's part,[189] in all such speaking yet another Heideggerian perspective is also in play. The existing of man, as early defined, continues to be presupposed, occasionally emerging, indeed, into immediate statement in some context descriptive of the existing that is human or of fully achieved human comportment.[190] Beyond this, as we have often specifically noted, throughout the historical portrayal whose changing configurations we have been surveying in accordance with the uniting yet diverse patternings present within them, in phenomena at once antecedent and intrinsically akin to the technology that now holds sway, human existing (*Dasein*) as such can be seen to be evidenced in various elements and structurings, as that existing is happened thus "historically" from out of Being;[191] and in Heidegger's portraying of modern technology also, specific elements in the peculiar existing characterizing man as man unquestionably evince themselves to us.[192] Among the early Greeks, the full possibilities belonging inherently to human existing so come to manifestation that the character of an age is decisively determined by them.[193] In succeeding ages human existing remains always in play, although now in a successive distorting that despoils that existing of any truly decisive happening in fullness and lets it carry itself out determinatively only in diminished guise.

In the deteriorative falling-away into inauthenticity that ever besets it man's existing, his openness-for-Being,[194] recoils from the awareness of the utter contingency of existing as such that is intrinsic to it[195] and

---

[189]But see as an exception HHF 35, HFH 100, where a reference to the everyday appears in a discussion of the "language machine." Cf. Ch. VI, pp. 237f.
[190]Cf. Ch. XII, p. 475, Ch. XIII, p. 521, and n. 118; and Ch. XIV, p. 545, and n. 22.
[191]Cf. Ch. VII, p. 268, n. 2, p. 269, n. 5, p. 270, n. 10 & p. 275, n. 22; Ch. VIII, p. 299, n. 23, p. 300, n. 24, p. 305, n. 42, p. 310, n. 61, and n. 63; Ch. IX, p. 334, and n. 11, p. 335, n. 15, p. 338, n. 35, pp. 340f, p. 341, and n. 49, pp. 343-345, 360, & 363f; and this chapter, p. 419, n. 65, p. 428, n. 92, and p. 453, n. 177. Cf. also Ch. I, p. 33, n. 36.
[192]See Ch. VI, p. 228, n. 10, and n. 12, and pp. 237f, 241, 244 & 264, and n. 107.
[193]Cf. Ch. IX, pp. 332-345.
[194]Cf. Ch. II, pp. 59f.
[195]Cf. Ch. II, pp. 74-76, 83f, & 85f.

from the receptive, restrained openness that must be, in its fullest happening, its own proper going-forth to meet the on-happening Twofold as the latter, in a mystery-fraught self-withholding in transiency, proffers itself to it.[196]   Recoiling, it blinds itself to the context of self-presentings in ever-shifting relatednesses—with its demand for a genuinely centered self-carrying-out in attentive, selective response—wherein it properly finds itself in its thrownness as Being-in-the-world.   Instead of venturing a demanding, wholly individual launching-forth to meet what ceaselessly meets it ever anew, it takes refuge in a secure comprehending of that with which it has to do that is undertaken from an unassailable stance of anonymity.   It understands everything, including itself, as something inherently isolable and manipulatable via thought or action, and it contents itself with the view of everything that can, without individual commitment to accountability, be held in common as the publically current opinion of the "they."   Transpiring thus, man's existing does not directly encounter, in a way genuinely its own, the components of its world in the fullness of the self-proffering that might be theirs.   Habitually, rather, in the crippling self-preoccupation that prompts it to trust itself to be able, through incessant searching about and incessant discussing and explaining, to provide for itself generally acceptable, entirely reliable knowledge regarding that which it encounters and regarding itself, it so discerns everything as to catch sight of it not in itself, but only as it appears to its own self-securing intending and on the basis of the concealedly anxious, knowing stance that carries the latter into play.[197]   And in all this, we can certainly discover readily an immediate parallel to the decisive comportment to which, ever increasingly, the Being of what-is, in its bringing of itself to pass historically, happens man away in happening him as "needed" partner to itself.

Although Heidegger does not discuss the point directly, that same inauthentic manner of human existing is to be seen as ubiquitously, if not determinatively, in play among the early Greeks.[198]   That same manner

---

[196]Cf. Ch. II, pp. 99-101, 105-108, & 91f.
[197]Cf. Ch. II, pp. 86-90.
[198]See, e.g., "*Moira*," VA 254, EGT 99 and EM 121, IM 132. See also EM 126, IM

of happening evinces itself and assumes a decisive role in Greek man's quailing before the utterly demanding, invincibly precarious onslaught of the Being of what-is happening as *physis*.[199] It is evident in the obtruding of a self-securing, humanly based grounding that, from its origin in the beginnings of philosophy,[200] intensifies and intensifies yet again with the ascendancy of *ratio* and with the new subjectivity of the modern age as established in metaphysics and carried into effect in science.[201] It exhibits itself finally and most clearly in the coming to open dominance of the technology that now does indeed exalt viewing-in-common, boldly assume the rightness of its evaluation of all that comes within its ken, and thoroughly overlook in its intrinsicness whatever it meets, as—in an extreme out-working of the ever garrulous, confident and curious busyness that marks human existing in its everyday transpiring in inauthenticity—it takes everything to be no more than an instrument available to serve its ends[202] and presses toward the fulfillment of its purposes and therewith toward the fulfillment of the purpose that tacitly underlies them, namely, that of achieving for those who ostensibly wield it, technology, a secure maintaining of themselves through the mastering of whatever is. Surely in the complacent tranquility that is yet driven into bustling activity by the very thrust of thrownness, of which Heidegger speaks in describing the manner in which inauthentic human existing takes place,[203] we can glimpse the counterpart of the assurance in which modern "technological" man maintains his stance in assumed dominion vis a vis whatever meets him and of the activity that is ever concomitant with it, i.e., the frenzied pushing-forward of technological undertakings, via which he so unquestioningly pursues his self-serving way.[204]

In one very significant passage in "On the Essence of Truth"—an

---

138, where Heidegger remarks that the "manner of Being" of the doer of *technē* is *"not* that of everyday." Cf. Ch. IX, pp. 335f & 341.
[199]Cf. Ch. IX, pp. 359f.
[200]Cf. Ch. X, pp. 371-374 & 388-392.
[201]Cf. Ch. VIII, pp. 302-304, and Ch. VII, pp. 268f.
[202]HHF 35; HFH 99f.
[203]Cf. Ch. II, p. 89.
[204]Cf. Ch. VI, pp. 245-247, 249-252, & 260-262, and Ch. VIII, pp. 319-321.

essay in which Heidegger, while remaining close to the depiction of
human existing given in *Being and Time*, had moved into the sphere of
overt concern with the happening of the Being of what-is that became
determinative for him in his later work—the link between the transpiring
of modern technology and the structures belonging to human existing as
such becomes overtly visible. Its presence goes unremarked, a fact which
itself attests that link's unquestioned place in Heidegger's thinking. In a
context in which he speaks at length regarding the disclosure of what-is by
way of the openness-for-Being lived out by man as man, as that openness
is made specific in particular human comportment, Heidegger says:

> Where for man what-is is little known and through science is
> scarcely and only imprecisely recognized, the manifestness
> (*Offenbarkeit*) of what-is as a whole can hold sway more essentially
> [more fundamentally and provisively (*wesentlicher*)[205]] than it can
> where what is known and what is at any time knowable has become too
> vast to be surveyed and is able to resist no further the industriousness
> of knowing, since technical[206] mastery (*technische Beherrschbarkeit*)
> of things conducts itself without limit. Precisely in the levelling and
> smoothing-down characteristic of omniscience and mere-knowing, the
> manifestness of what-is flattens out into the seeming Nothingness of
> that which is no longer even taken to be indifferent and all the same
> but is only forgotten.[207]

In thus adducing an example of non-disclosive disclosure to set
over against the genuine manifestness of what-is, Heidegger clearly

[205]For a passage from the same essay in which Heidegger takes *Wesen* to be provisive,
see "On the Essence of Truth," W 81, BW 125. On the meaning of *Wesen*, cf. Ch.
VI, pp. 252-257.
[206]Since the main reference here is to the knowing evinced by science and there is no
obvious reference to technology per se, this adjective, which could be translated
"technological" is here given its equally possible, less specific meaning. Cf. Ch. VI,
p. 225, n. 1.
[207]"On the Essence of Truth," W 88, BW 131.

alludes to the technical-scientific mode of knowing with which both he and his listeners are familiar. Here the parallel is plainly evident between his description of such knowing and his later portrayal of the purposeful, exploitative technological comportment that now ever more urgently and pervasively commandeers everything into its dominion, relegates it to the status of standing-reserve, cares only for its availability as thus provided, and leaves it quite unheeded in itself. And a like parallel is also discernible between that description and his characterization of the conduct definitive of modern science, which approaches everything by means of a preestablished, confident knowing and, seeing it from this perspective or that, gathers it into the purview of its own, ever specialized understanding in order, finally, that more and more phenomena—indeed all phenomena whatever—may, while unnoticed in their fullness, be made susceptible of such understanding and of the disposing control to which under the now decisive dominion of technology it can and will lead.[208]

In the context immediately successive to the depiction of the sort of technical-scientific knowing that is now in play, which has just been quoted, Heidegger, having returned to his original inclusive theme, speaks of the way in which the "as a whole (*im Ganzen*)" of what-is appears "in the field of vision of everyday calculating (*Rechnen*) and supplying (*Beschaffen*)," and this allusion, which manifestly refers back to the description of knowing previously given, with the use of the word "everyday" makes that description clearly exemplary of the transpiring of human openness-for-Being inauthentically in everydayness.[209]

In turning his attention to history, Heidegger is looking elsewhere and otherwise than when he inquired specifically into the structures of human existing. But the structures remain visible in his thinking, and many of their elements, albeit as variously transformed and augmented, remain to be seen in the historical portrayals that he presents. Not least is this true when the decisively ruling phenomenon that is modern machine technology is in question. Conversely, although Heidegger did not speak

---

[208]Cf. Ch. VII, pp. 278f.
[209]Cf. Ch. II, pp. 88f.

in such terms at the outset of his work in *Being and Time*, it seems fair to say that in the taking place of human existing he saw from the first a ruling happening of Being in which, as he would surely soon have said,[210] concealing and unconcealing, bringing themselves to pass via an inseparability of inauthentic and authentic existing, were ever underway. In his considering of the single Twofold, the Being of what-is, as unfolding historically by way of the self-carrying out of man, new perspectives disclosed new insights. But even as the happening of Being understood as self-unconcealing self-concealing, ruling as presencing, comes to the fore and all phenomena are seen first in its light,[211] inner continuity remains. In like manner, the originally seen structuring is remembered and its lineaments suggest themselves to us still.

▦   ▦   ▦

Thus it is that, for Heidegger, the roots and origins of the modern age lie far beyond our power to reach or to transform. Resolutely, even with assurance, we confront modern technology. Although it presses hard upon us, yet we trust and cherish our successes and there arises ever afresh our conviction that we shall grasp technology, understand it, channel it, and thoroughly bend it to our purpose. But this, according to Heidegger's understanding, we shall never do. Heidegger would have us see that the manner of holding sway of modern technology is rooted in that which is inaccessible to us. In carrying into play the temporally determined existing of man as such, it springs out of the past—not a past to be viewed simply through linear time and to be assessed as directly accessible to understanding and evaluation in our own terms, but a past that can be genuinely defined only in terms of the opening-up of time that is the self-opening, via successive self-withholdings, of the unitary self-

---

[210]Cf. "On the Essence of Truth," W 73ff, BW 117ff, where concealing and the revealing harboring forth that happens as an unconcealing are first made central, in a discussion of the role of man vis a vis what-is as disclosed via the happening of Being. On the relation of *Entbergen* and *Unverborgenheit* see Ch. I, p. 24, n. 12.
[211]Cf. Ch. I, pp. 24-26.

differentiating Happening named as the Being of what-is[212] and that therefore cannot be utilized or built upon, but can only be acknowledged and accorded with.

If we take our stand in this perspective, the accomplishments of Western man over the centuries can afford us no grounds for confidence. For we must see ourselves to have come not, as we suppose, to a time marked by laudable achievements, but to one wherein the unique happening of genuine self-disclosure is being decisively withheld. Seen thus, ours is not, as we prefer to think, a milieu wherein increasingly meaningful life is being both offered ever more widely and sought on behalf of more and more people. It is, instead, one that can boast, ultimately, only an obscure ambiance of interchangeability and of insidiously wasting having-in-common, wherein it is in fact nothing other than inauthentic concealment-ravaged human existing that is powerfully in play.

The Being of what-is, in its happening as the enframing summons that sets everything in place as supply, drives relentlessly forward. In it the most primal tendency of Being—its happening as self-concealing— brings itself more and more thoroughly to bear. Pervasively now throughout all our world, it governs us in our encountering of what-is. The prospect before us is bleak. It stands forth in even darker relief when viewed by way of the origin, the Greek manner of relating of Being and man that permitted genuine responsive openness and genuine unconcealing to take place. We are left, according to Heidegger's way of thinking, without self-reliant expectations of deliverance from our present plight that he so unsparingly depicts.[213] We pursue our way under the governance of Being. And only in that governance can lie the decisive source of whatever change of determining circumstance—and that means whatever change in the relating of the Being of what-is and man—is to come upon us from beyond the juncture wherein we find ourselves today.

---

[212]Cf. Ch. I, pp. 29-33, 37f, & 47-49, and Ch. XIII, p. 496.
[213]"Overcoming Metaphysics," VA 98f, EP 110; "Memorial Address," G 21, 22f; DT 51, 52.